New Perspectives for Energy Savings in Agriculture

Solar Energy R&D in the European Community

Series H:

Solar Energy in Agriculture and Industry

Volume 2

Publication arrangements: T. C. Jones

Solar Energy R&D
in the European Community

Series H   Volume 2
Solar Energy in Agriculture and Industry

New Perspectives for Energy Savings in Agriculture

Current Progress in
Solar Technologies

edited by

V. GOEDSEELS, E. VAN DER STUYFT, and U. AVERMAETE
Catholic University of Leuven, Belgium

and

H. BUIS and W. PALZ
D-G. XII, Commission of the European Communities,
Brussels, Belgium

D. Reidel Publishing Company

A MEMBER OF THE KLUWER ACADEMIC PUBLISHERS GROUP

Dordrecht / Boston / Lancaster / Tokyo

for the Commission of the European Communities

Library of Congress Cataloging in Publication Data
Main entry under title:

New perspectives for energy savings in agriculture.

    (Solar energy R. & D. in the European Community. Series H, Solar energy in agriculture and industry; v. 2)
    1. Solar energy in agriculture—European Economic Community countries.
2. Agriculture—European Economic Community countries—Energy conservation. I. Goedseels, Vic. II. Series.
TJ809.97.E85N49   1986   630   86–17657

ISBN-13:978-94-010-8607-3   e-ISBN-13:978-94-009-4740-5
DOI: 10.1007/978-94-009-4740-5

Publication arrangements by
Commission of the European Communities
Directorate-General Telecommunications, Information Industry and Innovation, Luxembourg

EUR 10649 EN
© 1986 ECSC, EEC, EAEC, Brussels and Luxembourg

LEGAL NOTICE
Neither the Commission of the European Communities nor any person acting on behalf of the Commission is responsible for the use which might be made of the following information.

Published by D. Reidel Publishing Company
P.O. Box 17, 3300 AA Dordrecht, Holland

Sold and distributed in the U.S.A. and Canada
by Kluwer Academic Publishers,
101 Philip Drive, Assinippi Park, Norwell, MA 02061, U.S.A.

In all other countries, sold and distributed
by Kluwer Academic Publishers Group,
P.O. Box 322, 3300 AH Dordrecht, Holland

PREFACE

In its continuing efforts to improve production and storage , European agriculture and food industry consumes ever more energy. Hence, as was the case for all European sectors founded on intensive energy consumption, agriculture was also severely affected by the shortages and price increases of conventional energy sources over the past decade.

The energy consciousness generated in this way led to a widespread consideration of the application of other, renewable energy sources. The potential applications in agriculture are, however, extremely diverse, and this explains to a great extent the fragmentary nature and even in some cases the mediocre level of current research.

The objective of this book is, therefore, to guide the reader in a systematic way through the apparent chaos of operational data which are currently available on the subject matter: thermal applications within agriculture of solar energy by means of solar collectors, passive designs and storage devices. While in volume I of this series H this area is assessed from a more general economic perspective, this tome II takes a more detailed technical approach.

The results of this book were produced in the form of a European concerted action under the leadership of the Commission in Brussels. All possible data were collected by national representatives in the EC member countries. Coordinators were Professor V. Goedseels and Mr E. Van der Stuyft from the Katholieke Universiteit Leuven and Professor G. Schepens from Facultés Universitaires Notre-Dame de la Paix, Namur. In particular, the coordinators had to produce a tremendous effort to achieve the results which we are now able to present in this book. I take this opportunity to thank them all for their efforts and their enthusiasm. The same is true for the support from Mr L. Crossby who served as editor.

For further reading, interested experts should refer to the various national reports which were also produced in the frame of this community action. They are published as EUR Reports and are available at the Commission's DG XIII in Luxembourg.

W. Palz
Head of the European R&D Programme
for Solar Energy

Summary

The opportunities for solar energy applications in European agriculture considered in this book are related to the direct thermal needs. A technical study is made of the potential of solar energy in three major agricultural areas : greenhouse heating, agricultural product drying, and hot water production in the animal sector, for digester heating, and aquaculture.

Given the intended interdisciplinary readership, the necessary background is first provided to each of the agricultural processes in which integration of solar energy is considered.

Next, a number of research projects and operating plants in the subject area are surveyed. Based on this survey some tendencies are reported and further research steps suggested.

In order to provide a more in depth view of the subject, the survey is supplemented by detailed technical reports on specific solar energy application projects.

In the greenhouse sector, a variety of measures can be taken for energy conservation purposes, many of which are situated at the limits of what can be termed solar energy applications.

The drying sector offers the brightest prospects.
In certain regions of Europes low temperature solar hay driers are already operating successfully; research indicates a clear potential for raisin drying, and in certain cases also for grain drying.
Other drying applications are currently being studied.

As far as hot water production is concerned, research is still limited and it is not yet possible to draw final conclusions. These installations are characterized by a relatively high complexity, as well as generally tight economic margins.
In this field, solar assisted milk preparation on calf rearing farms clearly ranks as the option with most potential.
Current research on floorheating techniques requiring lower water temperatures may bring about a significant increase in solar system efficiency and so lead to economically feasible solar installations.

Table of contents

Contributors

This survey of <u>solar systems</u> in European agriculture together with the <u>assessment study</u> on the same subject (Volume I of this series) were commissioned for the Directorate-General for Science, Research and Development (D.G. XII) of the Commission of the European Communities, within the framework of its Solar Energy R & D Programme.

The Contributors to this second volume were:

For the Commission of the European Communities

PALZ W., Head of the Solar Energy R & D Programme

BUIS H., Responsible for Project H: Solar Energy Applications in Agriculture

Acting as coordinators

Subcontractor

GOEDSEELS V., Project leader	Catholic University of Leuven
	Laboratory of Agricultural
AVERMAETE U.	Building Research
BERCKMANS D.	Kardinaal Mercierlaan 92
MERCKX J.	B - 3030 HEVERLEE
VAN DER STUYFT E.	

Prime contractor (in charge of Volume I)

SCHEPENS G., Project leader	Facultés Universitaires N-D
	de la Paix
BRIZZI A.	Faculté des Sciences Economiques
BUYDENS C.	Rempart de la Vierge 8
JACQUEMIN J-C.	B - 5000 NAMUR
LEBEAU P.	
MAHY D.	
MAJERUS J-P.	

Acting as action 2 national participants (cf. Introduction p. XXI)

COUNTRY	PARTICIPANTS	
Belgium & Luxembourg	The coordinating organisations mentioned above.	
F.R. Germany	MEUREN K.	Bayerische Landesanstalt für Landtechnik Vöttingerstrasse 36 D – 8050 FREISING–WEIHENSTEPHAN
Denmark	AMSEN M.G.	Institut for Vaeksthus Kulturer Kirstinebjergvej 10 DK – 5792 ÅRSLEV
Ireland	ROBINSON D.W.	Kinsealy Research Centre Malahide Road IRL – DUBLIN 5
France	MOLLE J.F.	CEMAGREF Parc de Tourvoie B.P. 121 F – 92164 ANTONY Cedex
Greece	KYRITSIS S.	Institut Agronomique d'Athènes Votanikos G – ATHINA 301
Italy	SCARPINI A.	Energia progettazione Corso d'Italia 19 I – 00198 ROMA
The Netherlands	LEIJENDECKERS P.H.H. BROUWER G.	RTB Van Heugten St. Annastraat 145 NL – 6524 EP NIJMEGEN
United Kingdom	CARPENTER J.L.	Engineering Department Seale Hayne College Newton Abbot GB – DEVON TQ12 6NQ

Acting as Action 3 participants (cf. Introduction, p. XXI)

1. Application of solar energy in greenhouses

PROJECT LEADER	ADDRESS	RESEARCH SUBJECT
AMIRANTE P.	Istituto di Meccanica Agraria Universita degli Studi Via Amendola 165/A I - 70126 BARI	Utilization of solar energy for heating of a greenhouse destined for horticultural cultivations
AMSEN M.	Research Centre for Horticulture Institute of Glass-house Crops Kirstinebjergvej 10 DK - 5792 ÅRSLEV	Biological effects of energy savings in greenhouses
BAILEY B.J.	British Society for Research in Agricul-tural Engineering Wrest Park Silsoe GB - BEDFORD MK45 4HS	A comparison of the light transmission of greenhouses covered with twin-walled rigid plastics
BERTRAND E.	Institut Technique Horticole de l'Etat 5 rue Verlaine B - 5800 GEMBLOUX	Evaluation of the energy savings by means of solar energy, compared with the total amount of energy required for greenhouse heating
DELTOUR J.	Faculté des Sciences Agronomiques de l'Etat 8 ave. de la Faculté B - 5800 GEMBLOUX	Evaluation of different covering materials for greenhouse production of strawberries
GRAFIADELIS M.	Agricultural Research of Northern Greece E. Georgiki Scholi G - THESSALONIKI	Development of a new solar system for heating green-houses, and root zone warming (RZW)
KYRITSIS Sp.	Agricultural College of Athens Iera Odos 75 Votanikos G - ATHENS	Greenhouse heating with solar energy and nutrient film techniques (NFT)
LETELLIER B.	Novelerg 12 Rue de la Baume F - 75008 PARIS	Monitoring the operation of a horticultural greenhouse with an economical heating system based on solar energy

RIGOPOULOS R.	Physics Laboratory II University of Patras G - PATRAS	Thermal storage of solar energy for a greenhouse
ROBINSON D.W.	Kinsealy Research Centre Malahide Road IRL - DUBLIN 5	Double-clad polythene greenhouses
ROBINSON D.W.	idem	Greenhouse energy saving through root zone warming (RZW) with reduced night air temperature

2. Applications of solar energy in drying processes

PROJECT LEADER	ADDRESS	RESEARCH SUBJECT
CASTELLI G.	Istituto di Ingegneria Agraria dell'Università degli Studi di Milano Via Celoria 2	The use of solar energy for drying agricultural products and washing of milking parlour installations
CLARK J.A.	Dept. of Physiology and Environmental Science University of Nottingham School of Agriculture University Park GB - NOTTINGHAM	Solar heating for barn hay drying
DAGUENET M.	Laboratoire de Thermo- dynamique et Energétique Université de Pergignan Ave. de Villeneuve F - 66025 PERGIGNAN Cédex	Development of a solar drier for wine lees
DEN OUÐEN C.	TPD - TNO - TH P.O. Box 155 NL - 2600 AD DELFT	The use of solar energy for bulb conditioning and soil heating
LASSERAN J.Cl.	Station expérimentale l'I.T.C.F. Boigneville F - 91720 MAISSE	Solar drying of grains

MEUREN K.   Technische Universität   Solar plant for drying of
                       München   spices
                       Bayerische Landesanstalt
                       für Landtechnik
                       Vöttingerstrasse 36
                       D - 8050 FREISING

PLETINCKX A.   Station de Génie Rural   Drying of forage and
                       Chaussée de Namur 146   cereals
                       B - 5800 GEMBLOUX

RIVA G.   Istituto di Ingegneria   Solar plant for forage
                       Agraria dell'Università   drying in mountainous
                       degli Studi di Milano   areas
                       Via Celoria 2
                       I - 20133 MILANO

SARAVACOS D.   Laboratory of Unit   Solar drying of agricultural
                       Operations   products
                       National Technical
                       University
                       28 is Oktovriou 42
                       G - ATHENS 147

3. Applications of solar energy for hot water production

PROJECT LEADER	ADDRESS	RESEARCH SUBJECT
CASTELLI G.	Istituto di Ingegneria Agraria dell'Università degli Studi di Milano Via Celoria 2 I - 20133 MILANO	The use of solar energy to accelerate the fermentation of cattle manure
GAUTIER E.	Sistemi Energia Sud S.p.A. Via Cuneo 20 I - 10152 TORINO	Bonafous Energy Farm
GOEDSEELS V.	Laboratory of Agricultural Building Research Fac. Agricultural Sc. Catholic University Leuven Kard. Mercierlaan 92 B - 3030 HEVERLEE	The use of solar energy for floor heating in a farrowing house
BROUWER G.	R.T.B. van Heugten St. Annastraat 145 NL - 6524 NIJMEGEN	The use of solar energy for heating circulation water for eel- and carp breeding

BROUWER G.	Idem	The use of solar energy for hot water production on a calf-rearing farm (2 different projects)
MEUREN K.	Technische Universität München Bayerische Landesanstalt für Landtechnik Vöttingerstrasse 36 D - 8050 FREISING	Solar heating of cleaning water and heating of farrowing house

Acknowledgements

First of all the coordinating team wishes to thank the Action 2 and 3 participants for procuring the background materials needed for the descriptions of projects given in parts I, II and III, and also for revising the descriptions in their adapted format.

Furthermore, the coordination team is greatly indebted to all those persons, not belonging to the above-mentioned group of contributors, whose dedicated help was a crucial asset in the preparation of this book.

Special mention should be made of the European researchers who kindly provided information on their experiments and investigations, so as to enable the Action 2 participants to develop a national inventory of research activities on solar energy applications in agriculture.

Those who generously gave professional advice in the context of preliminary drafts of this volume are too numerous to mention individually. Thanks are due in particular to Prof. F. Ollevier (aquaculture) and Prof. I. Impens (greenhouses).

Last but not least the team wishes to thank both Mr. L. Crossby (reviser of the text in English) as well as the people involved with the logistics of the editorial task (typing, drawing, lay-out) whose dedication and patience made the completion of this book possible.

READERS' GUIDE

1. Introduction

CEC project H, devoted to "Solar Energy in Agriculture and Related Industries", is part of a broad scale stimulation and coordination program of the Commission of the European Communities (CEC) in the area of Solar Energy Research and Development within the European Community (EC).

This project H involved 3 distinct steps:

- Action 1:
  The goal of this action, which is reported on in volume I of this series, was to determine the economically promising areas for solar applications in agriculture and related industries, based on:

  1. an assessment of the energy demands, as related to the climatic and agricultural environment.
  2. an assessment of the available solar technologies, as well as an evaluation of some of the existing applications.

- Action 2:
  In this action program, as precise and detailed an inventory as possible was made of the research and development within the EC countries in the given study area.
  The main purpose of this action step was to make possible an evaluation of the current state of the art based on the research activities and results, gathered for each EC country separately.
  This evaluation led to the identification of certain valid techniques, worthy of further development and/or of being spread in other regions and countries.

- Action 3:
  On the basis of the newly acquired knowledge concerning the at-first-sight economically promising applications of solar energy in agriculture (cf. Action 1), together with the identification of currently running significant research (Action 2), several projects were selected, most of which had started recently. In order to obtain extensive measuring data, the CEC provided these teams with financial support so that they could buy the equipment to ensure a high quality of research.

The present volume consists of a technical report of the work accomplished by the participants in Actions 2 and 3.

Research on the application of solar energy in agriculture has been running for five, and in some cases more years already, in varying circumstances; in universities, in counselling centres, and also on farms.

It is noteworthy that, in several instances, the farmers are ahead of the research. In this context, valuable technologies are often being developed, which are however usually lacking in the area of component optimization.

Furthermore, given the specific character of agriculture, the knowledge of component optimization in other sectors can seldom be applied. (E.g.: collector sophistication is investigated as a means of optimization outside the agricultural field, whereas in the case of agriculture cheap, easy to make collectors are often preferable).

An essential step in solar research therefore consists in the collection of broad agriculture-specific knowledge based on the presently scattered information. This is the precise aim of Project H : to take the first steps in the elaboration of an orderly survey and evaluation of current research, so as to provide the broadest possible picture of the agriculture-specific knowledge and perspectives in the area of solar energy.

It is hoped that this work will contribute significantly to accelerated and more effective research and development in the area of solar technologies adapted to agriculture in the future.

2. Methodology

The present volume is meant to guide an interdisciplinary readership through a long list of applications and techniques. The background to agricultural processes -- unfamiliar to some readers - is provided. Existing trends, the expected potential and known difficulties of projects and applications are pointed out. Consistent use is made of cross references, symbols, badges, standardized presentations, etc ... to facilitate consultation of this book.

The intended readership includes :
- Researchers or company executives, comprising both agronomists and "non agrarian" solar technology specialists, looking for the necessary background and up to date information in connection with the potentially successful further development of solar techniques in different application fields within the agricultural sector.
- Policy makers, people looking for fundamental documentation to help them decide which research activities and demonstration projects to support, in light of the development and expansion potential of the field in question.
- Consultants (within Ministries of Agriculture, agricultural organisations, co-operatives, ...), people acquiring information as a basis for promotional activities and advisory services to farmers.

The book consists of 3 major parts, each of which presents a discussion of one of the 3 main application areas for solar energy in agriculture :
- The greenhouse sector (Part I)
- Drying processes (Part II)
- Hot water production (Part III)

In each part, the first chapter, entitled "Process requirements-Technological options", describes the energy-requiring processes which are being considered as potential candidates for the integration of solar energy. In most cases, the description consists of an analysis of the parameters of the process studied, followed by a presentation of the main technologies - including solar technology - applied to meet the requirements of the process in question.

The second chapter of each part gives an overview of the main research programs within the EC on the given subject, drawing on information obtained from actions 2 and 3. It is important to note the provisional nature of the information presented here, as well as in the appendices.
Indeed, obtaining valid information on well established (non CEC subsidized) projects proved to be very difficult, while in many cases the CEC action 3 projects were still at an early stage and it was not possible to draw final conclusions from them.

Given these limitations, the information was combined and interpreted in such a way as to optimally bring about the objective of project H by drawing a clear global picture of the state of the art, as well as by pointing out some tendencies and prospects in the evolution of research and in the estimated feasibility of certain applications. The "General conclusions" and "Prospects" sections sum up the core findings and message of this book.

In the appendixes, a selection of the projects outlined in chapter 2 are described in detail.
The first part of these detailed descriptions gives general information, such as project identification and description, schematic representation and conclusions.
The second part is of a more technical nature, and, following a standardized presentation, it gives data on site and climate, plant, product, economic aspects and research results of the project in question.

Note : _Information on the reporting format of "Basic" and "Detailed descriptions" and the symbols employed in the diagrams can be found in an addendum._ _(see p. 467)_

3. Terminology

3.1. Main symbols

Symbol	definition	dimensions
C	cost factor	(money unit/product unit)
c_x	specific heat of material x	$J/(kg\ K)$
D	drying speed	$\dfrac{kg\ H_2O\ removed}{m^2\ prod.surf.s}$
E	Instantaneous energy flow (explanation of the indices is given under symbol Q)	Watt
e	Specific instantaneous energy flow (E/S) (explanation of the indices is given under symbol Q)	$Watt/m^2$
F_i	Flow meter i	
G_x	Mass of element x in a specific amount of air	kg
I_x	Instantaneous solar radiation on 1 m^2 of surface x For x = c : collector surface x = h : horizontal surface	W/m^2
I'_x	Solar radiation on 1 m^2 of surface x integrated over time (For explanation of indices, cf I_x)	$(M)J/(m^2.day\ or\ yr.)$
j_x	enthalpy of x	$J/(kg\ dry\ air)$
k	heat transfer coefficient	$W/(m^2\ K)$
K_i	value i	
M	mass	kg
$m.c._x$	moisture content in x Note : when not explicitly mentioned, the moisture content is given on a wet basis (w.b.) as opposed to a dry basis (d.b.)	decimal or %
P_a	atmospheric pressure	Pa; bar
P_i	Pump or ventilator i	
p_s	static air pressure	Pa; bar

symbol	definition	dimensions
P_{Tot}	total air pressure (sum of vapour and dry air pressure)	Pa; bar
p_v	partial vapour pressure	Pa; bar
p_v'	saturated vapour pressure	Pa; bar
Q	Energy flow (summation of E over time)	J; kJ; MJ; MJ/day

Note : * for x = c, s or h
- Q_{xi} refers to the energy flow at the input side of element x

    with x = c (collector) this means solar radiation on the collector

        = h (heat exchanger) this means energy input into heat exchanger

        = s (storage) this means energy input into storage

- Q_{xo} refers to the energy flow at the <u>output</u> of element x
- Q_{xl} refers to the energy losses in element x

    * for x = e, f, g
- Q_{ax} refers to the heat input of the auxiliary heater by means of fuel x

    x = e means electricity

      = f means oil

      = g means gas

    * in other cases
- Q_d refers to energy demand
- Q_e refers to energy input (in general) in the form of electricity
- Q_{no} refers to the net useful energy output of the heat producing system

symbol	definition	dimensions
q	specific energy flow (Q/S) (explanation of the indices given under symbol Q)	J/m^2; kJ/m^2; MJ/m^2; $MJ/(m^2 \cdot day)$
r	evaporation heat of water at a specific temperature	J/kg
r_o	evaporation heat of water at 0°C	J/kg
r.h.$_x$ $(R.H.)_x$	Relative humidity in x	%; (decimal)
S_x	Surface of element x	m^2
T	temperature	°C; K
t_i	time i	s; h
V	volume of storage tank	m^3; 1

symbol	definition	dimensions
$\overset{\circ}{V}$	flowrate	m^3/s; m^3/h
v	windspeed	m/s
$\overset{\circ}{v}$	specific flowrate through the collector	$\dfrac{m^3}{\text{(s or h)}.m^2}$ collector surface)
x	absolute humidity	kg vapour/ kg dry air or g vapour/ kg dry air
$< x >$	Average value of x	
ψ	relative humidity of the air	% or decimal
λ	Heat conductivity coefficient	$W/(m\ K)$
η_c	collector efficiency ($= Q_{co}/Q_{ci} \times 100$)	%; (decimal)
η_{system}	System efficiency ($= Q_{no}/Q_{ci} \times 100$)	%; (decimal)
ρ_x	specific mass of element x	kg/m^3

3.2. Indices

Note : Where the indices are used with a different meaning than the one given in this list, this is explicitly mentioned under the symbol in question.

X =

a   ambient, air in general (= dry air + vapour)

A   dry air

c   heat transporting medium in the collector; collector itself

d   in connection with time : daylight time
        in connection with temperature : desired temperature

dp   dry product

f   floorheating water

g   air

X =

h   the heat transporting medium in the heat exchanger; heat echanger itself

i   inside; at inlet

m   manure

o   outside, at outlet

p   product

s   the heat transporting medium in the storage tank; the storage tank itself

V   vapour

w   water

3.3. Abbreviations

Countries

B   Belgium/Luxembourg

D   F.R. Germany

Dk   Denmark

EIR   Ireland

F   France

Gr   Greece

I   Italy

Nl   Netherlands

UK   United Kingdom

Other

BD   Basic description of project (in chapters 2)

cop   coefficient of performance

D.I.Y. Do it yourself

EC	European Community	
m.c.	moisture content (cf. main symbols)	
N.A.	Not available	
NFT	Nutrient film technique	
N.O.T.E.	Editor's note	
P.A.R.	Photosynthetically active radiation	$(W/m^2$; Einstein/$(m^2.s)$
p.p.m.	parts per million	
r.h.	relative humidity	(%; (decimal))
r.p.m.	revolutions per minute	
RZW	Root zone warming	
T.O.E.	Tonne of oil equivalent	

PRELIMINARY NOTIONS

1. Fundamental criteria for a successful solar application

If one wishes to evaluate the potential of different solar applications, a
clear understanding of the fundamental criteria of success is obviously a
first requirement. In the context of this book these criteria have been
arranged under 5 headings.

Criterion I. ENERGY DEMAND

- The energy demand should be important enough, micro-economically
  speaking, to justify investment costs in a solar system for the
  individual farmer.
- At the same time, the macro-economic importance should be sufficient, in
  order to justify the development of solar systems.
  Note : In this context, not only the size of the energy need, but its
  role in the whole process should be adequately considered. If a small
  amount of energy has a tremendous influence on a larger process via a
  catalytic effect, it might be worth while to reduce the dependence on
  conventional energy sources by employing solar energy.

Criterion II. ENERGY SUPPLY

- Obviously the chances for the successful application of a solar system
  are greater in a climate with greater amounts of solar radiation.
- Since no practical solutions for long-term heat storage are available at
  present, this supply needs to be available at a time when there is a
  real heat demand.
  In the best case, supply and demand would evolve in phase with one
  another.

Criterion III. SOLAR SYSTEM EFFICIENCY

- High radiation intensity values and high outside temperatures greatly
  contribute to solar system efficiency. Thus the climate can have a
  significant influence on efficiency.
- Solar system operating temperatures closer to ambient temperatures yield
  higher efficiency rates.
- Compatibility problems between solar system and process can cause
  undesirable complications and efficiency reductions.
- An adequate solar system design, going from collector components,
  through insulation, to design of the system as a whole (optimization of
  dimensions, regulation mechanisms, etc...) will also influence the solar
  system efficiency.

Criterion IV. SOLAR SYSTEM COST

- The integration of solar collectors into a roof is one example of how
  the solar system cost can be reduced by spreading its use over two
  different functions (solar collector, roof cover).
- Spreading the use of a solar system over the whole year rather than over
  just a few months is an example of how the same system can be used to
  collect extra energy for comparable investment costs.
- In evaluating the solar system cost versus performance, its feasibility
  should be compared not only with the conventional system, but with all

the other **alternatives**. Indeed, a solar system that is competitive with the conventional system but not with a particular alternative is bound to disappear.

Criterion V. "SOCIO-ECONOMIC" OUTLOOK

Even if a solar system could technically speaking yield good results for acceptable costs, it might not penetrate successfully in certain socio-economic environments.
Different elements can play a crucial role here. E.g. :
- The **attitude of the farmers** towards this new technology (influenced by the quality and vision of farmers' unions, and advisory services, and by governmental support, subsidies, etc...).
- The size, and the degree of **intensity of the farms** (this factor is correlated with the amount of available **capital**, as well as with the level of **technical skill** of the farmers).

In the light of this last socio-economic element, it can be seen that the crucial requirement of small-scale farmers will often be simplicity in the solar installation.
E.g. : - Do it yourself easy to build collectors, to reduce investment costs and technical skill requirements.
       - Simple, reliable automatic controls, periodically checked by a travelling technician, who can at the same time provide necessary information to compensate for the farmers' possible lack of technical skills.

In more industrial large-scale agricultural environments on the other hand the integration of more complex solar technology is usually not so much of a problem.
- In this same socio-economic context it is appropriate to stress the widely recognized desirable aspects of solar systems from the **environmental** point of view. The application of solar energy does not contribute in any way to pollution, a typical problem attached to many of the other energy sources. This aspect will certainly have a significant influence on the success of solar technologies in countries or places where the **quality of the environment** is a high priority.

2. The (thermal) energy demand in agriculture

The opportunities for solar energy in agriculture in the EC considered in this book are related to the direct thermal needs.
The indirect energy input in agriculture (e.g. fertilizers, pesticides) and energy for mechanical power are excluded from this study. The present approach does not consider the opportunities for solar energy in rural dwelling houses or energy derived from biomasses, and solar energy products are also excluded as these are the subjects of other CEC activities.
Three principal agricultural sectors are distinguished in order to analyse the direct thermal energy consumption : greenhouses, animal sector (heating of animal housing, hot water for animals) *(1)* and drying (mainly

(1) The bulk of the solar projects in the animal sector are centered around the technique of hot water production. This same technique is also applied in solar heated aquaculture and biogas production plants. Given the technical focus of this book, these subjects were included in part III.

fodder, grain, fruits). The pattern of direct thermal energy consumption by country and by specific activity is shown in Table 1.

Table 1 : Direct thermal energy consumption   in different sectors of
         agriculture. (Source : project H action 1 reports - cf. p. XXI)
         (PJ/year)

	Greenhouses	Animal sector	Drying	Total
The Netherlands	125.0	5.5	0.3	130.8
W-Germany	70.0	8.9	6.1	85.0
France	31.2	16.7	33.6	81.5
United Kingdom	29.0	9.7	5.6	37.4
Denmark	18.7	9.2	0.7	28.7
Italy	6.2	11.1	5.8	23.1
Belgium	15.6	3.8	0.3	19.8
Ireland	2.6	2.1	1.9	6.6
Greece	0.7	0.6	0.5	1.8
Luxembourg	0.0	0.1	0.0	0.1
Total	299.1	60.7	55.0	414.8

This table shows clearly that, in EC agriculture, the greenhouse sector is the most important with respect to direct thermal needs.
In the northern countries (particularly in The Netherlands but also in northern Germany, the United Kingdom, Denmark and Belgium) over 75 % of the energy needs are related to greenhouses. In the Netherlands, the share of the greenhouse consumption in the total is 95 %.

In the southern countries of the EC (Italy, Greece, central and southern France, the south of West Germany), i.e. the countries belonging to agro-climatic zones I and II (see below, "Solar radiation in the EC"), the share of greenhouse consumption compared to the other sectors is less than one third. The zones in question represent only 15 % of the total EC greenhouse energy consumption.
The energy needs there are very temporary and occur only in the winter months. They are expected to increase as the demand for better quality products on the European market increases.

A few comments on the animal and drying sectors :

- Italy shows a relatively high energy need in the animal sector.
- In the agro-climatic zones I and II, a considerable amount of energy is needed to reduce livestock housing temperatures on particularly hot days.
- France spends over 40 % of its total energy expenses on the drying sector.

3. Solar radiation in the European Community   - the identification of agro-climatic zones

Solar radiation is one of the main climatic parameters : the maximal intensity of solar radiation on a horizontal surface is 1.2 kW per hour per square metre. This value can only be recorded at noon and under a clear sky.

The territory of the European Community lies between 35° and 58° north latitude. Generally, the solar radiation decreases with increasing latitude. The gradient of the isolines is weaker in summer than in winter. This is clearly indicated by a comparison of the daily solar radiation at 36° and 48° north latitude respectively, as shown in Fig. 1.

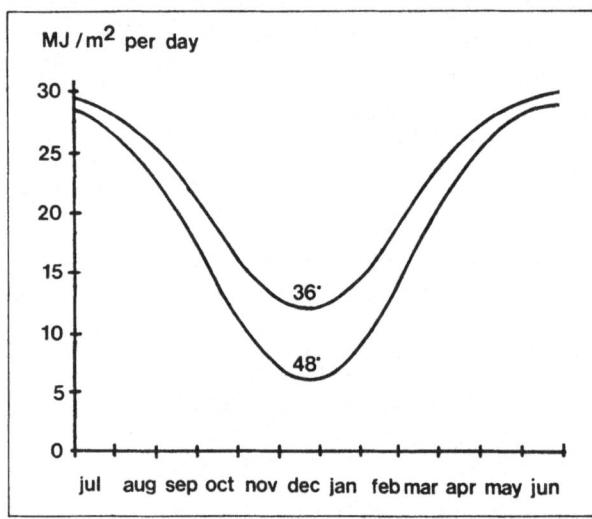

Fig. 1 : Variation of solar radiation over time for different latitudes
(1)

In addition to the effect of latitude, the isolines of solar radiation indicate an east-west gradient, especially at the coasts of Ireland, the United Kingdom, Denmark, Belgium and The Netherlands. Particularly during the summer months the coastal districts show a higher average radiation.
On the basis of those as well as a number of other considerations it is possible to identify homogeneous regions defined as "agro-climatic zones". These agro-climatic zones are shown in Fig. 2. A cumulated yearly global radiation representative of each of the four zones is shown in Fig. 3.
Zone I is roughly south of 45° N latitude, including the south of France, the south and centre of Italy and Greece. It is the sunniest part of Europe. The daily solar radiation varies between 5.4 MJ and 30.6 MJ per square metre per day. The global yearly radiation exceeds 5 400 MJ/m².
Zone II includes the greater part of France, southern and central Germany and the Po valley. The yearly radiation exceeds 3 950 MJ/m².
Zone III is the typical moisture region, including Belgium, The Netherlands, Denmark, northern France and northern Germany. The annual radiation is approximately 3 600 MJ/m² and the radiation during winter is poor.
Zone IV covers northern England, Scotland and the north-west of Ireland. Climatic conditions are generally cloudy with strong oceanic winds and high rainfall in most parts of the region. The solar use potential is rather low.

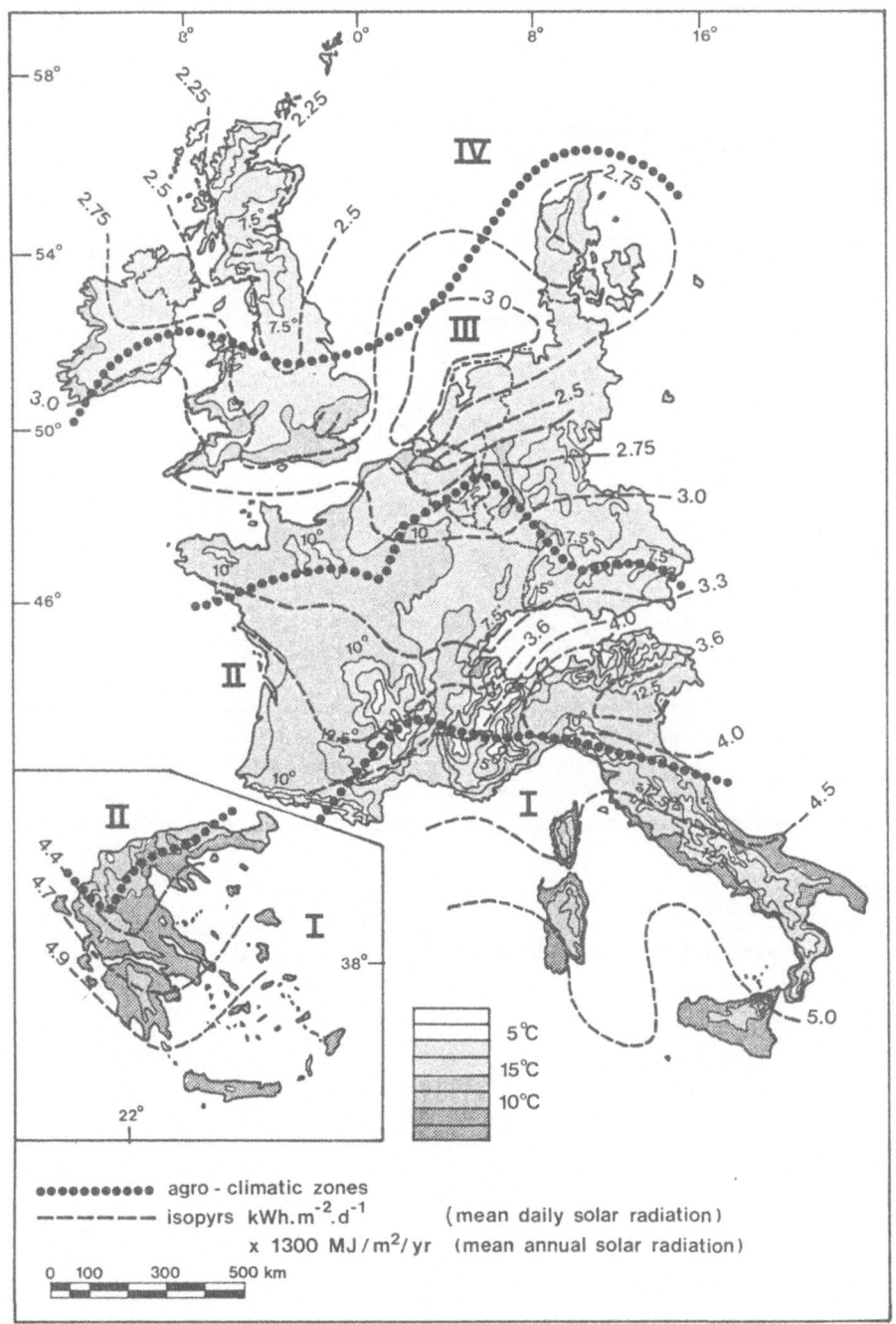

Fig. 2 : The agro-climatic zones within the E.C.
(from : CEC Project H, action 1 Working Group report, published
in Volume I of this series.)

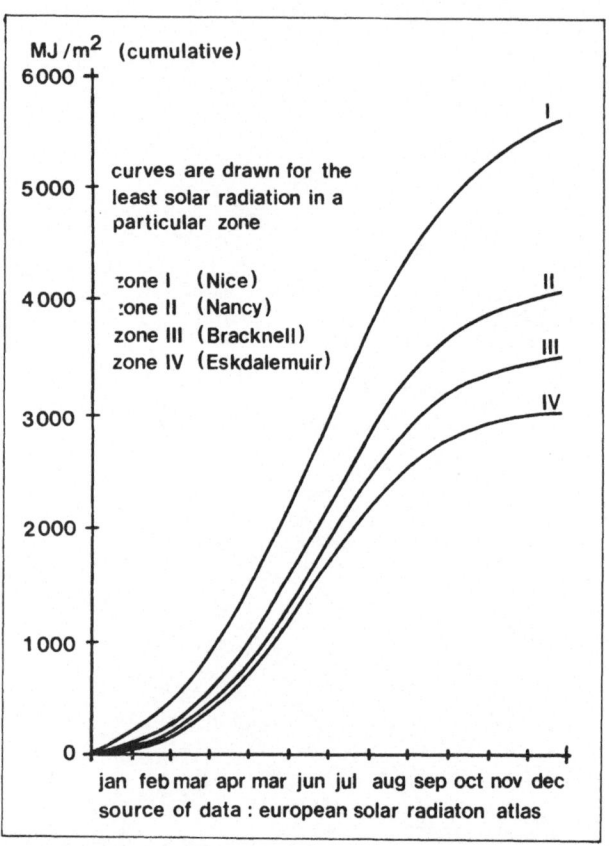

MJ/m^2 (cumulative)

6000

curves are drawn for the
least solar radiation in a
particular zone

5000

zone I   (Nice)
zone II   (Nancy)
zone III (Bracknell)
zone IV   (Eskdalemuir)

4000

I

II

III

3000

IV

2000

1000

0

jan feb mar apr mar jun jul aug sep oct nov dec
source of data : european solar radiaton atlas

Fig. 3 : Cumulated yearly global radiation in the four homogeneous zones.

References :

(1) Myers C. and C.J. Mackson : "L'Emploi de l'Energie solaire dans
       l'agriculture" FAO/ECE/AGRI/WP.2/R 68 : Septembre 1981.

PART I

Applications of solar energy in greenhouses

PART I

Applications of solar energy in greenhouses

CHAPTER I :
Process requirements - technological options

This chapter describes the general context of the application of solar energy in greenhouses. The purpose is to give a theoretical background of solar energy applications in greenhouses and to indicate the possible techniques for incorporating solar energy into the production process.
The incorporation of solar energy in greenhouses is bound to a number of plant physiological limits, depending on climatic conditions. Therefore a brief description of the major production parameters is given : photosynthesis, respiration, transpiration and soil nutrition. The reaction of greenhouse crops to a change of the production parameters indicates the limits of the introduction of energy-saving measures in general and of the solar energy applications in particular (8). Special attention is given to solar energy storage systems.

1. TRADITIONAL PROCESSES

1.1. Definition and importance of greenhouses

The different horticultural systems for protected cultivation can be divided into three main categories :

- low plastic tunnels and plastic covers, protecting the cultures against rain, wind and hail and improving the climatic conditions. These devices can be applied only to low growing plants and in the early stages of high growing plants (seedlings etc) ;

- high tunnels and greenhouses, with the added possibility for the grower of operating under sheltered conditions ; furthermore allowing the mechanization of the most important cultural operations (but high tunnels cannot withstand heavy snowfalls);

- heated greenhouses, possibly controlled by other climatic systems and achieving a more perfect control of the internal climatic factors.

Table 1 shows the most recently available data on the importance of these different horticultural systems for protected cultivation within the EC.

The energy crisis (1974 and later 1979) and the general stagnation of the economy resulted in a stagnation and decrease of the heated greenhouse area. The area of heated greenhouses has decreased in Denmark since 1973. In the United Kingdom the stagnation was observed since 1975, in Belgium since 1977, in Western Germany since 1978 and in The Netherlands since 1980.

In contrast to the other member countries of the EC, the investment activity in The Netherlands was extremely high in 1977, 1978 and 1979 (due to relatively low prices for natural gas).

Table 1 : Area under protected cultivation (1980).
(from : Volume I of this series, based mainly on O.E.C.D. figures).

Country	Total area under protected cultivation (ha)	Total area under glass greenhouses (ha)	Total area under plastic greenhouses and tunnels (ha)	Heated area under glass & plastic (ha)
D	3800	3400	400	3400
F	8560	3230	5330	2770
UK	4257	1835	2422	1149
B/L	1470	1470	0	1183
Dk	540	540	0	540
EIR	207	137	70	130
Nl	8434	8434	0	7476
I	27000	1577	25423	12500
Gr	8264	32	8232	425
EUR-10	62532	20655	41877	29573

Because of the increased demand for (out-of-season) horticultural products and the relatively high availability of solar energy in the southern countries of the EC, as opposed to the expensive fossil fuels required in the northern countries, a rapid growth of plastic greenhouses was observed in the southern part of the EC, particularly during the seventies (see table 1).

1.2. Production parameters

1.2.1. Photosynthesis (11), (15)

Photosynthesis can be summarized as the induction of chemical reactions by light which falls on the leaves, with the result of converting CO_2 and water to oxygen and organic compounds. Hydrogen atoms from water are transferred to carbon dioxide, the oxygen involved comes from the water and not from the CO_2. The process takes place in a watery solution, electrons are transferred from one molecule to another. In photosynthesis water is oxidized and loses electrons, carbon dioxide is reduced and receives electrons.
The primary requirement for the occurrence of photosynthesis is absorption of radiation by chloroplasts. The degree to which radiation is utilized depends on the chlorophyll concentration or, more precisely, on the concentration of photosynthetically active pigments. The photochemical process is initiated when the chloroplasts capture photosynthetically utilizable radiation.
The electron-transport in photosynthesis takes place against an energy potential of 1.2 Volts. It appears that this barrier is overcome by two light quanta.

The electromagnetic radiation is composed of quanta (photons) whose energy is
$$E = h.\upsilon \qquad h = 6.63 \times 10^{-34} \text{ J.s}$$
$$\upsilon = \text{frequency s}^{-1}$$

or
$$E = \frac{1.999\times10^{-25}}{\lambda} \text{ J}$$
λ = wavelength (m)

for λ = 400 nm (violet), E_{400} = 5×10^{-19} J.

λ = 700 nm (red), E_{700} = 2.86×10^{-19} J.

The primary processes of photosynthesis in green plants are driven by radiation in the range of wavelengths between 380 and 710 nm. This photosynthetically active radiation (PAR, often defined as the region 400-700 nm) is an important quantity in plant ecology. The photoreceptors involved in photosynthesis are the chlorophylls with absorption maxima in the red and blue.

The efficiency of the process is rather low. The leaf absorbs 75 % of the energy which it receives, the rest is reflected. Of this 75 % only a small part is used for photosynthesis (see fig. 1).

The total efficiency lies between 1 and 5 %, the rest of the energy is used for chlorovaporisation. An exception is the Clorella with an efficiency of about 55 %.

The quantity of light received by a plant should be measured either with an energy sensor with a flat response between 400 and 700 nm in $W.m^{-2}$ or preferably with a quantum sensor with wavelength-proportional response in Einstein (dimension quantity of matter) between 400 and 700 nm.

A common "lux-meter" is unsuitable since this instrument is adapted to the human eye which has its maximum sensitivity for daylight between 520 and 600 nm (green). Furthermore, the conversion of "Lux" to "Einstein" is impossible.

1.2.2. Respiration

Respiration is a biological phenomenon destined to furnish the necessary energy for the accomplishment of vital processes. (10)
Whereas anabolic processes bring about the synthesis of the substances that make up the plant, in catabolism substances are broken down to provide energy for the diverse metabolic functions of the cell. The substrate for these reactions is carbohydrate or fat; in the exergonic decomposition of these substances hydrogen is split off and energy released in a stepwise manner. Most of the energy is obtained at the step in which hydrogen is transferred to the final hydrogen acceptor. In respiration this acceptor is atmospheric oxygen, which receives the hydrogen from an electron-transport chain by way of the sequence of respiratory reactions in the mitochondria. In fermentation reducible organic compounds take up the hydrogen, and in anaerobic respiration the acceptors are inorganic ions such as nitrate and sulphate. Because the potential difference between hydrogen and oxygen is greater than those between hydrogen and other oxidizing agents, aerobic respiration provides much more energy than the other catabolic processes; it operates with an efficiency of 30-40 %.

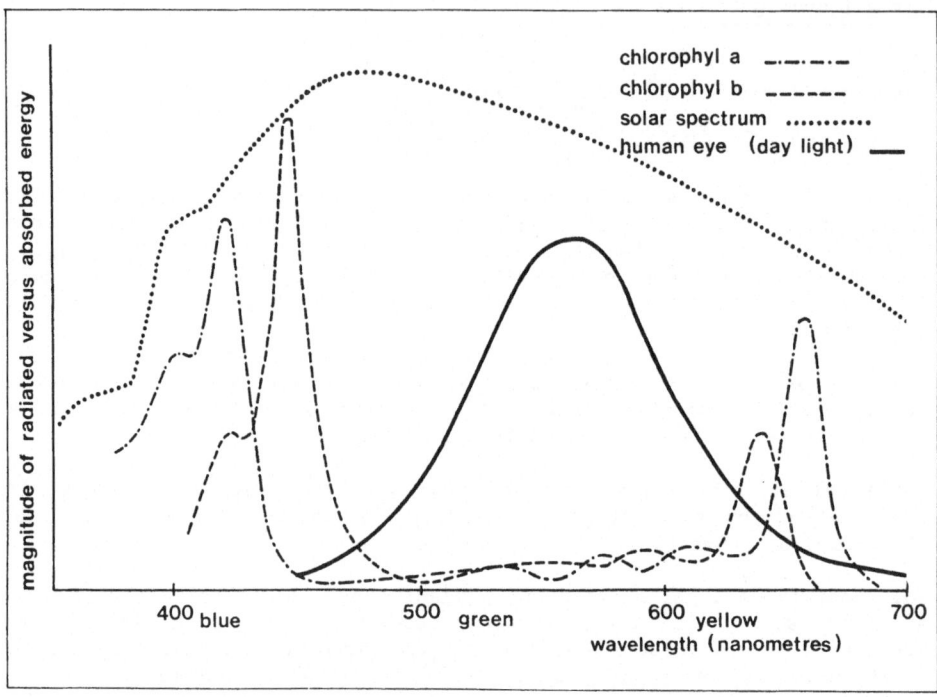

Fig. 1 : Absorption spectrum (source : Scientific American; Dec. 1974)

Temperature affects metabolic processes by means of its influence on the reaction kinetics of chemical events and on the effectiveness of the various enzymes involved in the photosynthesis. The temperature-dependence of growth results from the difference between the rate of photosynthetic CO_2 incorporation and the rate of respiration. The temperature limits for net photosynthesis are dependent on the genetic characteristics and plant species.

As long as radiation is the sole rate-limiting factor, the intensity of photosynthesis is represented by a saturating curve. In dim light this light dependence curve reflects a net release of CO_2 since more CO_2 is given off by respiration than is fixed by photosynthesis. At the compensation light intensity, photosynthesis fixes exactly as much CO_2 as is set free by respiration. Once the compensation point has been passed, CO_2 uptake increases rapidly. In the lower range of this increase (winter conditions in greenhouses), there is a strict proportionality between the yield of photosynthesis (growth) and the available radiation (fig. 2).

1.2.3. Transpiration

Transpiration is the simultaneous control of CO_2 entry into the leaves and the release of water vapour. The stomata are the most important regulators of the diffusion process by varying the width of the stomatal plant pores.

Opening and closing of the stomata is brought about by a turgor difference between the guard cells and the adjacent cells. If the turgor of the

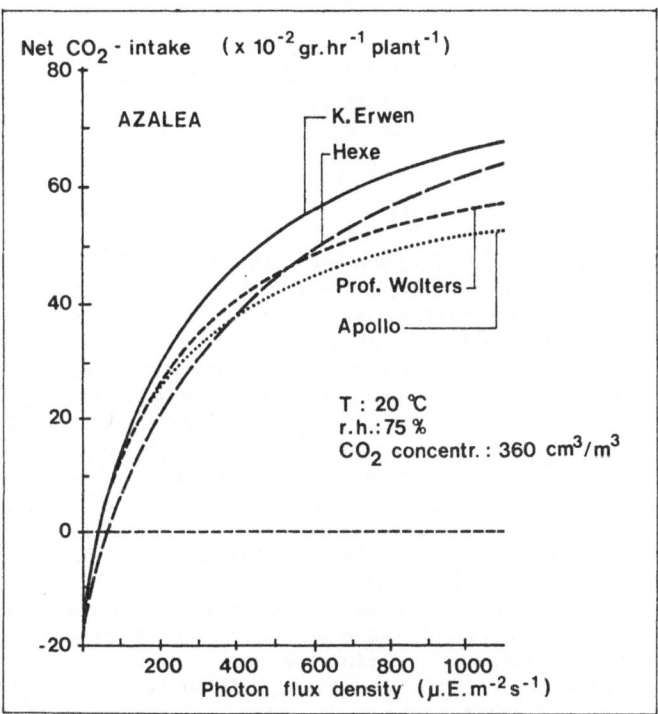

Fig. 2 : Net photosynthesis (CO_2-intake) of different types of Azalea-cultivars as a function of increasing light intensities (8).

guard cells becomes greater than that of the subsidiary cells the stomata open; when not under tension, they are closed. Increase in turgor is an osmoregulatory process associated with active transport of potassium ions from the adjacent cells into the guard cells.

Like carbon dioxide, water is used in the photosynthetic process, but it is not in this respect that water shortage can be a limiting factor, more important is the water necessary to maintain a high water potential in the protoplasm. The metabolic processes of the cell are critically dependent upon water in this sense.
The first effect of water deficiency upon vascular plants is on narrowing the stomata, which slows down CO_2 exchange. With increasing desiccation there is reduced hydration of the protoplasm in general, and thus reduced photosynthetic capacity. Normally CO_2 uptake is high only over a narrow range of the adequate water supply level; beyond this it begins to decline and eventually is entirely suspended. There are therefore two critical points in gas exchange versus water loss : the point of transition from full capacity to the limited region and the null point for gas exchange.

The first critical point comes at a level of water stress in which the stomata begin to close, causing the stomatal diffusion resistance to become greater than the residual resistance. If water is suppied after this first critical point has been passed, recovery is rapid.
The second critical point is determined by marked or complete closing of the stomata as well as by the direct effect of water shortage on the protoplasm.
Appreciable CO_2 uptake is no longer possible, though the CO_2 freed by

respiration can be bound again. Once this state has been reached, a renewed water supply does not lead to an immediate recovery of photosynthesis. Recovery is delayed, and after severe desiccation the original photosynthetic capacity may, under certain conditions, never be achieved again.

1.2.4. Nutrition

Nutrition concerns the influence of mineral nutrients on the carbon metabolism via synthesis of new tissue and growth. Direct effects upon photosynthesis and respiration result from the fact that the minerals either are incorporated in metabolites, enzymes and pigments or participate directly as activators of photosynthesis. In soils not seriously deficient in particular nutrients, the availability of minerals is less critical than the climatic factors. Nevertheless, it is almost always possible to enhance the yield of photosynthesis by the artificial provision of nutrients. In water cultures however, lack of minerals is very likely to be a limiting factor if not artificially added. Conversely, minerals in excess are also harmful; at too high concentrations certain minerals (heavy-metal ions in particular, as well as air pollutants) impair photosynthesis.

Manganese, for example, acts as an activator of photolysis, and potassium is involved in the electron-transport system. Nitrogen and magnesium are components of chlorophyll : various enzymes include iron, cobalt and copper, and phosphate is a component of nucleotides. The lack of minerals, as well as alterations in relative amounts of the elements taken up, can affect the chlorophyll content and the number, size and ultrastructure of the chloroplasts; this applies even if the elements in question, e.g. iron, are not themselves incorporated into the chlorophyll molecule. In conditions of nitrogen and iron deficiency, chloroses are observed, which cause a diminution of CO_2 uptake to less than 1/3. Lack of magnesium can have similar consequences. The chief result of insufficient chlorophyll is that the plants cannot make full use of intense light.

1.3. Parameter dependency and periodicity

The rate of absorption of radiant energy at various places on the earth's surface depends on their orientation with respect to the sun. This direction-dependence, because of the rotation and revolution of the earth, makes energy input a periodically varying environmental factor. It imposes dependency on periodicity upon all terrestrial phenomena. Particularly light and temperature show large seasonal and daily variations.

1.3.1. Light (13)

At low light intensity during the winter months, particularly in the northern countries of the EC, the resulting photosynthesis is low and growth possibilities are weak. Moreover, greenhouse structures cause an additional light loss, particularly by increasing the degree of insulation. A light reduction of 1 % causes a photosynthetic drop ranging between 0.6 % and 1.2 %. Generally speaking, foliage plants are far less sensitive to light reduction than are vegetables. This largely explains the expansion of foliage plants in Denmark.

Observation of single leaves could lead to the mistaken inference that

there tends to be a surplus of light. However, for the plant as a whole
and for stands of plants, this is not so. It is true that the individual
leaves of a plant are often arranged so as to favour interception of the
strongest average light. In the course of the day the leaves are struck
by light at many different angles. Moreover leaves shade one another.
In a stand of plants, the contribution of the various layers of foliage to
the overall photosynthetic yield is quite different, depending on the
arrangement and the amount of shading of one layer by another. In the
morning, the light compensation point in the top layer of the leaves
precedes by one to several hours this same phenomenon in the lower layer
of leaves.
It is important to note that young plants (relatively distant from each
other) do react quite differently to light reduction as compared to older
plants. This is clearly illustrated by a Dutch experiment on cucumber
plants. In Fig. 3, the vertical axis shows the percentage of growth
reduction per percent light reduction. The horizontal axis shows the
outside radiation conditions (winter-summer). During growing (average of
2.5 J per m^2 per day) 1 % light reduction results in 0.8 % growth
reduction for young cucumber plants. The reaction of growth to light
reduction is over 3 % in comparable winter circumstances for an older
cucumber crop (dry matter over 50 grammes per m^2).

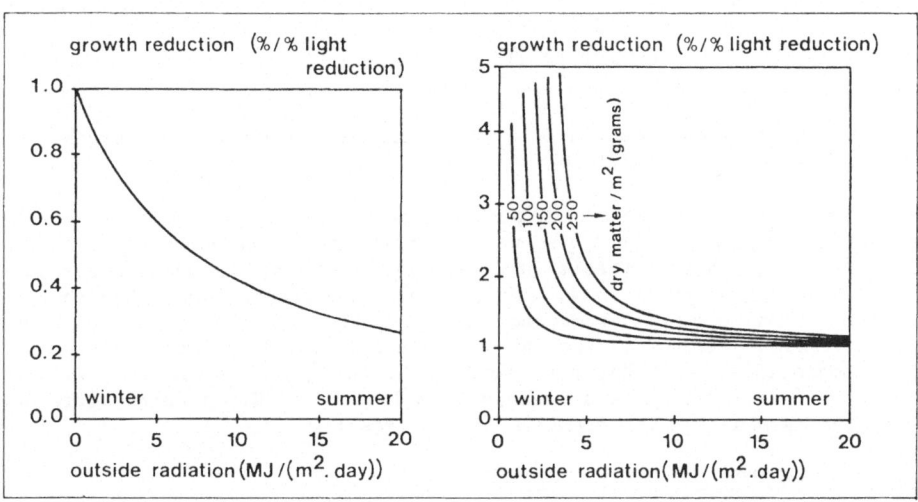

Fig. 3 : Young cucumber plants   Old cucumber plants

1.3.2. Temperature

Plants are poïkilothermal organisms, that is, their temperatures tend to
approach the temperature of their surroundings. Sufficient but not
excessive heat is a basic prerequisite for life. In the higher plants CO_2
uptake is blocked as soon as the assimilation organs begin to freeze.
Temperature affects metabolic processes by way of its influence on the
reaction kinetics of chemical events and on the effectiveness of the
various enzymes involved (secondary processes). A number of these
processes, e.g. respiration, rise exponentially with temperature
(Van't Hoff-reaction rate).
As temperature rises, the fixation and reduction of carbon dioxide by

means of photosynthesis also occurs with increasing speed, and this until a maximum value is reached : this rate is maintained over a broad range of temperatures. (11)

The energy available for plant growth results from the difference between photosynthesis and respiration (see Fig. 4).

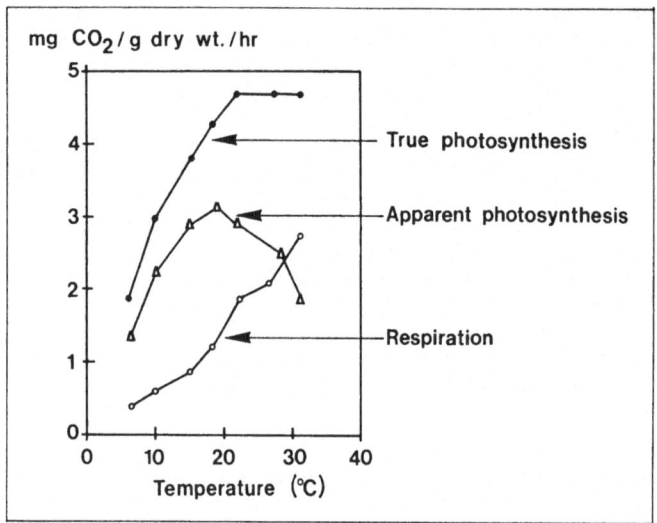

Fig. 4 : A classical example of temperature dependence of photosynthesis and respiration. (16)

The temperature dependence of photosynthesis can be illustrated by the number of bunches or buds per week, depending on the mean temperature. The rates of photosynthesis and respiration adapt to the temperature prevailing at a given time. Modulative temperature adaptation occurs within a few days, or sometimes a few hours, by shifts in substrate concentrations (hydro-culture) or other environmental circumstances.

For tomato plants, it has been found that the growth speed increases by about 50 % as temperature increases from 16°C to 21°C. A quite different reaction is observed for Bromelia plants (see Fig. 5).

1.4. Transparent covering materials

1.4.1. Greenhouse effect (3), (20), (27)

The advantage of greenhouses is the higher inside temperature : this occurs even when they are not artificially heated. Within the greenhouse, solar energy is transformed into radiation of a longer wavelength (infrared or heat radiation) the greater part of which is held inside by the greenhouse glass (or to a lesser degree plastic) cover. This is called the greenhouse effect.

A general solar and thermal energy balance for a greenhouse is illustrated by Figure 6 and involves the following factors : insolation, conduction heat loss, ventilation heat loss, evapotranspiration heat, heat loss to ground, thermal radiation, and thermal capacity of mass within the greenhouse.

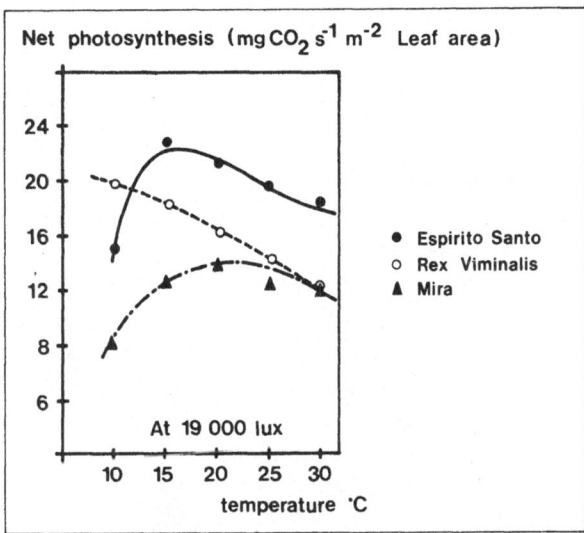

Fig. 5 : Net photosynthesis of different hybrids of VRIESEA (Bromelia) at
increasing temperatures (8).

The radiaton intensity inside a greenhouse is less than outside due to
reflection, absorption, and transmittance of the covering. The slope of
the covering and corresponding angle of incidence of the solar radiation
further affect the transmittance. Edlin and Willauer (1961) report only a
small decrease in transmittance for angles of incidence from 0 to 45
degrees but more rapid decreases for angles greater than 60 degrees.

Though portions of the roof or sidewalls in typical greenhouses may have
an angle of incidence to radiation greater than 45 degrees, it can be
generally assumed that the ground area within a south-facing greenhouse
with vertical walls and normal slopes is subject to a solar intensity
equal to the solar intensity outside the greenhouse multiplied by a
transmittance coefficient. This is particularly true during the hours of
maximum solar intensity.

In addition to the short-wave direct and diffuse radiation considered as
solar radiation, the atmosphere and objects radiate on the thermal, or
longwave, range. For glass-covered greenhouses, thermal radiation from
surfaces within the enclosure is not considered an important factor owing
to the virtual complete opaqueness of glass to radiation within these
wavelengths. However, many of the plastic materials do transmit varying
amounts of thermal radiation. Thus, heat losses from plastic covered
greenhouses on cold, clear nights can be significantly greater than by
conduction heat loss alone. The thermal radiation loss is appreciably
absorbed by a condensation film on the interior plastic surface.

An obvious way of energy saving is the maximal use of both the available
light and heat. For new structures this implies as much as possible an
east-west structure with a minimum of shadow-causing elements (e.g.
lowering of heat-tubes).

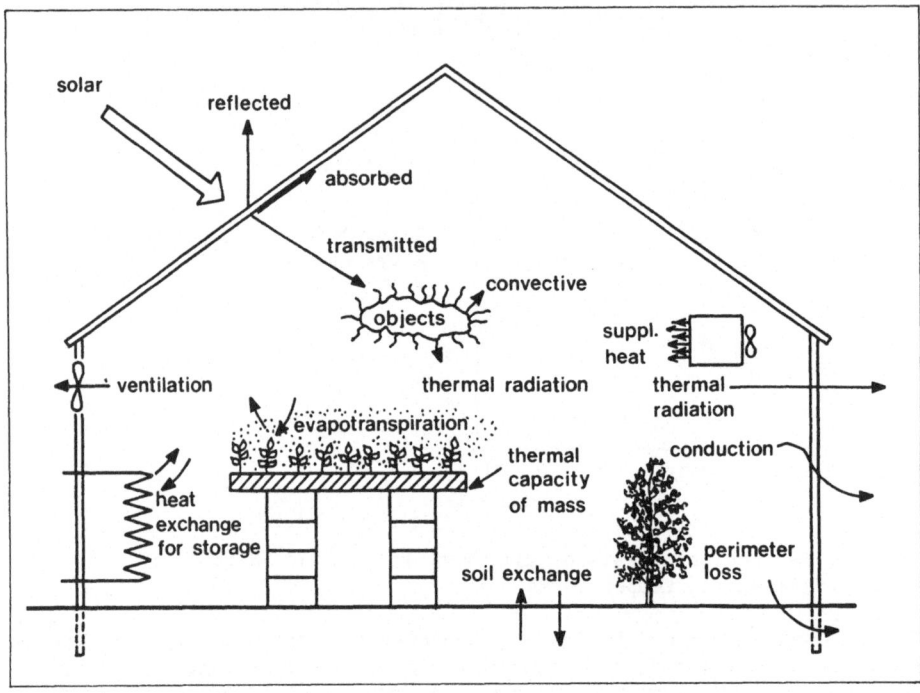

Fig. 6 : Scheme of primary energy flows in a greenhouse.

For existing greenhouses the maximal light transmission can be attained by carefully cleaning the cover. The glass (or cover) obscuration is relatively high, due to the chimmey emission and the overall increased pollution of the environment (particularly near industrial sites).

An additional option for energy savings consists in the choice of covering materials and screens which provide a maximal use of heat for a minimum of light reduction.

1.4.2. Covering materials (2), (5), (12), (14), (19)

Glass is commonly used in the northern EC-countries : its transmittance of solar radiation is extremely good. Moreover it is a stable and durable material. Coated glass is normal glass with a thin coating of metallic oxide (e.g. Hortiplus). This glass causes an additional light reduction of 10 % as compared to normal glass. However when the cover in question is dry an energy saving of 25 % is possible.
Plastic is available as a sheet or in panels. Polyethylene (PE) is most commonly used in thicknesses of 0.03 to 0.2 mm. PE shows a very weak greenhouse-effect. Polyvinylchloride (PVC), Ethylvinyl-aceton (EVA), Polyester, polymethylacryl (PMMA) and polycarbonate do offer a greenhouse-effect. For different covering materials, some values of light transmittance are given in table 2 and thermal heat loss coefficients in table 3.

The thermal heat loss coefficient of a number of twin covering materials is good as compared to double glazing.

Table 2 : Light transmittance of some covering materials (24)

	direct radiation %	diffuse radiation %
Single glazing	89-92	83
Double glazing	83-84	71
Acryl glazing (8-16 mm) ("Stegdoppelplatte")	83-84	75
Acryl glazing (32 mm) ("Stegdreidoppelplatte")	75	75
Polycarbonate	82-83	66-68
Hortiplus	79-81	68

Table 3 : Thermal heat loss coefficient (k) of some covering materials (24)

	$k(W/(m^2 °C))$
Single glazing	6.7
Double glazing	3.4
Double Acryl glazing (8-18 mm)	3.4
Triple Acryl glazing (32 mm)	2.3
Polycarbonate	4.5
Hortiplus	depending on r.h.

1.5. Screens

To further limit energy losses, at night times a film or fold is applied under the greenhouse cover. Some screens are also used as shade-blind to provide shade in periods of excessive insolation (particularly in the non-edible sector). A number of different materials are available.

1.5.1. Shadow effect - light reduction (17), (18)

Opening and closing of a thermal curtain at the beginning and end of each day is obviously based on a compromise between photosynthesis needs and energy conservation. During cloudy winter weather, the curtain may be open for only a few hours. During spring and autumn, the curtain may be open as long as it is closed.
A computer modelling study (13) concluded that photocell control is the preferred method. Sensing additional environmental parameters such as temperature did not improve the benefit sufficiently to justify a more complex system. The photocell method showed a significant benefit compared to a time switch even with frequent adjustments for changing day lengths. Even when fully open, the folding mechanism causes a shadow effect, which can be 5-15 % in comparison with a greenhouse with single glass.

1.5.2. Screen materials

Aluminized screens are reported to offer the highest energy savings, due to the reflective capability. Potential savings up to over 50 % are reported. Maximum savings of 30 to 40 % are more common in the industry. (see Table 4)

Table 4 : Energy savings with different kinds of thermal screens
(Source : A.F.M.E., Economie d'énergie dans les serres, 1982)

Screen type	Energy savings (%)	
	Calm weather	Strong wind
Plastic films		
Black P.E.	32	41
translucent P.E.	33	42
PVC	34	45
Aluminized screens		
P.E./alu/P.E.	54	63
Aluminized polyester	55	-
Aluminized tyvek	51	-
Woven or non-woven fabrics		
Reenay 2016	31	46
Fibuter	30	42
Tyvek (impervious)	57	-

Single layer, deployable thermal screens have been available to the greenhouse industry for several years.
Although a partial success, they have four troublesome problems associated with them. First, edge and end seals are often not complete. Warm, humid air is permitted to circulate from below the curtain to above which is both an energy loss and a source of serious amounts of condensation. Another problem arises from the relatively modest thermal insulation value of single layer fabrics. The underside of the curtain is thereby significantly cooler than the air beneath the curtain. Condensation on the fabric can occur, with water dripping on the plants, promoting disease or other quality loss problems. Retraction of the curtain into accordion folds promotes fabric wear and creates a region (albeit small) of shade during the day. A final problem is also of a mechanical nature. Usual deployment systems suspend the curtain from rollers or hangers which impose point loads on the fabric. The point loads and resulting stress concentrations accelerate fabric break-down.

Data of fuel savings, light reduction and expected yield losses of different energy saving options as compared to single glass are given in table 5.

1.6. Effect of airtightness

1.6.1. Humidity

The degree of turgor in the stomata of plants is influenced by the relative humidity of the surroundings. When greenhouses are made more airtight and the temperatures inside the greenhouses are lowered, the relative humidity increases. CO_2 uptake declines when the relative humidity reaches (very) high values.

The sensitivity of CO_2 exchange to high r.h. is to a large extent characteristic for each plant species. Lettuce for example withstands higher r.h. than cucumber or tomato-plants.

Table 5 : Basic data for economic comparison of energy saving measures in
The Netherlands (23)

	(comparison with a greenhouse with single glass)		
	Single screen	Double glazing	Twin covering -plastic materials-
- Fuel savings in % of total yearly fuel consumption	20	30	44
- Light reduction	5	12	12.5
- Resulting yield losses in light-sensitive crops (vegetables)	6	14.4	15
in less light-sensitive crops (potted plants)	3	7.2	7.5

Secondary effects of high r.h. are also extremely important. A sudden
increased turgor in cucumber and tomato-plants results in a cracking of
the fruit-skin and hence deterioration of the commercial value.

In lettuce, too high r.h. causes a complete film over the stomata and
results in so called waxiness.

The high r.h. and low transpiration causes a weak transport of CO_2,
resulting in a weak cell structure.
Increased ventilation and/or temperature increase can avoid these problems
(energy demand + energy loss).

1.6.2. CO_2 content

If one assumes that when photosynthesis is proceeding to completion, the
CO_2 in the chloroplasts is used up, and that during respiration in the
mitochondria the O_2 concentration there falls to zero, then the
concentration gradients of these two gases are determined by their
concentrations in the surroundings of the plant.

Carbon dioxide amounts to about 0.03 % of the atmosphere by volume. It is
readily soluble in water and the proportion of free CO_2 depends on the pH
of water, being high in the acid region. Above pH 9 only hydrogen
carbonates and carbonate ions are present : but these represent also CO_2
reserves for the plant. The used CO_2 is replenished under natural
conditions by movement along a slight concentration gradient. Inadequate
CO_2 supply is often a yield limiting factor for plants.

Energy savings may be achieved by reducing the losses associated with
ventilation. However, the reduced natural ventilation and the consumption
of CO_2 during the day time causes a reduction of the CO_2 concentration
inside the greenhouse (cf. fig. 7)

During the night (respiration) a certain amount of CO_2 is liberated, but
this soon appears insufficient for plant growth during the day.

Therefore in the more airtight greenhouses CO_2 supply is necessary in
order to maintain the same level of plant growth as in the other
greenhouses. (see Fig. 8 for the case of Azalea cultivars)

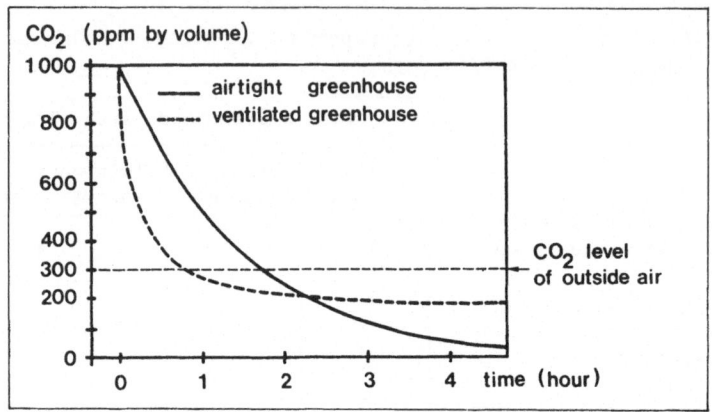

Fig. 7 : CO_2 concentration in an airtight greenhouse, compared with the CO_2 concentration in a well ventilated greenhouse.

When the CO_2 content of the air is artificially raised to 1 000-3 000 ppm by volume, plants are able to bind 2-3 times as much CO_2 as under natural conditions.
Research has indicated that a good growth is realised at a CO_2 concentration of 1 000 ppm. At increased concentrations up to 2 000 ppm growth increases by approximately 20-30 %. In tomatoes, a CO_2 increase from 300 ppm to 900 ppm resulted in a growth acceleration by 120 %. The possibility of raising the yield of photosynthesis by increasing the CO_2 concentration (CO_2 fertilization) is often applied by growers in greenhouses. By raising the plants in air containing 1 000 ppm CO_2 (by volume), the growth of tomatoes, cucumbers and leafy vegetables could be doubled. (4), (11)

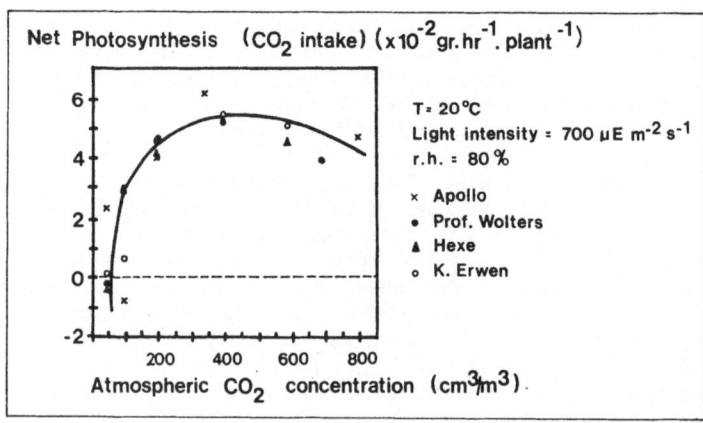

Fig. 8 : Effect of CO_2 concentration on photosynthesis of Azalea cultivars (8)

1.7. Cultivation techniques

1.7.1. Traditional techniques

The total greenhouse area showed a strong growth during the period 1950-1973. This expansion took place in the northern countries of the

European Community and was based on the availability of relatively cheap energy. A geographical survey shows that the glasshouse industry is mainly concentrated at short distances from the main oil importing ports (Rotterdam, Antwerp) : "Solar energy" was imported in the form of relatively cheap fossil energy.

Fuel costs for heated crops in the northern part of the EC have increased rapidly. The southern EC member countries and especially Spain have mild winter climates which allow for the profitable use of plastic greenhouses.

This relatively cheap production technique provides a better protection against low night temperatures, often a slight greenhouse effect during the day, it provides a good shelter against wind and rain, and consequently the quality of the products is improved. However, night frost could still cause serious damage as the outside temperature often drops below the physiologically supportable level of the plants. The products grown under plastic shelters have been readily accepted in different international markets. In France, one can observe the transition from glass greenhouses of the northern countries to the plastic greenhouses of the mediterranean area. In the mediterranean countries, protected culture is only possible during the winter season, generally from November to April. During the summer, the greenhouses cannot be cultivated because of the excessively high temperatures.

1.7.2. Nutrient Film Technique (NFT)

The continual culture of the same plants for several years in greenhouses causes accumulation of disease potential in the soil and results in a large number of primary and secondary disease infections.
Therefore, there is a permanent need for soil disinfection. But this technique is expensive (steam) and/or increases the residue-level of toxic elements in the final products (e.g. Bromides)
It is now possible to grow almost all commercial plants in water. In some cases an inert substrate (e.g. rockwool) is used for better fixing of the roots. Commonly, no substrate is used. This new technique is called Nutrient Film Technique (NFT).

1.7.3. Root-zone warming (RZW)

Research has indicated that a substantial energy saving could be achieved with tomato plants by allowing night temperatures to go down even to below 13°C. In order to keep the plants sufficiently active a careful regulation of the root temperature as a function of the air temperature is needed. The Nutrient Film Technique (NFT) allows for simple manipulation of the temperature of the water film in which the roots are bedded (26).

Thus NFT makes it possible to lower the night temperature substantially (energy saving) without affecting the vital activity of the plant very much. Moreover, the size of the fruit is often improved.

Although low night temperatures are possible for cultures on soil, the working limits are tighter (only part of the night cold; so called split-nights), and production lag times greater (cf. fig. 9).

It should be noted that the soil which has cooled overnight heats up slowly during the day.

This obviously entails a reduced root activity, whereas the relatively high day-time air temperature calls for intensive shoot activity and active water transpiration.
In extreme cases this disequilibrium can greatly weaken the plant.

Inversely, excessive root heating during cold nights by means of RZW may cause fall of flowers (e.g. freesias) or cracking of fruit (tomatoes).

2. ENERGY CONSERVATION

A brief overview (six approaches) of some energy conservation measures (other than the solar options described in section 3, p. 27) is given below. A distinction is made between biological and technical methods.

2.1. Biological methods

Research and development activities for greenhouse production have focused major attention on energy saving measures.
A first step was the increase of production by means of a higher production per unit of area and hence, a higher output per unit of energy. There are a number of indications that there is still potential for a further rise in crop production (micro-electronic control, cultivation methods).

A second step is the lowering of the heat demand. This implies the introduction and selection of less energy demanding varieties. This is a very promising and profitable approach. It is the task of research institutes and breeders to create cultivars which can be grown at lower temperatures and/or with lower light intensities, without incurring an economic loss of yield and quality (6).

For greenhouses in agro-climatic zone III an inside temperature reduction from initially 15°C to 14°C (Δ1K) would result in an energy saving of the order of 13 % of the originally required energy. The reduction by 1K of initial inside temperatures of 20°C and 25°C would in turn result in energy savings of 9 % and 6 % respectively.

The economic objectives are met if lower plant-temperatures are made possible without change of the time of planting. This is one of the objectives of the plant-geneticists in the creation of alternatives requiring less heat and possibly lower light intensity. The reduced interference with other growth parameters further increases the attractiveness of this solution.

2.2. Technical methods

A third option consists in the reduction of the heat losses. This option has an attractively wide field of research and experimental activities. A number of measures are discussed in sections 1.4 and 1.5 on pp. 15-19 as well as in section 3 on p. 27.

Improving the efficiency of the existing heating installation and heat transfer is a fourth general approach to energy saving.
A correctly functioning and efficient heat source is evidently a principal prerequisite. Significant energy saving has already been achieved in recent years.

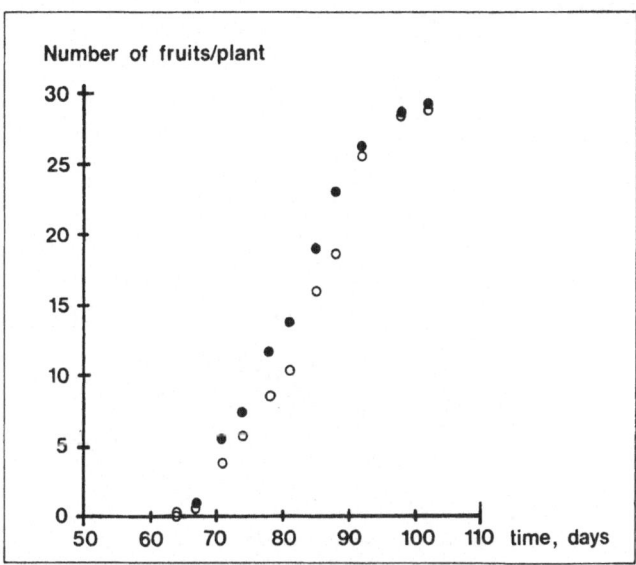

Fig. 9 : The average number of fruits of control (●) and split-night (○)
tomato plants grown in pots with soil. This figure shows that
control and split-night plants set fruit at the same rate, but
the split-night plants lagged several days : this lag became
smaller as plant development progressed.
(Source : Connecticut Agricultural Experiment Station, Bulletin
781; 11/1979)

An important advance in efficiency was achieved through the use of flue
gas condensers. In an optimal situation (natural gas), a saving of 16 %
can be obtained (Netherlands). An important prerequisite for the
introduction of flue gas condensers is the possibility of using low
temperature energy (30°-40°C). With a number of crops this can be
achieved e.g. by means of applying the root zone warming technique or by
the application of soil-heating (the so called fifth pipe) for which
however an additional investment is required.

In principle the use of total energy systems also offers an improvement of
the energy balance. The application of this system is however often
limited to exploitations with a relatively high consumption of electricity
for lighting - or cooling - purposes.

The utilization of a heat pump is another way in which energy can be
saved. As indicated in figure 10 an electrical heat pump can save nearly
30 % of primary energy when operating at a coefficient of performance
(cop) of 3.

Application of a heat pump does however require an investment of 3 to 5
times the cost of a conventional heating system. The rate at which the
reduction in fuel costs compensates for the extra investment costs is a
function of a combination of the level of the coefficient of performance
and the number of operating hours per year.

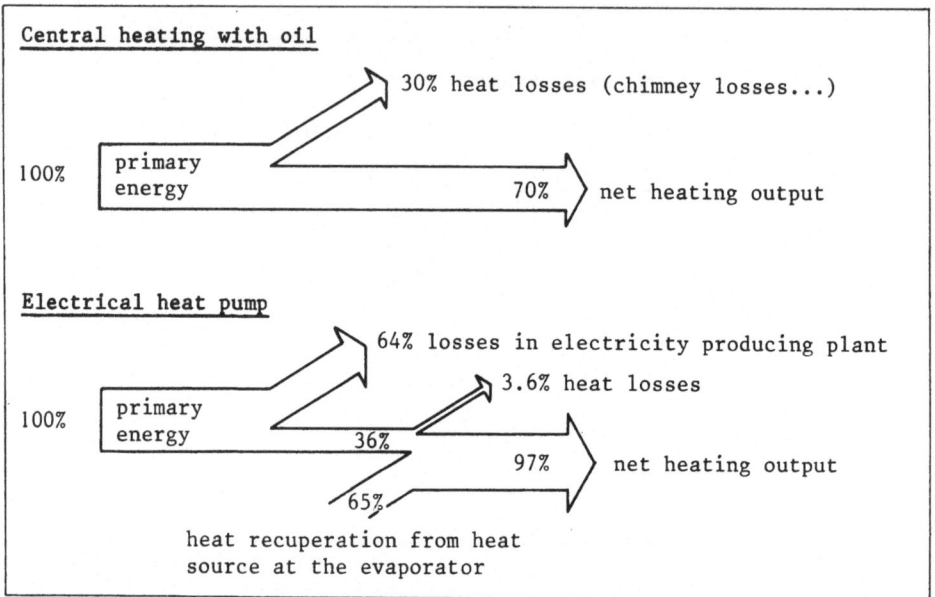

Fig. 10 : Comparison of energy flows in a conventional heating set-up and
an electrical heat pump with cop = 3.

Measurements in actual greenhouse heat pump set-ups in Belgium indicate
that at present and for the number of operating hours reached the obtained
cop is most of the time below the level required for economic feasibility.
The situation might be different for well thought out systems involving
large capacity heat pumps operating at least 4 000 to 4 500 hours per
year. For this purpose, the incorporation of the maximum number of
meaningful heat pump operation applications is a must. Examples are
drying of the moist greenhouse air to prevent condensation while
minimizing ventilation. This also involves the conversion of latent
(partly solar) energy into useful sensible heat. In many cases however,
operating hours can only be increased by the addition of costly and
complex control and buffering systems...

Where natural water bodies are used as a heat source, large quantities of
water at a relatively high temperature level need to be available in order
to obtain economically efficient installations.
In the case of ground water, the regulations on discharge of the used
water often constitute a prohibitive obstruction to its application as a
heat source.

The use of energy derived from industrial waste heat is the sixth general
approach to energy saving.
The heat discharge from industrial processes is in principle available
from two sources : (1) Cooling waters at a temperature level of about
15°-30°C or (2) Industrial waste water discharged at high temperatures,
e.g. 80°-90°C.

Cooling waters are available in huge quantities at electrical power
stations. Given certain internal adaptations within the power station
this heat source could be made available to potential users.

In this context several 5-30 ha large greenhouse projects using low temperature heating devices already exist in the neighbourhood of such power stations.

The use of heat in hot industrial waste waters and the combined production of heat and energy (e.g. total energy) needs a lot of technical work and also implies socio-economic consideration.

A lot of work has already been and still is being accomplished in this general area of energy savings. E.g. : for the Netherlands, per unit of production, an energy saving of 34 % has already been achieved between 1973 and 1980 and for the period 1980-1990 an additional energy saving of 50 % is expected (see Table 6) (6).

3. SOLAR ENERGY COLLECTION AND STORAGE

3.1. Short-time storage (typically day-night storage) with greenhouse as collector

3.1.1. Thermal mass

The thermal mass of the greenhouse is a central element in solar heating systems. However in greenhouses operated for commercial production, too much thermal mass is perhaps even more detrimental than no thermal mass at all. The blueprint for temperature control requires that the daytime temperature be reached early in the morning, long before there is excess solar heat. Thus much heating of the mass may be done by the conventional heating system.

Further work is required before a truly suitable thermal mass augmentation for commercial greenhouses is devised. Water mats have provided some benefit, but were found to be inconvenient.

3.1.2. Active short-term storage

Thermal mass may be considered as a passive means of short term storage of solar energy collected by the greenhouse. An active, more adjustable means consists in blowing hot greenhouse air through a heat storage at specific times. Thus, during periods of excessive solar radiation, the storage would cool the air; conversely at night, or at times with low insolation, the storage would warm up the air. As opposed to the case of thermal mass, here the storage medium forms an independent unit from the greenhouse. Consequently the heating system will not experience the abovementioned extra load when the greenhouse must be warmed from night to day time temperature. Different means of storage exist.
Soil is often available in great quantities below the greenhouses and it is logical to store heat in this natural reservoir. However, the soil heat capacity is relatively low and proportional to its moisture content. Evaporation losses are likewise not negligible. Therefore soil is not appropriate. The air-to-rock storage method is another simple approach but again, limited in performance capabilities due to small allowable air temperature differences. The air can circulate by natural or by forced convection through the empty spaces between the rocks. The warm air is best taken from the upper spaces of the greenhouse where the hot air stratifies.

Table 6 : Energy saving figures for The Netherlands. (6)

Measures	Obtained energy saving 1973–1980	Expected saving 1980–1990
1. Technical		
− Heating installation adaptation and maintenance retarders and flue gas	3 %	
condensers	5 %	8 %
− Greenhouse adaptations	2 %	8 %
insulation	2 %	3 %
thermal screen	1 %	15 %
2. Biological		
− Crop techniques existing crops	5 %	15 %
selection		10 %
3. Cumulative economy	17 %	45 %
4. Production increase per m^2	25 %	10 %
5. Economy per m^2	34 %	50 %

The circulation starts when the air temperature exceeds the desired level. Heat is extracted from the rock bed at night in the same manner. The rock beds may be located underneath the greenhouse, adjacent to it or as part of the greenhouse facility (built under benches, in the North wall, etc.). (Staged beds are sometimes proposed so as to more effectively use cooler air before and after the maximum air temperature is obtained.)

The air-to-water technique is likewise limited to small temperature differences. However it offers more versatility in flow control and storage but at a greater cost for heat exchangers and other associated equipment. Water systems are capable of approximately twice as much thermal energy storage per unit of volume as rock. Heat storage in water furthermore makes a solar contribution to root zone warming possible.

The crystallization of salt solutions at a given temperature level releases a large quantity of energy. The phase change of this solution (from solid to liquid) permits an energy storage of about 100 times more than the same volume of water. The first material used in greenhouses was paraffin (high fusion temperature). New phase change materials are being developed (cf. Solvay, Dow) and are investigated as to their potential in greenhouse climatisation (cf. also Basic Description no.14 in chapter 2, p. 52).

The phase change temperature of the salt solution should be sharply defined and should fall just above the daytime thermostat set point, and well below the ventilation set point.

3.2. Collectors other than the greenhouse itself

Flat plate collectors and concentrating collectors can produce higher temperatures than that of the greenhouse air. Hence the heat storage capacity for one and the same energy store is increased.

If the collector is placed within a greenhouse, interception of incoming solar radiation generally interferes with use of the same solar energy for normal plant growth or natural heating. Internal collectors cannot generally be located and oriented for optimum reception of insolation (facing south, tilted at an angle equal to the latitude plus 10 to 15 %) without obstructing valuable plant growing area and wasting expensive greenhouse space. Only the interior side of a north wall offers potential for a flat plate collector.

A collector placed outside the greenhouse has the advantage of not modifying the luminosity inside the greenhouse, as well as of not requiring adaptations within the greenhouse itself.

However for northern EC countries the application of collectors for greenhouse purposes does not have a very bright future because :
- the required area of collectors is very large;
- land value is very high and should preferably be devoted to cultivation rather than heating equipment;
- investments are still quite high.
For southern EC countries, more interest is shown :
- if very cheap collectors can be found;
- for non-heated greenhouses as a protection against frost in the winter;
- if enough space is available which is devoted to less intensive agricultural utilization.

Experiments in Greece have shown that the introduction of such equipment (1) gave a 30 to 50 % yield increase for tomatoes due to a temperature 6 to 8K higher than in traditional greenhouses.

3.3 Long-term storage

Given the excess of solar heat input in summer versus the shortage in winter, great interest has been shown all over the world in the collection (by means of the greenhouse itself or by means of another solar collector) and storage of excess heat over one to several weeks up to a whole season.

Owing to the significant heat losses, large storage volumes are necessary in order to ensure seasonal storage. Consequently, extremely high costs are to be expected for seasonal storage.

References

(1) Areskough M.: "Preparatory experiments on seasonal storage of heat energy in protected cultivation". Acta horticulturae, N° 115, 1981.

(2) Breuer J.J.G. en A.M.G. Van Den Kieboom: "Meer onderzoek naar hortiplus". Tuinderij. 20 (1980), 21: 34 - 35.

(3) Cesar G.: "Un exemple de climatisation sous serres plastiques". Revue horticole suisse". 42 (1969) 12: 357 - 364.

(4) Ceulemans R. en O.Y. Meckers: "De Nieuw-Zeeland Laurier: Gasuitwisseling bij verschillende temperaturen en belichtingsintensiteiten". De Belgische Tuinbouw. 1982, nr. 63.

(5) Gac A. et R. Bartoli: "Utilisation d'écrans réflecteurs pour réduire les dépenses de chauffage dans les serres". Comptes rendus de séances de l'Académie d'agriculture de France. 1974, <u>60</u>, 12: 952 - 958.

(6) Germing G. en D. Meijaard: "Vooruitzichten voor een betere energiebenutting in de glastuinbouw". Tuinbouwdagen, Wageningen. 1979.

(7) Harnette R.F.: "Rigid plastic still not a serious challenge to glass". Grower <u>78</u> (1972) 16: 786 - 788.

(8) Ceulemans R. en Impens I.: "Verdere informatie betreffende de fotosynthetische CO_2-opname en de groei van verschillende sier- en tuingewassen". Landbouwtijdschrift. 1983, vol. 36, juli-augustus, p. 1105 - 1124.

(9) Johnson C.B.: "Physiological processes limiting plant productivity". Butterworths. 1981, 390 pp.

(10) Klapwijk D.: "Kasklimaat, plantengroei en groeibeheersing onder glas". Elsevier, Amsterdam. 1971: 126 blz.

(11) Larcher W.: "Physiological plant ecology". Springer Verlag. 1981, 300 p.

(12) Mackroth K.: "Die Eigenschaften von Glas und Kunststoffen im Hinblick auf die Pflanzenkultur in klimatisierten Raümen". Gartenwelt. <u>71</u> (1971) 11: 256 - 259.

(13) Manbeck H.B. and R.A. Aldrich: "Analytical determination of direct visible solar energy transmitted by rigid plastic greenhouses". Special edition. Transactions of the ASAE. <u>10</u> (1967) 4: 564 - 567.

(14) Martin P., Gent N., Thorne J. and E. Aylor: "Split night temperatures in a greenhouse: the effects on the physiology and growth of plants". The Connecticut Agricultural Experiment Station, Bulletin N° 781. 1979.

(15) Odum E.: "Principles of plant ecology". Springer Verlag, 1979, 400 p.

(16) Stålfelt M.G.: "Der Gasaustausch der Moose". Planta 27. 30 - 60 (1937).

(17) Van Den Berg G.A.: "Het energiescherm in de glastuinbouw". Bedrijfsontwikkeling. <u>9</u> (1978) 3: 293 - 297.

(18) Van Den Berg G.A. en A.M.G. Van Den Kieboom: "Energiescherm in glastuinbouw". Vakblad voor de bloemisterij. <u>33</u> (1978) 25: 28, 29 en 31.

(19) Van den Kieboom A.M.G.: "Lichtdoorlatendheid van energiebesparende kasdekken". Vakblad voor de bloemisterij. <u>35</u> (1980) 14: 34 - 35.

31

(20) Van Huyzenbergh E.: "A history of greenhouses". The Institute of
     Agricultural Engineering, Wageningen. 1980.

(21) Vegter B.: "Kan zon een kas verwarmen". Vakblad voor de
     bloemisterij. 34 (1979) 41: 50, 51 and 53.

(22) Vegter B.: "75 % energiebesparing met: drievoudige bedekking,
     energiescherm, warmtepomp, warmteopslag?". Vakblad voor de
     bloemisterij. 35 (1980) 40: 42 - 45.

(23) Verhaegh: "Economische vergelijking van energiebesparende
     maatregelen". Landbouweconomisch Instituut, Den Haag. 1982.

(24) Von Zabeltitz: "Eigenschappen van bedekkingsmaterialen".
     K.VIV.-studiedag Steenokkerzeel. 1983.

(25) Von Zabeltitz: "Möglichkeiten zur Nutzung der Sonnenenergie für
     Gewächshäusern". Gartenbauwissenschaft, 41 (5), 1976.

(26) White J.W.: "Energy efficient growing structures for controlled
     environment agriculture". In: Horticultural reviews 1. 141 - 171
     (158 ref.). 1979.

(27) Zandbelt A.J.: "Kas als zonnecollector". Vakblad voor de
     bloemisterij. 34 (1979) 3: 24 - 27.

CHAPTER II :
Present research within the EC on the application of solar energy in greenhouses

The main objective of this chapter is to offer the reader a number of examples showing the incorporation of solar energy in greenhouses.

The first part of this chapter offers a discussion of the main criteria involved in projects with considerable potential for integration of solar energy in greenhouses. Furthermore a general classification of the projects is given, according to the discussion in the first chapter.

The second part consists of basic descriptions of research within the E.C. (their format is explained in the "Addendum", p. 467).

The third part gives the general conclusions based on the different projects. This discussion is ordered according to the fundamental criteria with respect to solar demand, solar supply, efficiency, economy and socio-economic acceptability. This third section summarizes the core message of this volume on the applicability of solar energy in the greenhouse sector.

1. CRITERIA (cf. p. 1)

The energy demand (first criterion) is generally high enough in greenhouses to justify the development of solar energy systems. In northern countries (Denmark, Germany, United Kingdom, the Netherlands) more than 75 % of the thermal energy consumption in agriculture is devoted to greenhouse heating (cf. p. 3). For The Netherlands, representing nearly 40 % of the total EC thermal energy consumption in agriculture, approximately 95 % is devoted to greenhouse heating. On the other hand southern EC countries (France, Italy, Greece), which benefit from sunnier climates, use less than 1/3 of their energy for greenhouses.

The energy supply in greenhouses is mainly characterised by the greenhouse effect. Obviously a number of measures are possible in order to take better advantage of the given energy supply (fossil or solar). This is the field of the passive solar energy systems: double glazing, thermal screen, thermal mass and greenhouse design. The possibilities of active solar systems are clearly more important in climates with a higher level of solar energy (agro-climatic zones 1 and 2, cf. p. 5). It follows that the possibilities of active solar systems are rather limited in the northern regions (agro-climatic zones 3 and 4) because of the lack of solar radiation at the time (winter, spring) when this energy could most profitably be used.

The efficiency of passive and active systems is a matter of great concern. In general, the thermal mass of a greenhouse is (very) low and the peak heat requirements are extremely high. This is especially important at the time of sunrise, when evidently the possible active solar energy supply in winter is nil. Therefore a number of alternatives are developed. One research approach consists of the use of energy at lower temperatures. This may be achieved by the nutrient film technique, in which the soil is replaced by a water substrate. Another interesting route is the use of heat exchangers for low temperature heating .

The economy of the balance between energy saving and production losses is an important consideration. The earliness of the crop is a prime criterion for economic profitability. The production loss due to light reduction in the greenhouses (thermal screens, double glazing) has to be balanced against the difference of the investment cost and the energy savings. In southern countries, a simple set up (e.g. plastic house without heating) is a rather cheap way of taking full advantage of the solar energy. The demand for products outside the traditional growing season explains the rapid growth of this production system in agro-climatic zone 1.

The socio-economic and technico-economic considerations are of minor importance in the evaluation of solar energy applications in greenhouses. Particularly in the northern countries, a positive attitude of greenhouse growers towards new technologies is observed.

2. OVERVIEW OF PROJECTS ON SOLAR ENERGY IN GREENHOUSES

2.1. Classification

Classification according to one criterion is hardly possible for the large number of projects included. In order to facilitate reference to the project descriptions, a broad classification is offered relating to the main research objective of each project (see page 35). A distinction is made between solar energy conservation projects and solar energy storage. The basic descriptions given in this chapter have been classified as follows:

- Solar energy conservation
  - Biological methods
  - Covering materials
  - Thermal screens

- Solar energy storage
  - Energy storage with greenhouse as solar collector
  - Energy storage with collector other than greenhouse itself.

The synoptic table 1 also shows the location of projects by country, the type of product (vegetables or flowers/fruit), and indicates whether researches conducted within a project whose major focus is scientific, or is conducted within an operational set-up.

Table 1 : Classification of the basic descriptions of projects
(The badges at the heads of columns are printed with each description, as oppropriate, to provide a visual index.)

Basic description number	page	Biological methods	covering materials	thermal screens	energy storage of heat collected by greenhouse — water	stone/soil	energy storage of heat collected by independent solar collector — water	stone/soil	Country	vegetables	flowers/fruit	major project focus = scientific	operational set-up	title of the project
1	38	●							Eir	●		●		Greenhouse energy saving through root zone warming with reduced night air temperature
2	39		●						B	●		●		Evaluation of different covering materials for greenhouse production of strawberries
3	40		●						Dk		●	●		The biological effects of energy savings in greenhouses
4	40		●						Eir	●		●		Comparison of double covered and single covered polythene greenhouse
5	41		●						Eir	●		●		Evaluation of double covered polythene greenhouses for early tomato production
6	41		●						Nl	●			●	Denar demonstration project
7	43		●						UK	●		●		A comparison of the light transmission of greenhouses covered with twinwalled sheets of polycarbonate and acrylic
8	43			●					D			●		Energy saving with mobile thermal screens
9	45			●					Eir		●	●		Estimation of the heat saved by using a removable internal lining in a glasshouse
10	46			●					Nl	●			●	Influence of a semi-movable screen on the energy consumption and the growth of tomatoes
11	47			●					UK	●		●		Evaluation of thermal screen in glasshouses on commercial nurseries
12	49				●				D			●		Greenhouse heating with solar energy
13	51					●			Gr	●		●	●	Development of a solar system for heating greenhouses
14	52						●		F	●	●	●		Solar energy storage in calcium chloride
15	54						●		F	●			●	A solar energy greenhouse (special case)
16	55						●		Gr	●		●	●	Greenhouse climatization by an earth-air heat exchanger
17	55							●	B			●	●	Heating of a greenhouse
18	56							●	D			●		Flexible plastic solar collector for low temperature heating

(Table 1, continued)

Basic description number	page	energy conservation			solar energy storage				Country	product		major project focus = scientific	operational set-up	title of the project
		Biological methods	covering materials	thermal screens	water — energy storage of heat collected by greenhouse	stone/soil	water — energy storage of heat collected by independent solar collector	stone/soil		vegetables	flowers/fruit			
		BIO												
19	58				•				D				•	Solar collectors and heat pumps for greenhouse heating
20	59				•				F	•		•		Solar energy greenhouse with selective glass filter
21	60				•				F		•		•	Heating of a greenhouse with solar energy
22	61				•				Gr	•		•	•	Heating a greenhouse with solar energy
23	63				•				I			•	•	Solar energy use for greenhouse heating
24	64				•				I			•	•	Greenhouse heating by solar collectors
25	64						•		B		•		•	Solar energy in grape-type greenhouses
26	65						•		D				•	Earth storage of thermal energy for greenhouse heating
27	66						•		F		•		•	Active and passive solar greenhouse climatization
28	67						•		I			•	•	Greenhouse heating by solar collectors

2.2. Location of the projects described

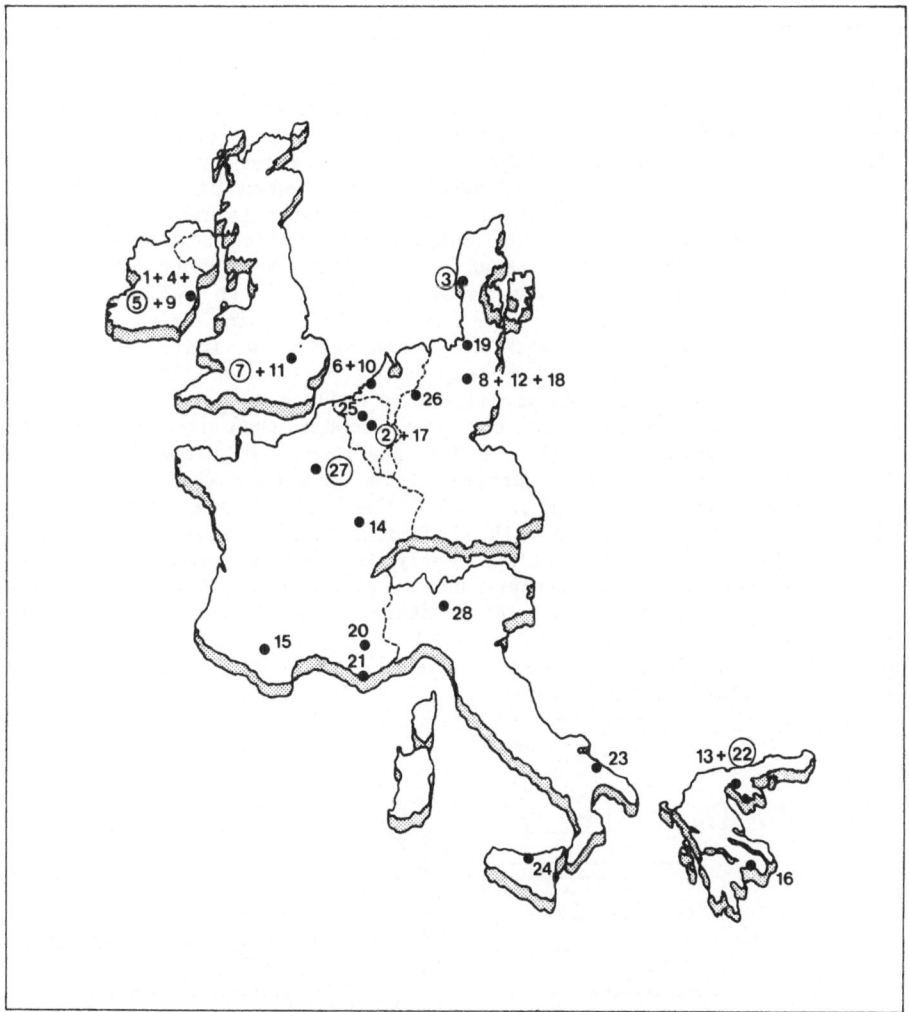

Fig. 1 : Location of the projects described
(The circled numbers refer to projects described in more detail
in the appendix to part I.)

2.3. Basic Descriptions

1.Greenhouse energy saving through root zone warming with reduced night air temperature

---

BIO

+ Project in Dublin - Ireland

+ Description :
Experimental work is in progress at Kinsealy Research Center in which tomato crops are being grown in a NFT (nutrient film technique) system with root zone warming and reduced night air temperature. In the NFT system, plants are grown in a flowing nutrient solution, i.e. water in which all plant nutrients are dissolved. The water is circulated to the plants from a tank in which the nutrients are added and in which the water can conveniently be heated. If plants are grown at reduced temperature without root zone warming, the loss in returns due to reduced yield far outweighs the saving in heating cost. However evidence from experimental work in recent years has indicated that if plant roots are maintained at a temperature above, say, 16°C, then the air temperature at night can be allowed to fall to as low as perhaps 5°C with much less ill-effect on crop development than if RZW were not employed. A single span 30 m x 9 m glasshouse at Kinsealy Research Centre is equipped with a nutrient film technique (NFT) system for tomato production. (Cf. fig. 1.1) The circulation tank for the nutrient solution is fitted with electrical heating elements, enabling the solution to be heated independently of the main heating system of the glasshouse, which is a piped hot-water system. Thus any selected combination of root and air temperatures can be maintained in the glasshouse, within the limits of capacity of the two heating systems.
An adjacent identical glasshouse is used for production of a tomato crop in peat at normal growing temperatures for comparison.

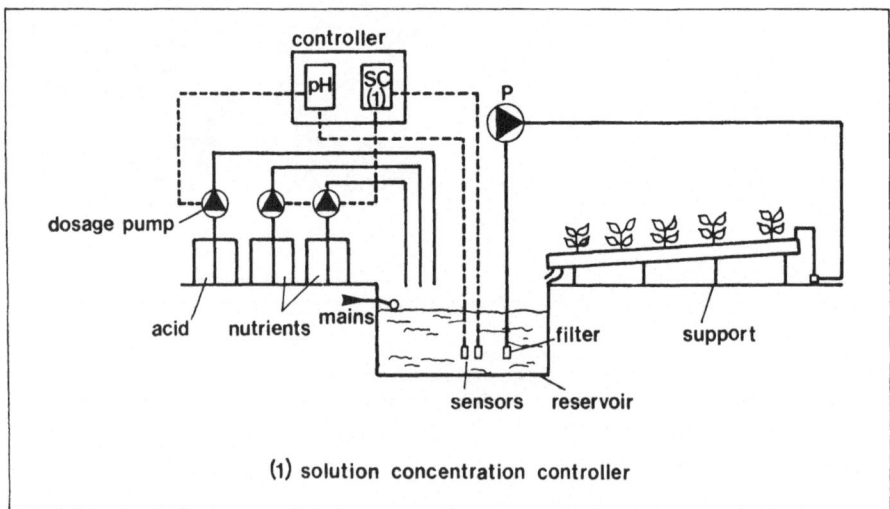

Fig. 1.1 : Scheme of the installation for the nutrient film technique.

+ Conclusions :
The use of a night temperature as low as 5°C results in substantial oil savings. On the basis of Irish meteorological records, these have been estimated as 13 litre/m^2 for a typical early tomato crop, or approximately 30% of normal oil consumption. Either in conjunction with NFT, or possibly through the use of other methods of root zone warming where a crop is grown in peat or soil, the technique is applicable to all existing glasshouses, and therefore in principle, its potential economic benefit is very large indeed.

+ Study Center : D. Robinson/M.J. Maher
            An Foras Taluntais
            Kinsealy Research Center
            Malahide Road
            Dublin 5
            Ireland

---

2. Evaluation of different covering materials for greenhouse production of strawberries

+ Project in Gembloux, Belgium (1979-)

+ Description :
Strawberries are cultivated in 9 separate greenhouses under different covering materials (single glass with and without polyethylene, double glass, hortiplus, polycarbonate, polyethylene, polyethylene infrared and ethyl vinyl acetate). An analysis is done of the behaviour, quality and yield of strawberries under different conditions. In an earlier study similar research was done for lettuce and tomatoes.

+ Conclusions :
- It is observed that quality and yield are not exclusively correlated with the light transmission coefficient of the different covering materials.
- Particularly the balance between light and temperature determines the final yield.
- The strawberry crop showed a low sensitivity to light and temperature as compared with the tomato crop.

+ Study Center : Faculté des Sciences Agronomiques
            de l' Etat
            8 Avenue de la Faculté
            B-5800 Gembloux

+ For a more detailed description, cf. p. 76.

3. The biological effects of energy savings in greenhouses

+ Project in Årslev, Denmark (1981).

+ Description :
An experimental installation was set up to investigate the effects of thermal screens and different coverings on the plant growth (single glass with and without aluminized thermal screen; double glass; double acryl). The project was used to evaluate the plant growth of different types of green foliage or flowering pot plants.

+ Conclusions :
- The energy saving of double acryl as compared to single glass amounted to 39% for the period November 1981 - April 1982.
- There has been a significantly higher relative humidity in the houses, covered with double covering materials.
- The light reduction in the permanently insulated greenhouses had only a limited effect on the dry matter production.
- Further use of better insulated greenhouses in the foliage sector of Danish horticulture might result in high energy savings.

+ Study centre : Danish Research Service for Plant and Soil Science
Research Centre for Horticulture
Institute of Glasshouse Crops
Kirstinebjergvej 10
DK-5792 Årslev
DENMARK

+ For a more detailed description, cf. p. 84.

4. Comparison of double covered and single covered polythene greenhouses

+ Project in Dublin, Ireland

+ Description :
The fuel consumption of identical double-covered and single-covered polythene greenhouses has been compared for three years, both heated to the temperature recommended for production of early tomatoes in Ireland in 1977, 1978 and 1979.

Three tunnel houses of identical dimensions, 30 m x 5.2 m, were used. Two of these houses had double polythene cladding, separation between the two coverings being maintained by air pressure. The third house had a single polythene cladding. All the houses were heated with warm air distributed through perforated polythene ducts from individual oil-fired heaters, and all had fan ventilation. Early tomato crops were grown in all three houses.

+ Conclusions :
In each of the three years, the fuel consumption of the double-clad tunnel was less (by ± 33%) than that of the comparable single-clad tunnel.
Comparison of the yields from the double-clad and single-clad tunnels shows that there was no difference between the two in 1978, while in

1979, the double-clad tunnel was slightly better. This result is surprising, as it would have been expected that better light transmission in the single-clad tunnel would have resulted in higher yields in this house.
A procedure for control of humidity in such a house was implemented in 1979, which gave rise to an increase in fuel consumption of no more than 2%.

+ Study Center : T. O'Flaherty
              An Foras Taluntais
              Kinsealy Research Centre
              Malahide Road
              Dublin 5
              Ireland

5. Evaluation of double-covered polythene greenhouses for early tomato production

+ Project in Dublin, Ireland (1977-1979)

+ Description :
A triple-span 30 m x 20 m double polythene greenhouse was built in order to develop a system for optimal control of air temperature, CO_2 concentration and air humidity. The energy consumption and crop production in the optimally controlled double polythene greenhouse was compared to the performance assessed in the previous project (BD nr. 4) of a single-covered polythene greenhouse.

+ Conclusions :
  - The fuel consumption of the double-clad polythene house was 20% lower than the reference single-clad polythene house.
  - The yield of tomatoes to June 30 in the double clad house was 40% higher than in the reference house.
  - Double polythene greenhouses have wide potential application in the northern countries of the European Community, because of the substantial reduction in energy requirement which they offer together with their low capital cost compared with traditional glasshouses.

+ Study Center : an Foras Taluntais
              Kinsealy Research Centre
              Malahide Road - Dublin 5
              Head of Centre : Dr. D.W. ROBINSON
              Project leader : Dr. T. O'FLAHERTY

+ For a more detailed description, cf. p. 95.

6. DENAR Demonstration project

+ Project in Delft, The Netherlands

+ Description :
A 10 000 m² greenhouse was built in order to demonstrate the possibilities of energy saving measures. The demonstration greenhouse is divided into three compartments : a normal standard greenhouse

(reference), an acryl greenhouse and a newly designed type (futuristic) greenhouse with characteristics of both the acryl greenhouse and the standard greenhouse. In each of the three types, two systems of cultivation are applied : crop on rockwool, and drip-irrigation in the soil. Experience is gathered over a period of 4 years with vegetable crops.

+ Conclusions :
In the reference greenhouse the energy consumption for a tomato crop is decreased from appr. 70 m^3 natural gas per m^2 in 1979 (second energy crisis) to 35 m^3 gas/m^2 recently. Due to the rockwool growing technique, another 5 m^3/m^2 is eliminated because in this case soil sterilization is unnecessary.

The drastic reduction of energy consumption in the reference greenhouse is caused by :
1) the use of double flue gas condensers;
2) the lowering of the heating pipes from the top to the bottom of the greenhouse;
3) the introduction of double glazing in the walls;
4) the better insulation of the greenhouses;
5) the process-controlling by means of a micro-computer;
6) the use of energy screens, which are closed during the night.

A major problem caused by the use of alternative covering-systems is the reduction of light and the increase of relative humidity. However, it was observed that 1% reduction of light caused approximately 1% of production loss. A compensation is possible by the use of process-controlling and the increase of the inside temperatures : this can entirely compensate for the disadvantages. Moreover, a reduction in CO_2-concentration is to be compensated by an additional source of CO_2-nutrition. In greenhouses with double glazing the water needed is approximately 30% of the reference greenhouse, but the nutritional concentration of the water has to be adapted.
The main findings of this project are illustrated in table 6.1.

Table 6.1 : Performance data on the crop of autumn cucumbers
(10 August - 30 November 1983)

	SOIL			ROCKWOOL		
	Acryl	Fut.	Ref.	Acryl	Fut.	Ref.
energy consumption (m^3 gas/ 100 m^2	580	668	868	606	659	924
production (kg/100 m^2)	1168	1228	1168	1251	1389	1356
average fruit weight (gram)	468	479	467	492	501	500
number of fruits/100 m^2	2496	2565	2500	2544	2775	2690
perishability after 14 days % (4 + 50% yellow)	5.3	5.1	4.8	5.4	5.7	5.6
percent inferior quality	0.8	2.1	0.7	0.7	0.8	0.5

+ Study Center : L. Koop
              Denar BV
              Sionsweg 2286
              KM Rijswijk
              The Netherlands

7. A comparison of the light transmission of greenhouses covered with twinwalled sheets of polycarbonate and acrylic

+ Project in Bedford, England

+ Description : The experiment comprised the comparison of the light transmission of twin-walled polycarbonate and acrylic clad greenhouses under both completely overcast and bright sunny weather.

+ Conclusion :
- The twin-walled acrylic was found to have transmissivity for PAR of 59%, both for diffuse and direct sunlight.
  This is 5% points better than for double layer polycarbonate.
- The acrylic greenhouse resisted a hailstorm which was fatal for the polycarbonate one.

+ Study Center : B.J. Bailey
                 National Institute of Agricultural
                 Engineering
                 Wrest Park
                 Silsoe, Bedford
                 England

+ For a more detailed description, cf. p. 104.

8. Energy saving with mobile thermal screens

+ Project in Hanover, West Germany

+ Description :
The effectiveness of a thermal screen is mainly influenced by the kind of screen material and the tightness of the system. In the Institute for Horticultural Engineering in Hanover a mechanical system has been designed (see fig. 8.2 & 8.3) to meet the requirements of
- diminution of light loss
- good tightness at sidewalls, gables and foundation
- small surface of the screen
- suitability for all screen materials and easy change of materials

In winter 1979/80 heat consumption measurements, each 20-30 days, have been compared with the open screen.

Screen acrylic shading cloth	Screen double black PE film
Transmissivity : light 60%;	Transmissivity : light 0%;
infrared 23%	infrared 4%

Screen black PE film 0.2 mm	Screen clear PE film
Transmissivity : light 0%;	Transmissivity : light 92%;
infrared 20%	infrared 75%

+ Conclusions :
Because of the good tightness of the system very good energy savings could be obtained. Figure 8.1 shows how the heat loss coefficient of the greenhouse is dependent on the screen material and the wind speed.

Calculated for a wind speed of 4 m/sec which is the average wind speed in northern Germany, and with the normal heat loss coefficient of a greenhouse of 7.6 W/(m²K), table 8.1 shows the percentage energy saving at night and the average yearly energy saving with different greenhouse temperatures calculated with average climate conditions and different heat consumption coefficients for day and night.

The horizontal temperature distribution under the screen is improved by the screen because of the lack of leakages. The temperature difference above and below the screen was up to 16 K (double PE film). This difference leads to a steep temperature decrease when the screen is opened in the morning. This problem can be overcome by the control system.

The use of air-tight materials increased the absolute humidity under the screen and can cause condensation on the material. Whether problems with too high humidity will occur depends on the kind of irrigation system and crop, which influence the rate of evapotranspiration.

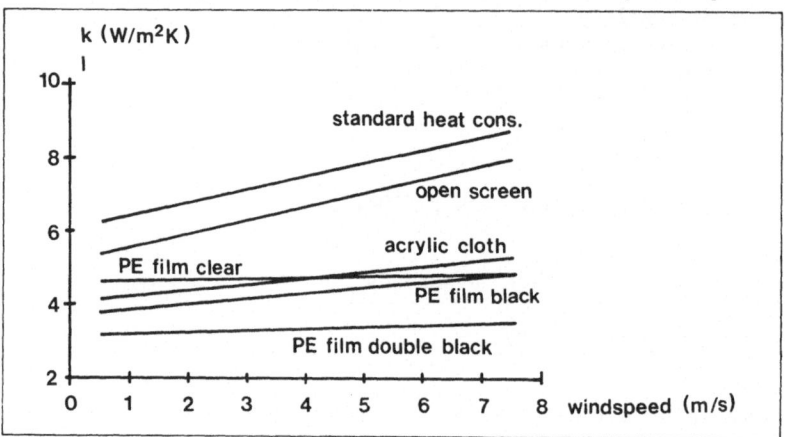

Fig. 8.1 : Regression curves of heat loss coefficient versus wind speed.

Table 8.1 : Impact of different screen materials on energy spendings

	k(W/(m²K)) at 4 m/sec	energy savings over nighttime (%)	Overall energy savings (added over day and night) (%) for different inside temperatures		
			8°	16°	20°
standard heat loss	7.6				
screen open	6.7				
acrylic shading cloth	4.7	38.2	30.6	27.6	26.8
PE film, clear	4.7	38.2	30.6	27.6	26.8
PE film, black	4.3	43.4	34.7	31.6	30.3
PE film double, black	3.5	53.9	43.1	39.3	37.7

+ Study Center : J. Meyer
                 Institute for horticultural and agricultural engineering
                 Universität Hanover
                 Herrenhäuserstrasse 2
                 Hanover 21
                 W. Germany

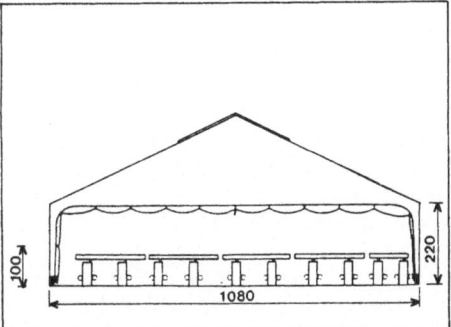
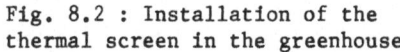
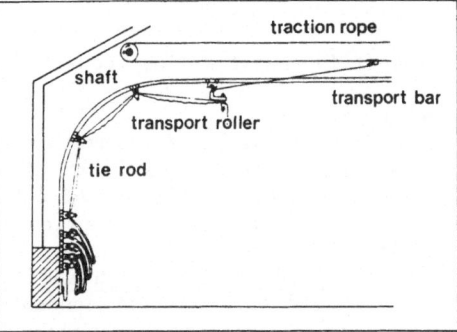

Fig. 8.2 : Installation of the
thermal screen in the greenhouse

Fig. 8.3 : Details of the mechanical
system

9. Estimation of the heat saved by using a removable internal lining in a glasshouse

+ Project in Dublin, Ireland

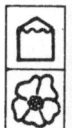

+ Description :
  Heat loss was measured in two adjacent 0.4 ha glasshouses in a Dublin
  nursery in March 1974. Each glasshouse was equipped with a removable
  internal lining, of the type used commonly as an automatic black-out
  system in the production of all-year-round chrysanthemums. By comparing
  the heat requirements of the two houses on a number of nights, with and
  without the linings in use, an estimate was obtained of the reduction in
  heat loss attributable to the lining.

  The dimensions of the greenhouses used in the experiment were identical.
  The layout of the heating sub-mains in the houses was identical also and
  the pipe loops selected for temperature measurement were located at
  corresponding positions in the two houses.

+ Conclusions :
  Over four nights, the heat requirement of the lined house was on the
  average 47% less than that of the unlined one.
  The percentage heat saving attributable to the glasshouse lining was
  substantial on all the nights that the lining was used. However, the
  results varied considerably from night to night. It is not possible to
  say from the data available to what extent this variation was due to
  change in the relative heat requirements of the glasshouses themselves,
  or to change in the effects of the linings.

  Notwithstanding this, however, the results obtained are clear evidence
  that substantial fuel savings can be achieved by using an internal
  lining in a glasshouse at night.

+ Study Center : T. O'Flaherty
                 An Foras Taluntais
                 Kinsealy Research Centre
                 Malahide Road
                 Dublin 5
                 Ireland

10. Influence of a semi-movable screen on the energy consumption and the growth of tomatoes

+ Project in Westland, The Netherlands

+ Description :
The first six weeks of a tomato crop require, depending on the planting date approximately 8 to 12 m^3 of natural gas per m^2 ground surface of the greenhouse. By the use of a polyethylene screen, it is possible to reduce this energy consumption substantially. In three different holdings a semi-movable screen was installed. <u>Planting dates were December 23rd, January 26th and February 9th</u> respectively. The screen was provided with an anti condensation sheet of 0.05 mm polyethylene. A comparable section of the greenhouse without the screen was used as reference.
The screen consisted of a very simple system of wires on which a folio was installed. It was possible, by a manual intervention to make openings in the screen. (see fig. 10.1)
After the initial period of 5 to 6 weeks, the screen was entirely removed.

+ Conclusions :
There were no visible disadvantages in the use of the screen. During the period that the screens were in use, the light reduction was 12%. Averaged over the whole period from planting to first harvest, the light reduction was 2 to 4%. A change in the vertical temperature distribution was noted : warmer at the soil level and colder in the top of the greenhouse. It was possible to manage the relative humidity by the use of the openings in the screen. In total, the energy saving amounted from 4 to 4.5 m^3 natural gas per m^2. In all holdings the harvest time slightly advanced. (see table 10.1)

Table 10.1 : Comparison of the harvested quantities as a function of time between holdings with and without screen.

Harvest (kg/m^2) before	Holding 1		Holding 2		Holding 3	
	screen	control	screen	control	screen	control
30 April	6.23	5.75	2.34	2.28	0.04	0.05
28 May	11.68	11.32	8.37	8.28	3.10	2.98
25 June	16.40	16.35	14.36	14.42	8.33	9.44

It was concluded that the screen offers an economically usable alternative for energy saving. It could even be used as a second screen in combination with a movable screen.

+ Study Center : C.P.A. Van Holsteijn
        Glasshouse Crops Research and Experiment Station
        Naaldwijk
        The Netherlands

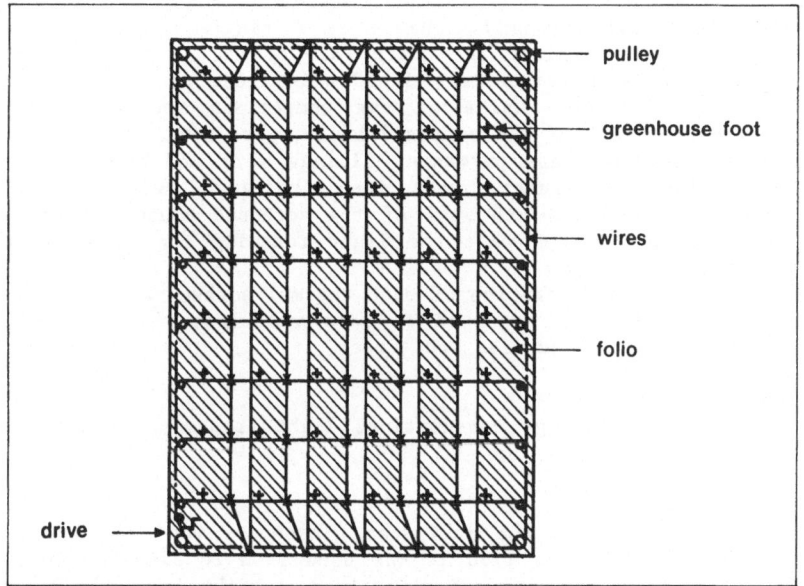

Fig. 10.1 : Schematic representation of the semi-movable screen.

11. Evaluation of thermal screens in glasshouses on commercial nurseries

+ Project in Bedfordshire, England

+ Description :
The project on Thermal Screens for tomatoes took place on two nurseries : one multi-span glasshouse and one single span glasshouse.

Multi-span glasshouse
A glasshouse block, consisting of 10 spans of 10.7 m each and 78 m long, was divided into four 0.21 ha compartments, one was fitted with the thermal screen and another used for comparison. The screen was made from Tyvek, a spun bonded polyolefin material aluminised on the upper surface and white on the underside. The material was punctured with a large number of small holes which gave a permeability of 0.85 x 10^{-3} m³/(m²s) when a temperature difference of 5K existed across it. The thermal screen was supported horizontally in the glasshouse on wires at a height of 3 m. Seven sections were required for the block, each covering a distance of 5.6 m. In its withdrawn position the screen coincided with the base of the roof trusses. The external walls of the screened block were lined with clear polyethylene film; there was no insulation on the partition walls. At the end of the first season this lining was replaced by withdrawable covers of Tyvek. The tomatoes were grown in the border soil.

Single span glasshouse
The screened glasshouse was the southernmost of four east-west houses each with a span of 22.3 m and a length of 100.5 m. The thermal screen was made from Peritherm, an impermeable film of aluminised polyester laminated to black polyethylene (is used with the aluminised surface uppermost). For infrared radiation, the emissivity of the aluminised and black surfaces were 0.07 and 0.91 respectively (Bailey). The screen

was installed to move across the house as a single sheet, and was withdrawn to a position along the north wall of the glasshouse. All walls of the house were permanently lined with clear polyethylene and the north, east and west walls of the comparison house, situated immediately to the north of the screened house, were similarly treated. For the second season these permanent linings were modified so they could be lowered. Peritherm, with the aluminised surface facing inwards, was used on the north wall of the screened house. This was raised on sunny days to increase the light reaching the crop in the vicinity of the north wall.

In both greenhouses the operation of each thermal screen was controlled by a timeswitch which was adjusted to open and close the screens at times which were close to sunrise and sunset.

+ Conclusions :
Results are given for energy consumption, light transmission and crop response.

Energy consumption
During 1979 the screen in the multispan glasshouse was used from 10 January until 15 May and in 1980 it was used from 26 January until 6 June. The average air temperatures maintained at night during 1979 were 13.6 and 13.3°C in the screened and non-screened houses, while in 1980 the values were 16.3 and 14.4°C respectively. The lower saving in 1980 was a reflection of the milder climatic conditions which prevailed; during 1979 the average ambient temperature was 4.1°C while in 1980 it was 7.0°C.

During the periods the screens were in use they reduced the heat loss at night by 55-60%. Over the complete season they gave savings of between 25 and 30%.

Light transmission
The transmittance of PAR by the screened and non-screened houses at both nurseries was measured when the sky was completely overcast with low cloud.

The measurements in the multispan glasshouse were made along a traverse beneath a ridge beginning at the external gable wall. The sensor was mounted 1 m above the soil.

The average reduction in the transmittance of PAR by the screened house was 3.0%. Part of this difference can be attributed to the 80 mm deep steel members placed under the gutters for support because angle braces were removed to accommodate the screen.

In the single span glasshouse the traverses were made across the complete width of the houses at a height of 1.5 m.

The screen influenced the light received at a height of 1.5 m for a distance of 3 m from the midpoint of the screen. When averaged over the whole house this reduction in light was equivalent to 1.5%.

Crop response
The yield of tomatoes from the screened and non-screened houses for 1979 and 1980 are given in table 11.1. It can be seen that in 1979 there were some losses from the screened house. In the multispan house there were fewer flowers and fruits on the first six trusses, and the plants were a little taller. At the other nursery the fruits from the screened house were rather smaller. The low early yields experienced in 1979

were not repeated in 1980, although the total yield from the screened houses was again less than from the unscreened houses.

Table 11.1 : Yield of tomatoes from screened and non-screened houses.

Crop yield (kg)	1979		1980	
	30 Apr	30 Sep	30 Apr	31 Aug
multispan screened	4.2	51.9	6.9	46.5
non-screened	6.0	57.0	7.0	50.3
single span screened	7.2	55.4	8.6	50.2
non-screened	8.2	57.2	8.2	51.8

+ Study Center : B.J. Bailey
　　　　　　　　National Institute of Agricultural Engineering
　　　　　　　　Silsoe, Bedfordshire
　　　　　　　　England

12. Greenhouse heating with solar energy

+ Project in Hanover, W. Germany.

+ Description :
In 1976 a plant for heating a greenhouse by means of solar energy was built in Hanover. (See fig. 12.2). The principle is to use the greenhouse itself as solar collector. The greenhouse is heated by solar energy from April to October, except in longer bad weather periods.
At daytime cold water of about 6°C from the cold water storage flows through two heat exchangers in the greenhouse. The cold water will be warmed up by the air, itself heated by the greenhouse-effect. The greenhouse is closed. The air cooled in the heat exchanger passes through perforated plastic tubes back into the greenhouse air volume. Between the cold- and warm- water storage is a heat pump. This heat pump brings the low temperature energy stored in the cold water storage to a higher temperature-level which is sufficient for heating. As the heat pump works independently of the greenhouse heating and cooling, the operation time can be up to 24 hours and the heat pump can be relatively small. In the warm water storage an additional heating system from the boiler is installed. Instead of the water store one can use other stores too (latent stores for example). For cooling and heating two different water temperatures (6°C for cooling, 45°C for heating) are necessary. These temperature levels are achieved by a heat pump, cooling the cold water tank back to 6°C and warming up the hot water tank to 45°C. The heat pump is automatically stopped, if one of the temperature levels is reached.
Because the greenhouse was especially tightened it needs an additional forced ventilation to control the humidity of the air and the CO_2 concentration. This ventilation system with an air exchange rate of about twice the greenhouse volume per hour is combined with a heat exchanger which recovers energy between the inside and outside air flow. In the summer when there is a very large amount of solar energy input into the greenhouse, the cooling system is not sufficient to cool the greenhouse because it has been designed for spring and autumn

conditions. Therefore a second forced ventilation system with 2 ventilators is needed. But instead the normal ventilation system of the greenhouse could be used.

+ Conclusions :

For the heat pump between the cold and warm water storage a power of about 10 Watt for one m^2 greenhouse ground area is necessary for the outside weather conditions of Hanover. Considering the storage-problem a distinction is made between the short-time-storage from day to night and the long-time-storage from summer to winter. It was found that the short-term-storage will be more economic. For the long-term-storage huge volumes of 12 to 20 m^3 per m^2 greenhouse ground area are necessary. The storage volume for a short-term-storage from day to night is about 0.2 m^3 water per m^2 greenhouse ground area for the conditions in Hanover.

Figure 12.1 shows the results of the measurements and calculations of the energies in the different months of the year. The different curves represent : q_d the average daily heat consumption in MJ per m^2 greenhouse ground area for double glass, 18°C inside temperature and the mean meteorological conditions in Hanover. There also is a heat requirement in summertime. q_s is the heat energy per m^2 which can be stored at daytime from the greenhouse air. The dashed area under the two curves q_d and q_s shows the amount of heat energy saved by using solar energy. These measurements have shown that the greenhouse can be heated from May until September with solar energy. One can partly use the solar energy for greenhouse heating from March till October. q_s shows the curves for the gain per m^2 greenhouse of heat energy out of separate solar collectors near the greenhouse depending on the percentage of collector surface area to greenhouse ground area Sc/Sg. If the saved energy is summed up, one can find that the greenhouse can produce the same amount of heat energy as a collector area of about 60% of the greenhouse ground area. For the first system, the greenhouse as collector, a heat pump is needed; for the second system the separate collectors are necessary.

The calculations have been made for a 10 000 m^2 greenhouse area divided in 4 sections with different temperature sequences for flower production over the year. For conventional heating 1.4 GJ per m^2 and year are needed. This means 1.75 GJ/m^2 of oil consumption in the case of a heating efficiency of 80%. In the solar energy system with the heat pump in addition 1.0 GJ/m^2 are needed.

Given a heating efficiency of 80 percent 1.25 GJ/m^2 are required for additional heating. The electricity power for the heat pump is 0.11 GJ/m^2. The large amount of electricity used here is a disadvantage. The oil saving in this case is about 29% but the requirement of electricity power is too high.

The aim of further research must be to increase the efficiency of the whole system by decreasing the use of electricity especially for the heat exchangers and increasing the amount of storable solar energy.

+ Study Center : J. Damrath/Chr. von Zabeltitz
                Institute for Horticultural Engineering
                University of Hanover
                Herrenhäuser Str. 2
                3000 Hanover
                West-Germany

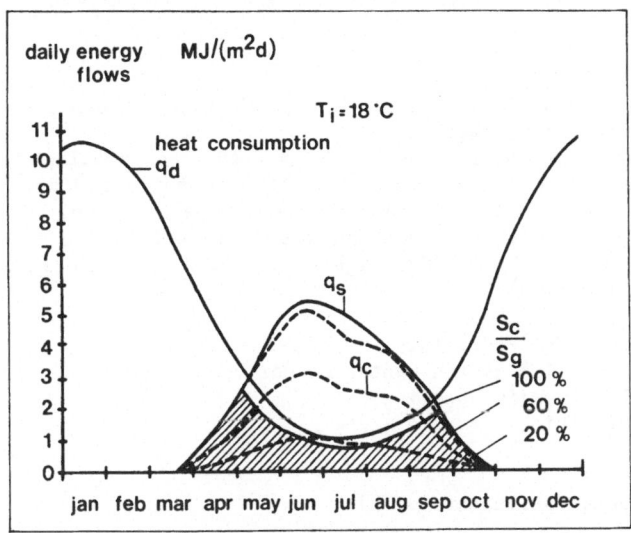

Fig. 12.1 : Results of measurements and calculations of the energies in
the different months of the year
q_d : Average daily heat consumption
q_s : Storable heat energy out of the greenhouse
q_c : Storable heat energy from collectors

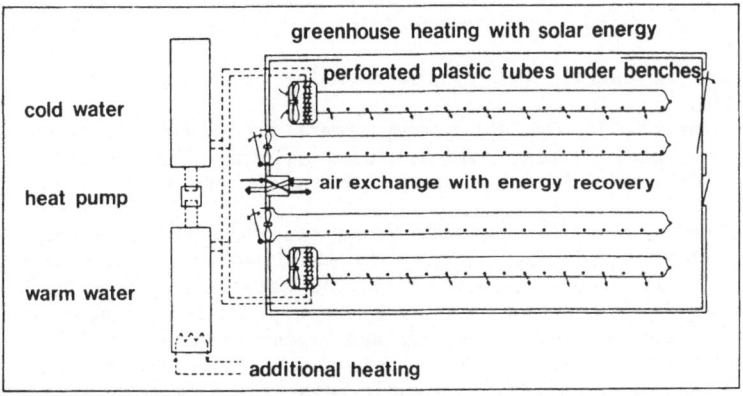

Fig. 12.2 : Principle of the solar greenhouse in Hanover

---

13. Development of a solar system for heating greenhouses

+ Project in Thessaloniki, Greece (1981-'83)

+ Description :
Air from the greenhouse is blown through a heat exchanger, consisting of
two plastic tubes with a simultaneous water circulating system. On
moments with a heat excess during day time the water is heated by the
greenhouse air. This heated water is then stored in an 11 m³ storage
tank. When the greenhouse air temperature drops below 12°C the system

is activated again. In this case, to heat the greenhouse air with the
stored water.

+ Conclusions :
  - This simple and low cost solar system is suitable as a temperature
    regulation device for Greek greenhouses.
  - Over a few measuring days in April, an energy recuperation was
    registered of around 20 % of the solar radiation measured inside the
    greenhouse. This amount would otherwise be lost by ventilation.
  - The earliness of the vegetable production was increased by 2-3 weeks
    whereas the yield was considerably higher as compared with a non
    heated greenhouse (+ Δ 10 %).

+ Study Center : Dr. Mavr. Grafiadellis,
                 Agricultural Research Center of Northern Greece
                 Elliniki Georgiki Scoli
                 Thessaloniki
                 Greece

+ For a more detailed description, cf. p. 131.

| 14. Solar energy storage in calcium chloride |

+ Project in La Gaude, France.

+ Description :
  The solar compartments of the studied greenhouse of 500 m^2 are
  characterized by a low heat loss coefficient through the walls and a
  high inertia.
  The insulation is carried out by the doubling of the glass wall by a
  polythene film.
  The thermal inertia in the solar compartment is obtained through the
  storage of heat under latent form in fused calcium chloride salt (see
  fig. 14.1). This salt is set in buried tunnels, regularly distributed
  in the greenhouse.
  The heat is transferred to the salt by an air circulation system (see
  fig. 14.2); the supplementary energy is supplied by a classical air
  blown heater. The fusion temperature of the calcium chloride is 28°C.
  The fusion temperature is decreased for the application in greenhouses
  by the simple addition of water, to approximately 15°C to 25°C.   In
  these conditions the latent energy of this material is 104.3 kJ/kg.
  The calcium chloride is packed in plastic bags of 1.5 kg each and placed
  in a tunnel; the hot greenhouse air is blown through those tunnels.

+ Conclusions :
  The insulation of the greenhouse has reduced the energy needs by 50%.
  The system of energy storage supplied 60% of the resulting demand.
  In comparison with a reference greenhouse, the energy need of the solar
  system was only 20%, increased by 5% for the energy needs of the
  fan-system.
  The production of roses in the solar energy greenhouse was one week
  earlier as compared with the reference greenhouse (harvest February,
  March).
  But the total production decreased by approximately 12%, due to the 20%
  lower transmission of light in the solar energy greenhouse.

+ Study Center : GREAT Agricultural Chamber
                 Domaine de la Baronne
                 RN 209
                 06610 La Gaude
                 France

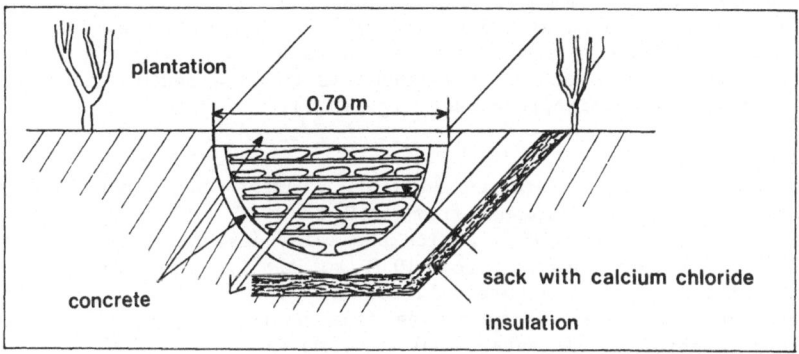

Fig. 14.1 : Scheme of an element of a tunnel equipped with calcium
            chloride.

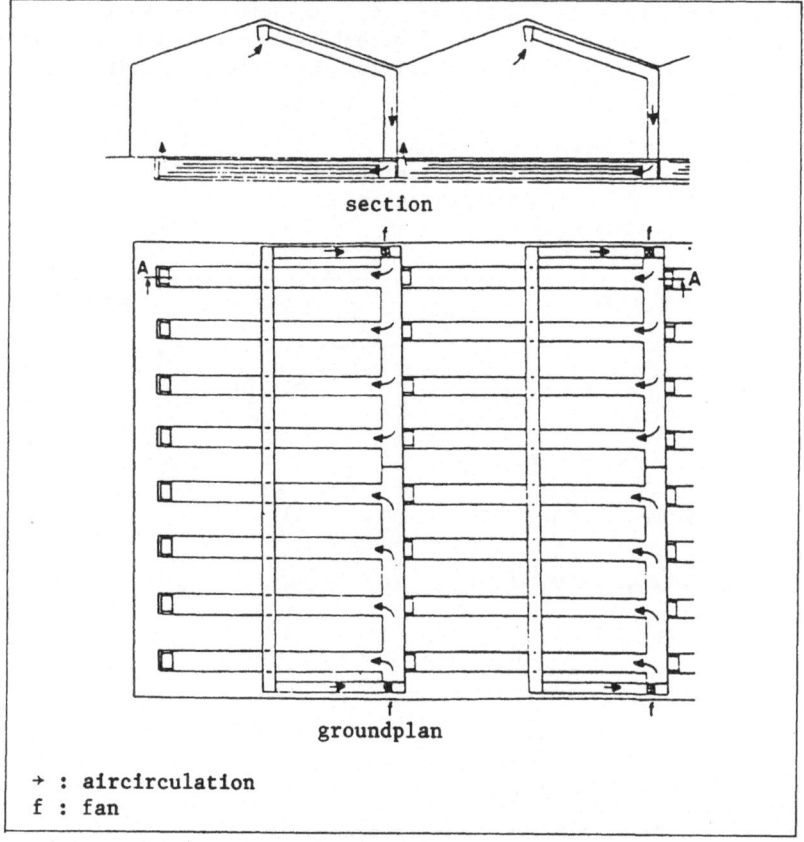

Fig. 14.2. : Distribution of the air in the heating tunnels.

15. A solar energy greenhouse (special case)

+ Project in Tarn, France.

+ Description :
This greenhouse occupies a surface of 60 m² and is especially built in
order to make maximum use of the solar energy. The north-east side of
the greenhouse is an existing rock with an inclination of 23%.
The solar energy is partly stocked in water and partly in a stone bed.
The principal aim of this greenhouse is to take maximum advantage of a
bioclimatical architecture (see fig. 15.1).

+ Conclusions :
The observations relate to 1980/81. During the winter the average
outside temperature ranged from -2°C to 15°C with a drop to -4°C in
January 1981. During this period the solar designed greenhouse operated
at temperatures ranging from +9°C to +28°C. The minimum was 6°C,
without damage for the horticultural crops.
The consumption of energy during this winter by the solar greenhouse
was : - February - November 1980 : 4500 MJ
      - Nov. 1980 - March 1981   : 10500 MJ
      - Dec. 1981 - November 1982 : 6850 MJ
During the representative season 1980/81 the fossil energy consumption
was approximately 6 1 oil per m². In order to maintain the same
climatic conditions in a normal greenhouse it was estimated that 20 1
oil would be necessary. Therefore, the energy saving amounted to
approximately 70%.

+ Study Center : Laurence and Jerome
                ARES (application of energy research)
                29, rue croix Baragnon
                Toulouse
                France

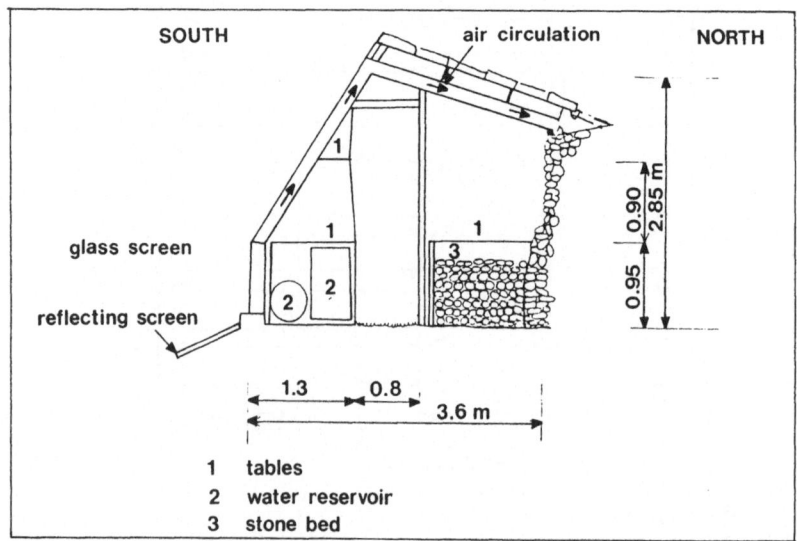

Fig. 15.1 : Passive solar energy greenhouses

16. Greenhouse climatization by an earth-air heat exchanger

+ Project in Athens, Greece (1982-83)

+ Description :
  An earth-air heat exchanger is built under a greenhouse. The greenhouse
  air is blown through the heat exchanger pipes whenever the greenhouse
  temperature exceeds 28°C (excess heat storage) or falls under 12°C
  (consumption of stored heat). The project was used to determine the
  benefits of such a system in energy efficient temperature control of the
  greenhouse.

+ Conclusions :
  - The method resulted in energy savings of 70% for the greenhouse
    heating in Athenian conditions.
  - It is a method that could offer energy savings in other spaces,
    needing heating or cooling.
  - It is expected that the results would be even better in more southern
    parts of Greece.

+ Study Center : Prof. S. Kyritsis and Prof. G. Mavrogianopoulos
                 National University
                 Agricultural College of Athens
                 Iera odos 75
                 Votanikos Athens 301
                 Greece

+ For a more detailed description, cf. p. 122.

17. Heating of a greenhouse

+ Project in Grand Manil (Gembloux) - Belgium.

+ Description :
  Construction of a small experimental greenhouse (50 m^2), based on 3
  fundamental principles :
    - maximal thermal insulation without hindering the luminous
      transmission;
    - a storage system for solar energy during the day with restitution to
      the culture at night;
    - a heating system to replace and support the solar energy with an
      air-water heat pump generator.
  The transmittance of the greenhouse cover to PAR-radiation has to be
  high. The cover must furthermore be weather-proof, keep its properties
  and have a small heat loss coefficient (its mean value should be 3 to
  3.5 W/(m^2 K)).
  This value is obtained with a polycarbonate cover made of alveolic
  plates (k = 3 to 3.5 W/(m^2 °C)).

  Aiming to use the maximal solar energy recovery possibilities, a surface
  of 10 m^2 of solar panels is used.
  A tank of 1 m^3 of water is used as storage. The heated and stored water
  will be used to feed radiant panels, utilising principles of low
  temperature central heating.
  An electrical immersion heater is set inside the storage tank, powered

by a windmill (Canadian type of 3 kW - 220 V direct current).
A water wall has been set on the north side to absorb and store the
energy. The wall is made of metallic tanks, filled with water (capacity
= 4 m³; surface 20 m²) and presents a non reflecting surface.

A recovery system for 5 to 8 greenhouse volumes per hour of the hot air
is fitted in the top of the greenhouse. The hot air is driven to the
recovery system, made of a heat pump. The calories produced by the
heater are stored in :
  - a tank of 1 m³ mentioned above;
  - a second tank of 1 m³ in parallel with the first one;
  - the water wall (capacity = 4 m³).

+ Conclusion :
Not available.

+ Study Center : E. Bertrand - J. Bazier
                 Institut technique horticole de l'Etat
                 4, rue Verlaine
                 5800 Gembloux
                 Belgium

18. Flexible plastic solar collector for low temperature heating

+ Project in Hanover, W. Germany.

+ Description :
The requirements that static heating elements for solar heat purposes
have to meet are :
- optimal adaption to the greenhouse,
- maximum energy output,
- minimal loss of light,
- low costs for installation of heating elements.
For the use as flexible static heating systems plastic heat exchangers
seemed to be suitable. A thin layer solar absorber of Hoechst AG,
Frankfurt/M was chosen. This solar absorber consists of a 3 layer
texture with outside PVC-coating. Enclosed fibers form a stable
distance of 2 to 4 mm. This structure is rather pressure stable and can
be filled with water. The element runs at maximum with 0.5 bar pressure
and up to 60°C. The maximum size is now 1.5 m width and any desired
length (see figures 18.1 & 18.2).
Two single elements are connected to form a unit suitable to fit into
the German Standard Greenhouse construction.
Hydraulically the 2 elements have a series connection with the inlet and
the outlet at the bottom and a connection between the two elements on
the top sides. In that way there is an upward stream of the heating
water in one element and a downward stream in the other.

A winding shaft is placed in the greenhouse parallel to the gutter and
the elements are held up by cable ropes. At night, the elements are
lowered down to a height corresponding to the inside greenhouse
structures. In that way the light loss at daytime can be minimized, but
the elements can contribute to the daytime heat demand.

+ Conclusions :
The investigation showed that the flexible solar collectors are suitable as a low temperature greenhouse heating system for flow line temperatures of more than 40°C. If they are used in well insulated greenhouses, they can cover a high percentage of the total annual heat demand.
The mechanical resistance to folding damage has proved to be sufficient for the use in greenhouses; till now no results could be obtained in respect of long time durability of the material.
The day time light loss can be minimized by lowering the elements down to the plant level.
For an economic use, the elements are too expensive at the moment, but prices will fall when mass production is possible.
In general, the insulation of the greenhouse should be the first step when low temperature heating systems are used.
Future efforts will be made to reduce heat losses by radiation from the elements to the side walls by using reflecting surfaces. The size of the units possibly will be increased to reduce installation costs.

+ Study Center : K. Schockert
                       Institute for Horticultural and Agricultural Engineering
                       University of Hanover
                       Hanover
                       West Germany

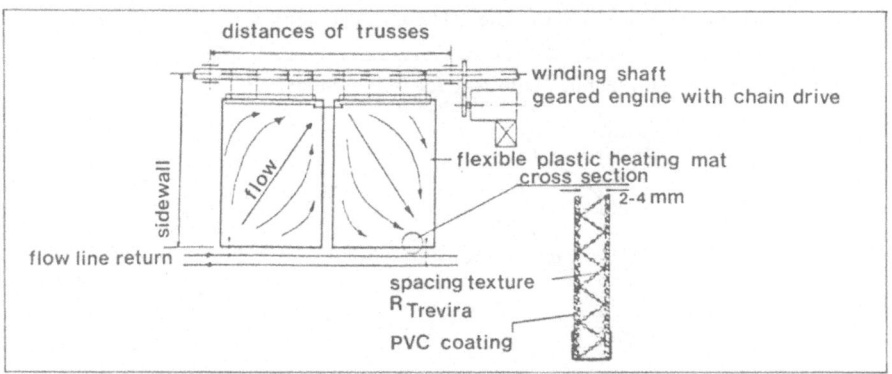

Fig. 18.1 : Principle of the flexible heating mat attachment at the sidewall and installation scheme.

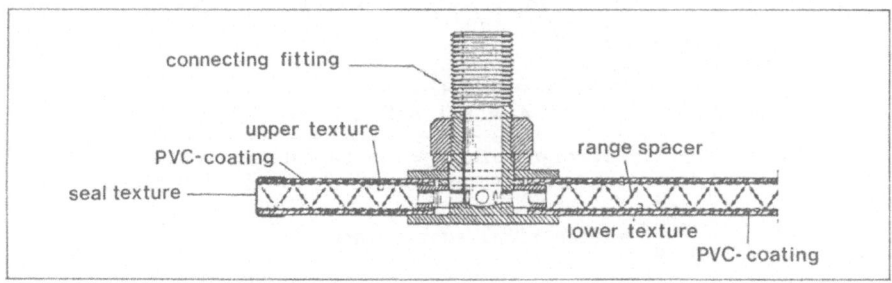

Fig. 18.2 : Structure of the flexible heating element.
               Ref. : Hoechst AG

19. Solar collectors and heat pumps for greenhouse heating

+ Project in Fünfhausen, West Germany.

+ Description :
This plant is a combination of a solar collector and a heat pump. For heating, the solar system has priority. The heat pumps work when the solar energy is lower than the energy demand. The plant is regulated by the greenhouse thermostat, so that there is no difference with a conventional heating system.
The 104 collectors with an area of 208 m² are installed outside the greenhouse and deliver the energy to a greenhouse (ground area of 600 m²).
The energy can be used directly or can be stored in 3 tanks with a capacity of 15 m³ each.
The two heat pumps with a power of 11 kW each use the solar storages and pump water as heat source. To reduce the heat load, the greenhouse is double glazed and provided with a low temperature heating system.

+ Conclusions :
The stationary collector curve was measured for a single collector and for the collector field. For the collector η_c = 0.75 – 6.4 x RP (1), for the collector field η_c = 0.7 – 7.7 x RP (1). This reduction is caused by the piping system.
In fig. 19.1 the monthly insolation Ic, the useful energy Qno and the electrical energy input for the circulating pumps (Q_e) are shown.

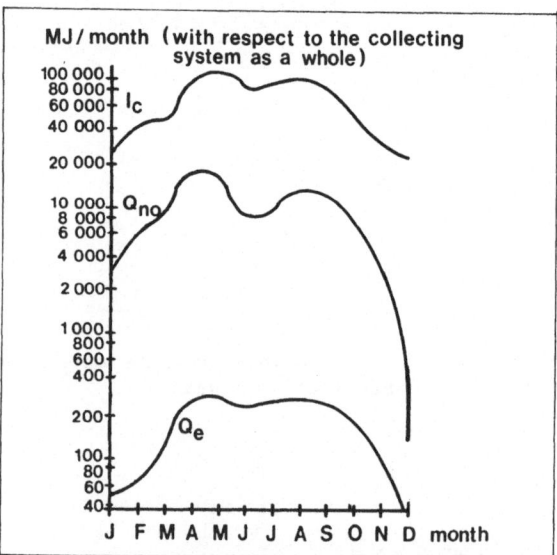

Fig. 19.1 : Monthly average solar system performance data
I_c = monthly isolation on the collector field
Q_{no} = useful energy
Q_e = electrical energy input

(1) RP = $\dfrac{Tc - Ta}{Ic}$ with   *Tc collector fluid temperature (°C)*
     Ta ambient air temperature (°C)
     Ic solar radiation on collector (W/m²)

During the measuring period from 1980 to 1983 the solar fraction of the collector-field was 13.4% and lower than expected. Given the fact that on sunny days the time period of heat surplus in the greenhouse corresponds with that of the operation of the solar system, it is the relatively small size of the storage volume which represents the limiting factor.
The electric energy consumption for the circulating pumps was 2% of the thermal output of the collector-field and is negligible. The water/water heat pump provided the remaining 86.6% of the heat demand.
Economic calculations of this solar plant, considering the low solar fraction, show that the collector will not be paid back within its lifetime. For the heat pump pay back times of about 15 years are expected.

+ Study Center : U. SCHMOLDT
                Hamburgische Gartenbau-Versuchsanstalt
                Fünfhausen

---

20. Solar energy greenhouse with selective glass filter

+ Project in Montfavet, France.

+ Description :
A 200 m^2 fluid-roof solar greenhouse was constructed in 1980 at the INRA Research Center in Montfavet (France). The greenhouse (INRA-CEA patent) is PAR-transparent and absorbs the near-infrared wavelength range. The roof is made of a double glass : a normal glass for the upper part, and a special one for the lower part. Water is allowed to circulate between the two glasses. The special glass (BOUSSOIS S.A.), is a selective filter.
A selective glass filter on the roof of the greenhouse lets the visible light pass. The collected infra-red rays (absorbed by the lower wall) heat up water which circulates in the double wall of the roof. This water is stored and used to heat up the greenhouse. In summer, a (water-water) exchanger on ground water allows cooling of the greenhouse.
The circulation fluid is normal water. The volume of water storage is 25 litre per m^2 of greenhouse surface. The interior temperature of the greenhouse is the regulating indicator of the solar greenhouse. A pump starts circulating the water at a temperature above 23°C. During the night, the water circulation starts at greenhouse temperature below 12°C (see figure 20.1).

+ Conclusions :
A set of experiments was carried out on tomato plants between March and July, 1980. The energy saving in heating requirement is nearly 100% during this period, but the planting dates (March, 5th) may have been too late for the plant to face high heating needs.
The experimental results are in good agreement with theoretical conclusions of VAN BAVEL and DAMAGNEZ (DAMAGNEZ et al., 1980) based on a dynamic simulation of the greenhouse energy balance.
During the night, the leaf surface temperature under the solar greenhouse is 2 K higher than under the normal glass. For the same temperature set point, the microclimatic conditions are better in the solar greenhouse.

During daytime, the needs of ventilation are considerably reduced. When exchanging heat with the underground water table, it is possible to save 100% energy in ventilation needs. But during daytime, the CO_2 concentration under the solar greenhouse is lower than under the normal glass, which may explain the reduction of earliness and yield under the solar greenhouse.

+ Study Center : J.P. Chiapale
I.N.R.A. Station of Bioclimatology
84140 Montfavet
France

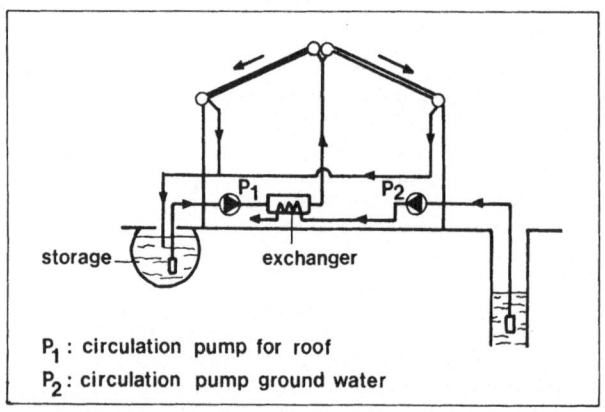

P_1 : circulation pump for roof
P_2 : circulation pump ground water

Fig. 20.1 : Principle of solar energy greenhouse

21. Heating of a greenhouse with solar energy

+ Project in Hyeres, France.

+ Description :
The solar energy project is part of a demonstration unit.
The floor area of the solar greenhouse is 525 m². A reference greenhouse is heated with fossil energy. In both greenhouses, roses were grown during the observation period in 1981/82. The roses were planted in February 1980.
The absorber area of the solar collecting system is 150 m², with an inclination of 60°. The fluid is water and the dimension of the heat storage is 10 m³. The investment in the solar greenhouse equals approximately 2.5 times the equivalent investment for the reference greenhouse.

+ Conclusions :
The average air temperature in the solar greenhouse was 1.9 K lower than the reference greenhouse. This was mainly due to the type of heating : air heating in the reference greenhouse versus soil heating in the solar greenhouse. Hence, the average soil temperature of the solar greenhouse exceeded the soil temperature in the reference greenhouse. The crop results of the solar greenhouse were slightly better.

Table 21.1 : Energetic balance of the greenhouses.

Month	Energy used in Ref. greenh. MJ/month	Energy collected by collectors (Qco) MJ/month (1)	Energy used in solar greenhouse MJ/month		% solar energy in solar greenh.
			solar(Q_{no})	non-solar	
Nov 81	35 300	15 550	14 500	10 800	57
Dec 81	63 550	13 350	11 550	16 000	25
Jan 82	60 500	21 000	19 550	22 350	46
Feb 82	52 150	13 950	9 900	42 950	19
Mar 82	50 100	27 150	25 100	50 350	33
Apr 82	27 150	24 000	21 800	8 250	73

The energetic balance of the two greenhouses is shown in table 21.1. The share of the energy delivered by the solar system varies between 19% and 83% (2). The data for December and February are unreliable due to faulty regulation system of the greenhouse.

+ Study Center : M.F. Cory
                 Lycée Agricole et Horticole
                 Quartier des Grés
                 83400   Hyeres
                 France

(1) Difference between Qco and Qno lies in duct and storage losses.
(2) Average performance between 31/10/81 and 21/11/81.

---

22. Heating a greenhouse with solar energy

+ Project in Thessaloniki, Greece.

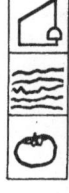

+ Description (see fig. 22.1) :
  - Low cost solar collector (water heating) of 68 m^2
  - storage of water in underground reservoir of 17 m^3
  - vegetable culture in N.F.T. system during the winter
  - space heating of plastic cover greenhouse of 160 m^2

The system consists of three major elements. These are : (i) the external solar energy collector (see fig. 22.2), (ii) an underground water reservoir and (iii) plastic tubes of black polyethylene through which warm water is circulated to heat the greenhouse (see fig. 22.3). Beside these three major elements, water pumps, thermostats and water pipes are needed. The water reservoir and solar energy collector were designed to meet the heat requirements of the greenhouses for three days.

+ Conclusions :
The results obtained indicate :
  - In cold weather conditions, when the air temperature in non-heated polyethylene covered greenhouses is as low as 2-3°C, the solar heated greenhouse had an air temperature 6.5 K higher and also 6-8 K higher soil temperature. In warmer weather conditions when the minimum air temperature was as high as 10°C, the solar heated greenhouse had only 2-3 K higher air temperature in comparison to the non-heated control

greenhouse.

- Heat gained from the 60 m² solar plastic collector during 100 days of use in spring time was 46 GJ (equivalent to 1 100 l of oil).

- The cost of the system with external solar energy collector of 300 m² surface, to heat a greenhouse of 1 000 m² is estimated to be around 3 200 ECU.

- It was found that tomatoes can grow satisfactorily in the solar heated greenhouses in which the air temperature was above 8°C and the soil temperature above 18°C. Plants grew even more vigorously in the solar heated greenhouses than in the fuel heated greenhouse where the air temperature was higher by 4-5 K. In the non-heated control greenhouse, the tomato plants took on a purplish appearance usually associated with nutrient deficiency resulting from cold soil. These plants did however recover at a later stage as days lengthened and ambient temperatures increased.

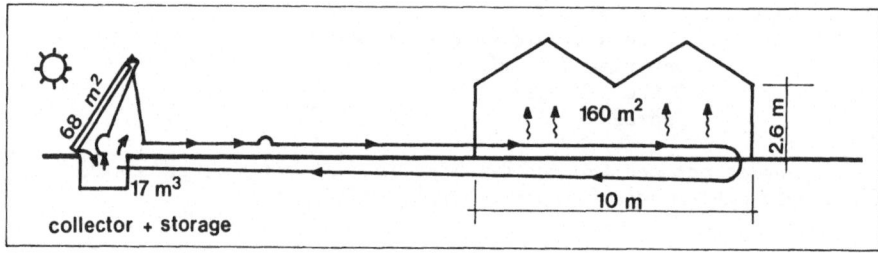

Fig. 22.1 : Scheme of the solar collecting system.

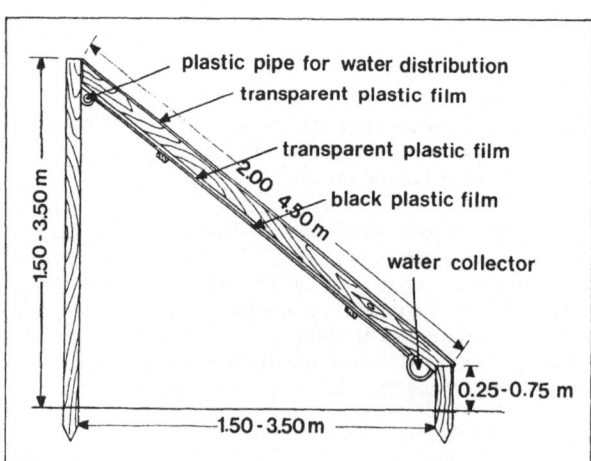

Fig. 22.2 : Components of the solar energy collector.

+ Study Center : -M. Grafiadellis
                    Agricultural Research Center of North Greece
                    Thessaloniki

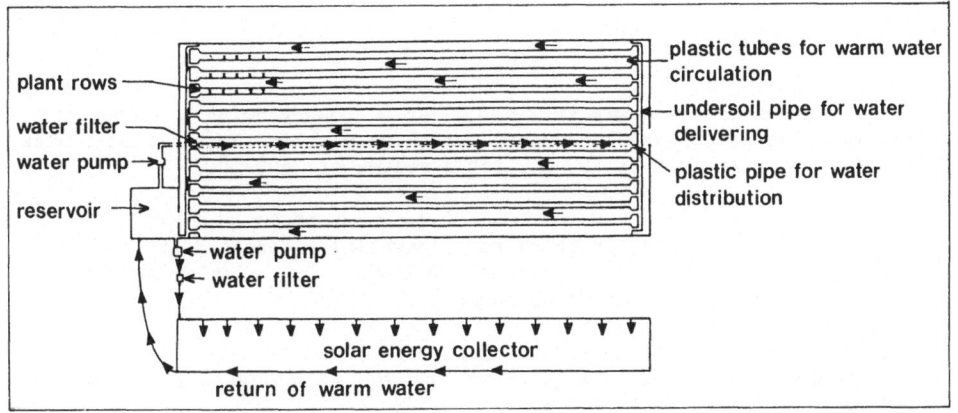

Fig. 22.3 : Arrangement of water circulation system to heat a greenhouse
with solar energy.
(The greenhouse heating system is linked with a water
reservoir and monitored by a water pump.)

23. Solar energy use for greenhouse heating

+ Project in Bari, Italy.

+ Description :
The solar greenhouse in this project is of the asymmetric type, having a
slope southwards leaning in such a way as to permit the optimal entry of
solar rays. The north front is fitted with protections against frost
and heat losses in winter, and against excess heat in summer. The solar
installation is mainly composed of :
- Five solar collectors of a total area of 15.45 m^2,
- a system for medium heating,
- a support system of air heating,
- an automatic humidification plant.
The size of the greenhouse is 96 m^2 (8 x 12 m^2) and it is covered with
alveolated polycarbonate of 10 mm thickness.

+ Conclusions :
The everyday conditioning in winter is done by heating the substratum on
soil with the storage hot water; and in summer by circulating inducted
air in the greenhouse through a controlled system of motorized opening
and closing of the windows.
The flat-plate solar collectors, installed through the primary flow,
accumulate heat in the storage tanks, which supply hot water for
possible zootechnical use in summer or in the hottest periods. An
auxiliary boiler protects the culture from low temperature peaks.

+ Study Center : P. Amirante
                 Institute of Agricultural Mechanics
                 165/A, via Amendola
                 70126 Bari
                 Italy

24. Greenhouse heating by solar collectors

+ Project in Sicily, Italy

+ Description :
Solar system for the winter heating of a rose-growing greenhouse in Sicily. The greenhouse is composed of three plastic tunnels placed side by side : its surface is 500 m², for a volume of 1400 m³.

The heating system is composed of three main parts : a set of solar collectors (flat-plate, in three rows, parallel - connected by insulated ducts : a hot water store of 15 m³), hot water circulation system with water-air exchanger and feeding pump.

+ Results :
The average energy available from the 80 m² collectors during the months of utilization is 565 MJ/day.
The average energy required for heating the greenhouse is 712 MJ/day.
The solar energy provides 79% of the energy needs.

+ Study Centre : Cattaneo e C.
                 S.A.S.
                 Via Baluardo Quintino Sella 24
                 28100 Novara
                 Italy

25. Solar energy in grape-type greenhouse

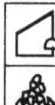

+ Project in Overijse, Belgium (1982).

+ Description :
A solar collector was installed in the top of a traditional type of greenhouse. Surplus solar energy is stored in a rockbed by circulating greenhouse air through the storage bed. The greenhouse itself is used as the solar collecting system and the heat transfer is improved by means of a DIY solar collector in the top of the greenhouse roof. When the temperature in the collecting area exceeds a setpoint value, the air is cooled by means of a fan.
During the night, the same ventilation system delivers the heat inside the greenhouse. The stone bed is constructed under the floor of the greenhouse, permitting a maximum use of the stored solar energy during night (see figure 25.1).

+ Conclusions :
 - The system resulted in energy savings of 26 % during the month of May and 100 % during the month of June, as compared to the energy need in a traditional greenhouse.
 - The installation costs were kept to the minimum (± 1 000 ECU for a greenhouse of 20 m by 7 m).
 - The light reduction because of the installed solar collector inside the greenhouse may cause prohibitive light reduction for light sensitive cultures.
 - The increased average temperature, due to better use of solar energy, resulted in a better growth and earlier flower production in the culture of orchids.

+ Study Center : Steenstra and Van Loey
                 Vrije Universiteit Brussel
                 Pleinlaan 2
                 B-1050   Brussel
                 BELGIUM

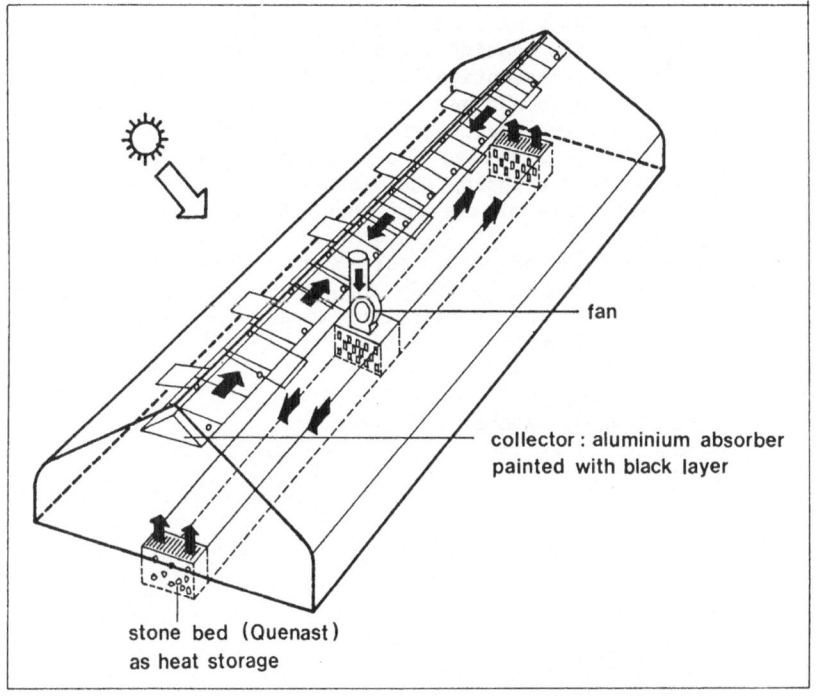

fan

collector : aluminium absorber
painted with black layer

stone bed (Quenast)
as heat storage

Fig. 25.1 : Solar greenhouse set-up.

26. Earth storage of thermal energy for greenhouse heating

+ Project in Nordrhein-Westfalen (Büttgen), West Germany.

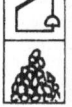

+ Description :
  In a single glazed greenhouse a new type of collector was installed to
  recover heat surplus and store it in the ground.
  The installation consists of two systems of tubes (see figure 26.1); one
  system, made up of absorbing tubes is fixed below the roof of the
  greenhouse, the other is installed in vertical holes (ϕ 64 mm) of 4.5 m
  depth in the ground below the greenhouse. In each hole there are two
  tubes (PVC); in the outer one water is pumped down and in the inner one
  it returns back to the absorber. The two systems are connected through
  a pump.
  During the day, a transfer of heat surplus in the greenhouse is achieved
  by circulating water into the ground.
  During the night, as the temperature in the greenhouse decreases, the
  flow direction is reversed.
  Both processes are controlled automatically. In the holes a special
  mass of clay which holds a great amount of water was used in order to
  improve the contact between the soil and the tube.

The holes are above the ground-water level, because otherwise the flowing ground-water would carry away the heat.

+ Conclusions :
  During one year, from June 82 to May 83, 28 200 MJ were stored in the ground, and in the same time 26 650 MJ were regained for heating the greenhouse. That means losses of about 5.5%.
  During that year the pump was running 4 175 hours, consuming 900 MJ.
  The plant reduces the oil consumption by 5 to 10% or a saving of 3 to 6 1 oil per m^2 greenhouse area and per year.
  The costs of electrical energy reduce the savings by 10%.
  The temperatures of the soil varied from 14°C to 24°C; also at low temperatures in the greenhouse during the winter the soil was relatively warm. The higher temperature of the soil has positive effects on the plant.
  The investment is between 13 and 15 ECU per m^2 greenhouse area.
  The costs of investment are relatively high but can be reduced by simpler tube construction. The amortisation time is 20 years for the used system, if some of the work is done "do it yourself".

+ Study Center : V. Musil
                 Landwirtschaftskammer Rheinland
                 Edenicher Allee 360
                 5300   Bonn
                 West Germany

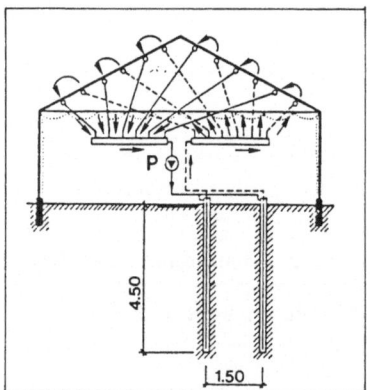

Fig. 26.1 : Collector and storage tube system.

27. Active and passive solar greenhouse climatization

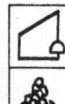

+ Project in Auteuil, France.

+ Description :
  The project is part of the renovation of PARIS municipal greenhouses located in Auteuil, which includes the testing of a new solar heating system.

  In the first place the performance of "passive solar collecting" adaptations were tested (double glazing, automated shading and aeration systems).

Secondly an active solar system is being built in combination with heat storage by means of "helioblocs".

+ Conclusions :
  - End 1983, the measurements indicated that a 33% energy-saving could be achieved by the passive solar energy system, (25% due to double glazing; 8% due to air-tightness).
  - The average relative humidity increased from 84% to 89%; average temperature from 22°C to 23°C.
  - The vertical stratification of air was estimated as favourable for the future functioning of the active solar energy system.
  - The total energy savings, to be realized by both active and passive solar energy systems are estimated at ± 50%.

+ Study Center : J.P. Hauet and B. Letellier
                 Novelerg
                 (Subsidiary of the C.G.E. Group)
                 Research and engineering company
                 12 rue de la Baume
                 75008 PARIS
                 FRANCE

+ For a more detailed description, cf. p. 111.

---

28. Greenhouse heating by solar collectors

+ Project in Milan, Italy.

+ Description :
  Development of solar active system for heating a greenhouse of appropriate geometry with heat storage system.
  The greenhouse has been designed to obtain the maximum utilization of the incident solar radiation.

  The heating is supplied by 30 m² of air collectors, with an average efficiency of 45%.

  The storage is of the pebble-bed type, with a capacity of 46 m³.
  The specific thermal capacity is 1.74 kJ/m³, and the expected time of utilization is 1 - 5 days.

+ Conclusions :
  The energy required for heating the greenhouse is 76.5 GJ/year.
  The solar plant provides 22.1 GJ/year (28.7%) and the auxiliary heater 54.5 GJ/year (74.3%).
  The efficiency of the system is 30%.

+ Study Center : G. Castelli
                 C.N.R. Institute for Agricultural Engineering
                 Università di Milano
                 Via Celoria 2
                 Milano
                 Italy

3. GENERAL CONCLUSIONS FOR THE GREENHOUSE SECTOR

3.1. Discussion of the listed research projects

3.1.1. Phytotechnical aspects

Considerable fuel savings were obtained by the incorporation of root zone warming in tomato production.(B.D. nr. 1) *(1)* The biological effect on plant growth of lower night temperature (even 5°C) is rather limited as long as the root temperature remains at 15° - 16°C: this can be realized only by the use of nutrient film technique (NFT).
The phytotechnical and biological effects of covering materials on plants are discussed in section 3.1.2.

3.1.2. Covering materials

Single glass is used as a reference basis in order to determine the characteristics of a number of alternative covering materials.

The reduced ease of thermal transport through a number of alternative covering materials results in important energy savings.(see Table 2).

Table 2 : Heat loss coefficient of some covering materials.
(source: Von Zabeltitz, Hanover)

	(K in $\frac{W}{m^2K}$)
Single glazing	6.7
Double glazing	3.4
Double Acryl glazing (8 - 18 mm)	3.4
Triple Acryl glazing (32 mm)	2.3
Polycarbonate	4.5

Reported <u>energy savings</u> range from 20 % to 38 % as compared with the reference solution. The Danish project (B.D.nr. 3) reports energy savings of 32 % (double glazing) and 39 % (double acryl) as compared to a reference solution (single glass). In the Denar experiment (B.D. nr. 6) energy savings of the acryl greenhouses are reported as 34 % (soil culture) and 35 % (rockwool culture), compared to the reference solution. During a period of three years average savings between 31 % and 35 % are reported for double covered and single covered greenhouses. Also according to B.D. nr. 5, the energy savings of a triple-span double clad polythene greenhouse were reported to be over 20 % compared to a single clad polythene greenhouse.
The effect of <u>light reduction</u> (cf. p. 14) differs according to the plant characteristics. The study described in B.D. nr. 2 indicated that the quality and the quantity of the yield of strawberries are not exclusively correlated with light transmission of the alternative coverings. Particularly, the balance between light and temperature determines the final yield. This explains the 40 % higher yield of a tomato crop grown in a double clad house as compared to the reference solution (B.D. nr. 5). The <u>biological effect of different coverings</u> on ornamental plants is rather limited. The Danish experiment (B.D. nr. 3) indicated that light

(1) See section 2.3 (B.D.nr. = basic description number).

reduction in permanently insulated greenhouses had only a limited effect on the dry matter production of seven selected pot plants. According to results of other experiments, the biological effect of light reduction on vegetable plants is significantly different from the observation for pot plants. In a Dutch study the relative importance of fuel savings, light reduction and production losses, were estimated as shown in Table 3 :

Table 3 : Basic data for economic comparison of energy saving measures in the Netherlands (source: Verhaegh, 1982).

	Double glazing	Twin covering - plastic materials -
- Fuel savings in % of total yearly fuel consumption	30 %	44 %
- Light reduction in comparison with a greenhouse with single glass	12 %	12.5 %
- Resulting yield losses in light-sensitive crops (vegetables)	14.4 %	15 %
in less light-sensitive crops (potted plants)	7.2 %	7.5 %

The relative humidity can be controlled by ventilation. High relative humidity may cause important problems such as fruit setting of vegetable crops and increase the disease infection level. In the Danish experiment (B.D. nr. 3) ornamental plants were not significantly affected by high r.h. values and for this application better insulation of greenhouses was suggested as a promising option. In this same context can be situated the research on the application of heat pumps for greenhouse air drying, involving both a positive effect on the air relative humidity and the energy balance of the greenhouse.

3.1.3. Thermal screens

The effectiveness of thermal screens depends on the types of screen materials and the tightness of the system. In B.D. nr. 8 the Institute for Horticultural Engineering in Hanover reported energy savings ranging from 38 % to 54 %, depending on the screen material. This study indicated that the temperature distribution under the screen is greatly improved as air movement in the greenhouse decreases. The temperature difference above and below the screen (up to 16 K) causes a steep temperature decrease when the screen is opened in the morning. This problem can be overcome partly by the use of control processes. As for double covering materials, the use of airtight materials results in an increase of the absolute humidity under the screen and can cause condensation on the plants. In addition, the use of thermal screens causes a light reduction when the screens are open as well as when they are closed. On average, a light reduction of 5 % is reported for open screens.

Two projects on the use of screens for tomatoes are reported (B.D. nr. 10; 11). B.D. nr. 11 confirms the above findings. During the periods that the screens were in use they reduced the heat loss at night by 55 - 60 %. Over the entire tomato growing season, a fuel saving of 25 to 30%

was reported. The average reduction of photosynthetic active radiation (PAR) transmittance of the screened house was 3 % (see also table 2 on p. 19).

Results of the application of thermal screens on chysanthemums are also described (B.D. nr. 9). Over a period of four nights, the heat requirement of the screened house was on average 47 % less than the non-screened house. The results varied considerably from night to night, but there was clear evidence of substantial fuel savings.

3.1.4. Energy storage with greenhouse as collector

An example of passive energy storage by means of thermal mass is given in B.D. nr. 15. The north side of this greenhouse consists of an existing rock, and a stone bed offers additional storage capacity. As compared to traditional greenhouses, the energy savings amounted to 70 %. Due to the building design and the geographical site, the application of this system is rather limited.

As to active means of energy storage, a distinction is made between storage in earth, storage in water, storage in stones and the use of phase change materials.
Reports on earth storage are available in different places (B.D. nr. 16; 26). The project described in B.D. nr. 26 indicated the limited possiblities of this application of solar energy in greenhouses in the colder climatic circumstances. Of course, storage of solar energy in the soil caused higher soil temperatures and hence an improved growth. But the economic evaluation indicated an estimated amortisation period of twenty years. The experiment described in B.D. nr. 16 indicated an energy saving of 70 % for greenhouses in Athenian conditions. Economic results should be even brighter in more southern parts (E.g. : Kreta) .

Reports on water storage are available for Greece (B.D. nr. 13), France (B.D. nr. 20), Germany (B.D. nr.12). Again, the German study indicates the limited possibilities of water storage in northern parts of the EC.
There are some prospects for the use of a heat pump and the short-term storage from day to night. For the Hanoverian conditions, a water volume of 0.2 m³ per m² ground area was found sufficient. For long term storage, water volumes of 12-20 m³ per m² would be required. The high demand for electricity (heat pump) is reported as a great disadvantage. The possible fuel savings are not large enough to allow for the required investments. The Greek project however offers broader perspectives because of the limited investment cost and the short-term use of solar energy (day-night temperature cycles). Moreover, it was possible to use a part of the greenhouse solar energy, which otherwise would be lost through ventilation. In the French project, the use of a fluid roof solar greenhouse and the water storage of the collected energy also substantially reduced the need for ventilation.
Between March and July, the energy savings in heating requirement are nearly 100 %. However, a relatively high investment is needed so that the economic prospects are rather limited.

Storage of solar energy in a stone bed is reported for Belgium (B.D. nr. 25). This solar system was built in an old type of traditional grape-type greenhouse. Considerable energy savings are reported for the period from

March to September. Application is bound to the particular greenhouse-type and hence limited.
The use of phase change materials is rather new and in an experimental stage. The system in France (B.D. nr. 14) decreased the energy consumption by 75 %.

3.1.5. Energy storage with collectors other than the greenhouse

Economic prospects for energy storage with such a collector are very limited in northern parts of the EC (B.D. nr. 17; 19). Economic calculations for a solar plant in northern climatic conditions indicate that the collector will not be paid back within its lifetime.

Applications in southern parts of the EC are more promising. The investment for a solar greenhouse in France (B.D. nr. 21) was 2.5 times the equivalent investment for the reference greenhouse. The share of the energy delivered by the solar system varied between 19 % and 83 %.

Similar results were reported for Italy (B.D. nr. 23; 24; 28). A calculation of the pay-back period was not available. In most cases the reported relative savings are high but the absolute amount of energy savings is generally too low for an economic application.

3.2. General evaluation

The E.C. horticulture is mainly located in the agro climatic zones I and III, very different regions as far as energy requirements are concerned.

In Fig.2a, the diagram for agro-climatic zone I reveals an excess solar supply, often resulting in excessively high temperatures during the summer months.
Light is not a limiting factor, so that active as well as passive solar systems combined with heat storage elements could help transfer the heat excess of the day to cover the heat demand of the cold nights, mainly during the months of February to April and September to November. However, a prerequisite for these and other applications is the development of new rational designs as well as innovative materials to replace the poorly constructed existing greenhouses.
As for agro climatic zone III, (cf. figure 2b) the major part of the heat demand coincides with the winter season where availability of light to the plant is a limiting factor, the value-in-use of which far exceeds the value-in-use of any form of thermal energy at that time of the year. Consequently, the primary solutions are sought in energy-saving devices, most often not explicitly solar ones, or even solar at all :
- Maximum boiler efficiency ;
- Application of crop varieties and cropping systems which have a lower temperature requirement ;
- New designs and adaptations of greenhouses to permit maximum light entry while minimizing heat loss (thermal screens, cladding, other passive forms of solar energy applications) ;
- Use of Nutrient film techniques and root zone warming, (applying temperature regulation procedures in which solar energy could possibly play a role ;
- Even though chances of success are smaller, here too, studies are being made of the potential use of storage media (combined with active collecting systems, or using the greenhouse itself as a collector)

which transfer the heat excess of the day to the cold nights during the period May to September.

AGRO CLIMATIC ZONE I

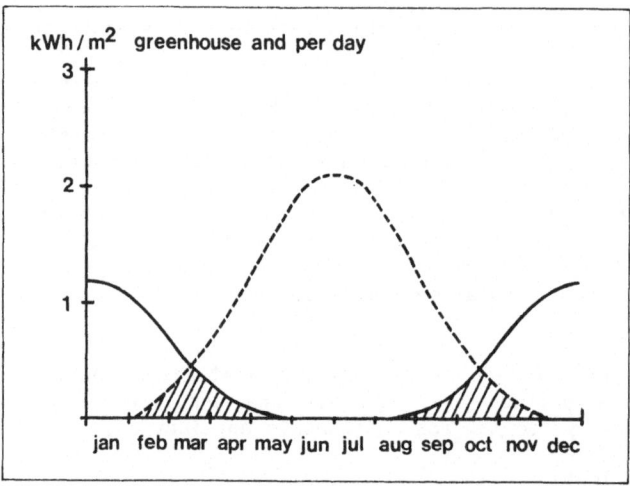

Fig. 2a : Comparison of energy demand (——) and maximum amount of solar heat that can be stored by a greenhouse in agro climatic zone 1 (- - - -) (▨ Potential solar energy input) (1)

The economic feasibility of the more explicitly solar systems - operative 4 (Agro climatic zone III) to 6 (Agro climatic zone I) months per year - with temperature level requirements near to ambient temperature - still remains to be established by means of more data *(1)*. Feasibility will depend greatly on the development of simple low cost solutions, yielding reasonable efficiency values. For agro climatic zone I, the "simplicity" factor is also expected to be essential from the socio-economic viewpoint.

In any case, only a general approach, analyzing the diverse influences of the different techniques (solar as well as others), and optimizing their integration in the plant production process, can have a far-reaching impact on the energy efficiency in both climatic regions.
This approach - involving computer controlled data collection and automatic regulation systems - is in full development in agro climatic

(1) *The estimated pay-back period is a good indication for this economic feasibility. However, it is extremely difficult to judge the production results of an experimental set-up. Indeed, a number of reasons, entirely outside the field of solar energy application may cause extremely high variations in the obtained results.*
Moreover, it seems even more difficult to judge the results of the experimental trials in money terms because of the random influence of a number of exogenous variables (bargaining power, distance from city, etc.).
The evaluation of the fuel saving, particularly if a reference solution is available, offers a valuable indication of the possible economy.

zone III, this in contrast with zone I.

A survey study group *(1)* estimated that over the decade 1980–1990 application throughout the EC of this general approach could result in energy savings of up to 1.5 x 10^6 t.o.e./year (25% of the estimated direct thermal energy consumption).

They gauged the maximum possible contribution of solar energy devices at about 10% of this figure, its main impact being expected in agro climatic zone I.

AGRO CLIMATIC ZONE III

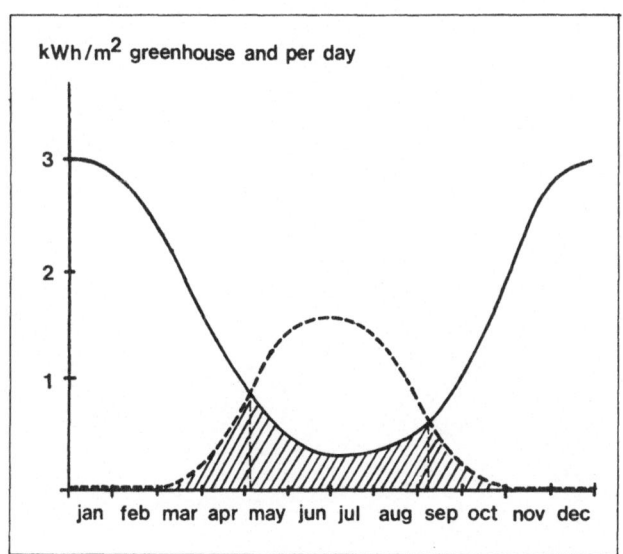

Fig. 2b : Comparison of energy demand (————) and maximum amount of solar heat that can be stored by a greenhouse (—————) in agro climatic zone 3
(////// potential solar energy input) (1)

(1) Participants in action 1 (cf. "Readers' guide", p. XXI).

References :

(1) Damrath J. and Von Zabeltitz Chr. : "Greenhouse heating with sun energy" Acta Horticulturae 115, 1981, p. 542 ; (Combined with personal calculations)

APPENDIX TO PART I :
Some specific research projects on greenhouses

In this appendix, seven projects in the area of solar energy applications in the greenhouse sector are described in detail.

1. EVALUATION OF DIFFERENT COVERING MATERIALS FOR GREENHOUSE PRODUCTION OF STRAWBERRIES (Belgian project, cf. p. 76)

2. THE BIOLOGICAL EFFECTS OF ENERGY SAVINGS IN GREENHOUSES (Danish project, cf. p. 84)

3. EVALUATION OF DOUBLE-COVERED POLYTHENE GREENHOUSES FOR EARLY TOMATO PRODUCTION (Irish project, cf. p. 95)

4. COMPARISON OF THE LIGHT TRANSMISSION OF GREENHOUSES COVERED WITH TWINWALLED SHEETS OF POLYCARBONATE AND ACRYLIC (English project, cf. p. 104)

5. GREENHOUSE CLIMATIZATION BY MEANS OF HELIOBLOCS (French project, cf. p. 111)

6. GREENHOUSE CLIMATIZATION BY AN EARTH-AIR HEAT EXCHANGER (Greek project, cf. p. 122)

7. DEVELOPMENT OF A SOLAR SYSTEM FOR HEATING GREENHOUSES (Greek project, cf. p. 131)

The reporting formats used to compile the detailed descriptions are explained in the ADDENDUM, p. 467. The numbering of the subheadings in the detailed descriptions corresponds to the reporting format.

1. Evaluation of different covering materials for greenhouse production of strawberries

Project location

GEMBLOUX, BELGIUM

Study Center :
  - Project Leader : Prof. J. Deltour
  - Institution : Faculté des Sciences Agronomiques de l'Etat
    State University
    8 Avenue de la Faculté
    5800 Gembloux
    Belgium

Running since 1979.

Project description

The strawberry crop is grown in different greenhouses of similar size (18 m²-22 m²). This research is part of a program of similar experiments with other vegetable crops (lettuce and tomatoes).
The main objective is to evaluate the relation between the plant production and different covering materials.
The research was also aimed to analyse the reaction on yield and quality of strawberries when energy saving covering materials are employed.

Scheme

The site plant of the 9 greenhouses is as follows :

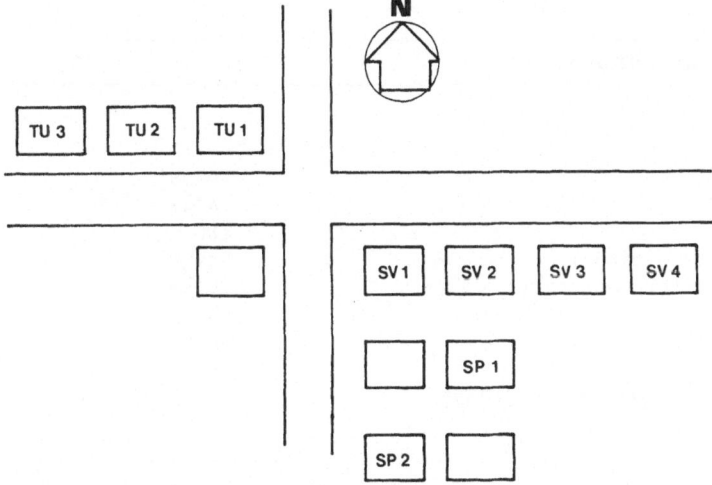

SV 1 = hortiplus covered greenhouse
SV 2 = glass doubled greenhouse with a polyethylene film
SV 3 = reference greenhouse
SV 4 = double glass greenhouse
SP 1 = polycarbonate covered greenhouse
SP 2 = acryl covered greenhouse
TU 1 = Ethylene Vinyl Acetate covered tunnel
TU 2 = polyethylene "Infrared" covered tunnel
TU 3 = polyethylene covered tunnel

Conclusions

- The rate of air renewal differs greatly between the various types of greenhouses and depends very much on the wind speed.
- The strawberry crop is not very sensitive to variations of illumination and temperature conditions, as compared with lettuce and tomatoes.
- However, it is observed that the quality and the yield of strawberries, as found earlier with tomatoes and lettuce, are not exclusively correlated with light transmission.
- The final yield is mainly determined by the balance between temperature and light.
- Due to the limited dimensions of the greenhouse, the observed results have a very relative significance.

PROJECT ANALYSIS

1. Site and climate

1.1. Site

Latitude :	50°35' N
Longitude :	4°41'14" E
Altitude :	180 m
Nearest main city :	Namur
Distance from main city:	20 km
Direction from main city :	NW
Obstructions :	None

1.3. Annual long-term averages

(Over 30 years : 1949-1979)

			Project
1. Prevailing wind direction	Winter		SSW
	Summer		SSW
2. Average wind speed	W	m/s	3.3
	S	m/s	2.1
3. Total precipitation		mm/yr	779
4. Relative humidity	W	%	88
	S	%	79
5. Global irradiation on horizontal plane		MJ/(m^2.yr)	3 452
6. Diffuse proportion		%	52
8. Hours sunshine		hrs/yr	1 553
9. Ambient temperature		C	9.1
10. Average max. temp. (Jul/Aug)		C	26.2
11. Average min. temp. (Jan)		C	-4.9

12. Meteorological station :   Gembloux (at distance of 2 km from site)

2. Plant

2.1. Greenhouses

SV 1 : Hortiplus (low emissivity glass) covered greenhouse
SV 2 : Greenhouse covered with glass, doubled with a polyethylene film
SV 3 : Reference greenhouse : glass cover
SV 4 : Double glass covered greenhouses
       - Floor area :   18 m^2
       - Volume :   35.1 m^3
       - Glazing area :   50.5 m^2 (roof 21 m^2; sides 29.5 m^2)
       - Ventilation system :   1 fan

SP 1 : Acryl (PMMA) covered greenhouse
SP 2 : Polycarbonate (8 mm) covered greenhouse
  - Floor area :   21.6 m^2
  - Volume :   51.8 m^3
  - Glazing area :   64.9 m^2 (roof 24 m^2; sides 40.9 m^2)
  - Ventilation system :   1 fan

TU 1 : Ethylene Vinyl Acetate covered tunnel
TU 2 : Polyethylene "Infrared" covered tunnel
TU 3 : Polyethylene covered tunnel
  - Foor area :   21.6 m^2
  - Volume :   34.9 m^3
  - Glazing area :   47.6 m^2

2.5. Installation parts and measuring devices

2. Description of the sensors
   - Temperature. NTC resistor IN 914 or Pt 100, giving an accuracy of 0.3°C in the range - 20°C + 90°C. Calibration is realised at 5°C and 50°C. The sensors are mounted in a black integrating sphere (radiation temp.) and in a reflecting cylinder (air temp.).
   - Relative humidity. Rotronic RHTT with an accuracy of 2% in the range 0 - 100%; calibration at 80%. The sensor is mounted in a reflecting cylinder.
   - Photosynthetic photon flux density. LICOR 190 S or MACAM SD IOI Q, giving an accuracy of 1% in the range 0 - 2000 uE/(m^2.s). The Macam sensors are calibrated against the Licor which is calibrated by the National Bureau of Standards. The sensors are mounted on horizontal levelling.
   - Energy consumption. Standard kWh meters.
   - Wind speed. Standard cupola anemometer.

3. Scheme of measurement points (cf. figure 2.5.3.)

3. Products

3.1. Culture in greenhouse

(for : SV 1, SV 2, SV 3, SV 4, SP 1, SP 2, SP 3.)

1. Type of crop : Strawberry.
2. Date of sowing / planting : 01/08
                harvesting : 15/04 - 15/06
3. Growing system : Conventional

4. Required temp. regime : min. temp.   15/01   15/02   15/03
                              night °C   5   7   10
                              day °C   8   10   15
                           max. temp. : 35°C
7. How sensitive is the crop to light ? Low sensitivity.

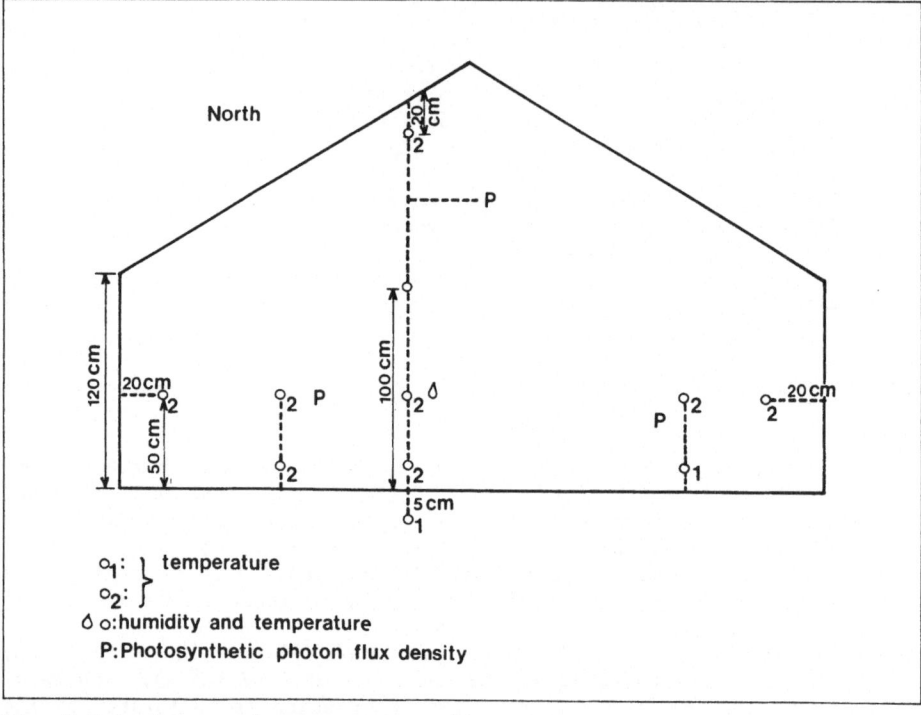

Fig. 2.5.3. Scheme of measurement points.

3.2. Culture in tunnels

(for : TU 1, TU 2, TU 3)

1. Type of crop : Strawberry
2. Date of sowing / planting : 01/08
   harvesting : 15/04 - 15/06
3. Growing system : Conventional
4. Required temp. regime : min. temp. : 5°C
   max. temp. : 35° C
7. How sensitive is the crop to light ? Low sensitivity.

---

4. Economic aspects *(1)*

4.1. Economic environment

1. Energy price of oil (ECU/1)   0.33
2. Energy price of electricity (ECU/kWh) 0.11
3. Energy price evolution prospects
   (real terms) $\overline{+}$ 1%/yr
5. Interest rate of loans for equipment   12%/yr

(1) presumably 1983 prices (N.O.T.E.)

4.2. Investment expenses

Price of the different covering materials in ECU/m².

1.	Hortiplus (low emissivity plan)	6 - 7	ECU/m²
2.	Glass doubled with a polyethylene film	3.9 + 2.3	ECU/m²
3.	Normal horticultural glass	2.3	ECU/m²
4.	Double glazing	17 - 18	ECU/m²
5.	PMMA-acryl	30	ECU/m²
6.	Polycarbonate	11 - 14	FCU/m²
7.	Ethyl Vinyl Acetate	0.45	ECU/m²
8.	Polyethylene "Infrared"	0.45	ECU/m²
9.	Polyethylene	0.33	ECU/m²

5. Results

5.2. Weather conditions and energy

For the period 15/01 - 30/05 the average ambient temperature was 6.44°C, the average wind speed 2.83 m/s and the average solar radiation 9.5 MJ/m². The electricity consumption for the different houses, in MJ, was as follows :

SV 1	9 360
SV 2	4 586
SV 3	9 450
SV 4	8 440
SP 1	4 530
SP 2	8 640
TU 1	8 506
TU 2	5 072
TU 3	12 376

5.3. Inside climate of greenhouse or tunnel

1.	ambient												from : 15/01 PERIOD
2.	outside	SV1	SV2	SV3	SV4	SP1	SP2	SP3	TU1	TU2	TU3		till : 30/05
3.	10.0	5.0	5.1	4.9	5.2	5.1	4.9	5.0	5.0	5.3	5.1	°C	min.
4.	6.44	15.4	15.4	16.7	13.9	14.4	15.8	14.0	12.5	12.1	12.2	°C	average ⟩ TEMPERATURE
5.	16.0	34.9	37.2	35.3	32.6	39 '	35.1	30.0	32.6	33.4	30.2	°C	max.
6.	38.5	36.7	34.6	39.8	39.2	31.0	23.1	29.1	50.3	47.8	56.4	%	min.
7.	78.8	71.0	86.2	88.9	85.0	72.7	66.8	66.6	82.8	75.0	74.0	%	average ⟩ HUMIDITY
8.	100	81.3	99.8	100	97.6	100	100	81.3	96.6	85.3	78.9	%	max.
10.	100	42.9	45.4	53.3	48.6	53.2	40.7	23.9	60.3	56	51.4	%	PAR (% of outside)

5.4. Crop results

	SV1	SV2	SV3	SV4
1. Date of sowing	01/08	01/08	01/08	01/08
2. Harvesting Date :	15/04-15/06	15/04-15/06	15/04-15/06	15/04-15/06
4. Flower number/plant (24/3)	6	4	4	6
5. Fruit number/plant	22.1	21.0	22.5	19.4
6. Leaf area measure- ment/plant (m^2)	23.97×10^{-4}	19.44×10^{-4}	19.54×10^{-4}	22.80×10^{-4}
7. Production obtained/ plant (gr)	304.7	294.6	363.5	301.6
8. Other differences : yield ECU/pl.	11.64	10.14	12.72	10.88
9. Half harvesting date	08/05	15/05	12/05	07/05

	SP1	SP2
1. Date of sowing	01/08	01/08
2. Harvesting date :	15/04-15/06	15/04-15/06
5. Fruit number/plant	14.0	13.0
6. Leaf area measure- ment/plant (m^2)	15.53×10^{-4}	11.80×10^{-4}
7. Production obtained/ plant (gr)	243.9	204.7
8. Other differences : yield ECU/pl.	8.49	7.43
9. Half harvesting date	04/05	12/05

	TU1	TU2	TU3
1. Date of sowing	01/08	01/08	01/08
2. Harvesting date :	15/04-15/06	15/04-15/06	15/04-15/06
5. Fruit number/plant	12.7	16.8	13.1
6. Leaf area measure- ment/plant (m^2)	16.12×10^{-4}	16.84×10^{-4}	16.77×10^{-4}
7. Production obtained/ plant (gr)	205.5	236.8	224.3
8. Other differences : yield ECU/pl.	6.48	7.40	7.02
9. Half harvesting date	13/05	19/05	12/05

5.7. An example of some measured data

A number of data are available for a central computer system.
Below an example is given of the average greenhouse temperature (sensors 0-9), the soil temperature and the average transmissivity of radiation for the different greenhouses. The data are the average on March 8th 1983 from 7.30 - 18.30.

	Average green-house temp.	Average soil temperature	Average transmissivity
SV1	14.0°C	11°C	0.48%
SV2	18.4°C	12.5°C	0.43%
SV3	19.9°C	–	0.60%
SV4	20.8°C	–	0.52%
SP1	17.4°C	15.2°C	0.61%
SP2	16.9°C	13.2°C	0.55%

2. The biological effects of energy savings in greenhouses

Project location

AARSLEV, DENMARK (since 1979)

Study Center :
- Project Leader : M.G. AMSEN
                   O. FRØSIG NIELSEN
- Institution :   Danish Research Service for Plant and Soil Science
                   Research Centre for Horticulture
                   Institute of Glasshouse Crops
                   Department of horticultural engineering
                   DK-5792 Aarslev, Kirstinebjergvej 10
                   DENMARK

Project description

In four greenhouses plant growth and use of energy are recorded.
The four houses are identical with respect to dimensioning and technical
installation, except for the following differences :

House 1 Covered with single glass (reference solution)
House 2 Covered with double glass (Sedoglass 2.4)
House 3 Covered with double acryl (Stegdoppeltplatte 16 mm)
House 4 Covered with single glass, and equipped with aluminized thermal
        screens, which are activated at night.

A selection of 7 different pot plants are grown in the houses.
In houses 1 and 4 shading curtains are installed and activated also during
night.

The main objective is to investigate how thermal screens and different
coverings influence the heat loss from greenhouses and the plant growth in
these houses.

The glasshouses are built according to a normal Danish wide span house
equipped with rolling benches for maximum space utilization.

The heating system is dimensioned to keep a maximum temperature difference
of 30 K between inside and outside.

The heating system is divided into top- and wall heating, floor heating
(below the benches), and bench heating in the benches insulated with
styropor downwards.

The three heating systems are controlled individually and energy
consumption is measured separately for each system.

Furthermore, the total energy consumption for the whole house is measured
independently.

The benches are equipped with a capillary watering and feeding system and
covered with a matting which is kept wet constantly (as common in Danish
pot plant nurseries).

Maximum air humidity is controlled by a hygrostat and mutual activation of
the heating- and ventilation system.

Carbon dioxide is supplied at a steady flow rate which allows 1 200 ppm at
an outside wind velocity of 4 m/s.

Scheme

The air temperature in the four greenhouses is controlled by a set point
min. temperature, which is 17°C when the light intensity is below
300 Lux (1) outside and 20°C when the irradiation is higher than 300 Lux
outside.

The bench temperature is controlled around a min. of 19°C.

The shading curtains are drawn when the irradiation is higher than
200 W/m^2.

Both the shading curtains and the thermal screens are drawn when the
irradiation is lower than 100 Lux outside.

The air ventilators are opened when the air temperature inside the
greenhouse is higher than 28°C, or if the air humidity is higher than 92%
R.H.

(1) "Lux" is a measure of the light intensity as the human eye perceives
it. Cf. chapter 1, p. 11. (N.O.T.E.)

The benches are watered with 3 mm of water, whenever 2 mm are evaporated from an evaporimeter.

<u>Greenhouse no.</u>

1. House covered with single glass and shading curtains acting as a reference house.
2. House covered with double glass (sedoglass 2.4).
3. House covered with double acryl (stegdobbeltplatte 16 mm).
4. House covered with single glass, shading curtains and equipped with thermal screens, which are activated at night.

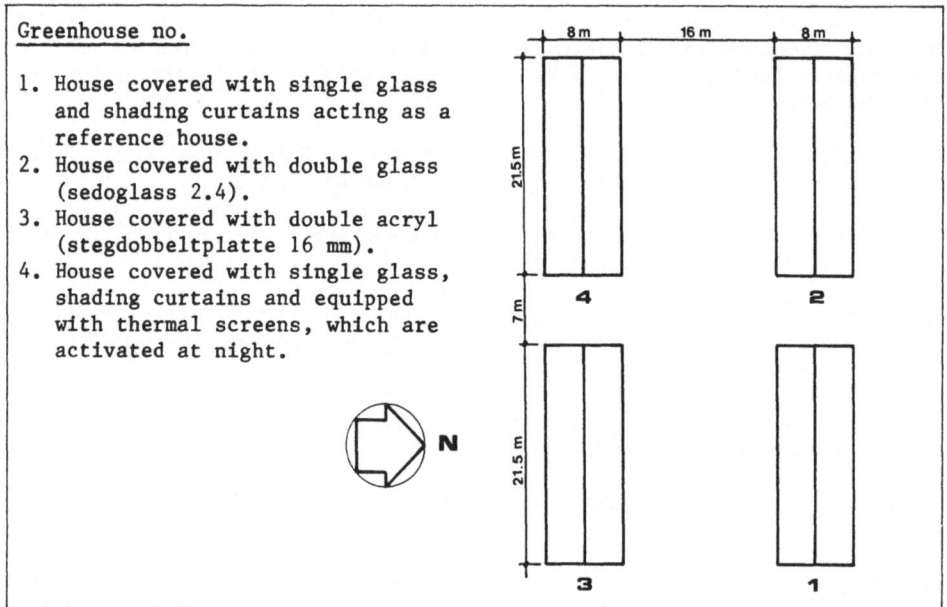

Conclusions

By using double glass or double acryl for covering greenhouses or by installing aluminized thermal screens in greenhouses much energy for heating can be saved, as compared to a single glass covered house without thermal screens.

Results for the period : 1/11/1981 - 30/4/1982 :

House	Heating GJ	energy rel.	Average room temp. °C	Average Rel. humidity %	Average light inside house W/m²
Thermal screens.	338	76	18.5	77	105
Double glass.	302	68	18.8	84	109
Double acryl.	273	61	19.1	83	105
Single glass (reference)	446	100	18.7	80	107

The effect on plant growth of the different insulating materials has proved to be rather small.
In some cases the use of insulating materials resulted in a shorter growth period. This means that even more energy may be saved per produced pot plant, as compared to production in a single glass covered house.

In the houses with double covering materials there has been a significantly higher relative humidity. However, no effect of this on plant growth has been observed.

It has been observed that the max. Rh was significantly higher in the houses with twin walled materials in spite of air humidity control set at 90% rh. Due to installation of shading curtains the amount of light inside the four houses has been very similar, even though the cover materials have different light reductions. There is however still a difference between the houses when light intensity is low (shading curtains not used).
In houses with twin walled covers, the heating system may be reduced significantly as compared to normal standards.

PROJECT ANALYSIS

1. Site and climate

1.1. Site

Latitude :	55°18'29" N
Longitude :	10°26'53" E
Altitude :	10 m
Nearest main city :	Odense
Distance from main city :	12 km
Direction from main city :	South
Obstructions :	None

1.3. Annual long-term averages

			Project	Country in general
1. Prevailing wind direction			SW W	SW W
2. Average wind speed	W	ms^{-1}		3.6
	S	ms^{-1}		3.0
3. Total precipitation		mm/yr	631	664
4. Relative humidity	W	%		85
	S	%		73
5. Global irradiation on horizontal plane		$MJ/(m^2\ yr)$		3 635
7. Degree days (Base temp = 17°C)		°C days	3 191	3 209
8. Hours sunshine		hrs/yr	1 777	1 739
9. Ambient temperature		°C	7.9	7.9
10. Average max. temp. (July/Aug)		°C		20.8
11. Average min. temp. (Jan)		°C		- 2.4

Note : W = Winter
       S = Summer

1.4. Monthly long-term averages

Month	Solar irradiation on horizontal plane MJ/(m².month)	Sunshine hours Hours/month	Average ambient temp. °C
January	47	39	−0.2
February	108	64	−0.4
March	251	126	1.7
April	411	176	6.4
May	577	258	11.3
June	633	268	14.7
July	569	259	16.6
August	485	239	16.3
September	299	170	13.1
October	153	102	8.6
November	66	47	4.7
December	36	29	2.1

2. Plant

2.1. Greenhouse

1. Dimensions : Floor area :   172 m²
             Volume :   504 m³
             Glazing area : 322 m²
             Wall area :   47 m² (concrete)

4. Ventilation systems :
Ridge ventilators activated by a control systemat 28°C.

2.5. Installation parts and measuring devices

1. <u>Technical description of the installation parts</u> :

Capacity of the heating system in each house

Bench heating system :   9.3 kW
Floor heating system :   22.5 kW
Top- and wall heating system : 27.1 kW − 51.1 kW

2. <u>Description of the sensors</u>

- Temperature measurements
Type : Platin − resistance Pt 100
Dimension : Ø 8 x 115 mm
Precision :+ 0.05°C

- Air humidity
Dry-wet
Type : Platin-resistance Pt 100 DIN 43760

- Waterflow
  Indication : Magnetic flowmeter
  Precision :+ 1%
  Fabr. : Fischer & Porter

- Solar irradiation on horizontal plane (Kipp solarimeter)
  Wavelength Range : 300 ηm to 2.5 μm
  Accuracy within 1%; linearity < 1%; temperature coefficient
  0.15%/K

- Infrared measurement instrument
  Objective : K
  Range : 0-100°C
  Emission : 0.95
  Fabr. : Heimann.

3. <u>Measurement points</u> (cf. figure 2.5.3.)

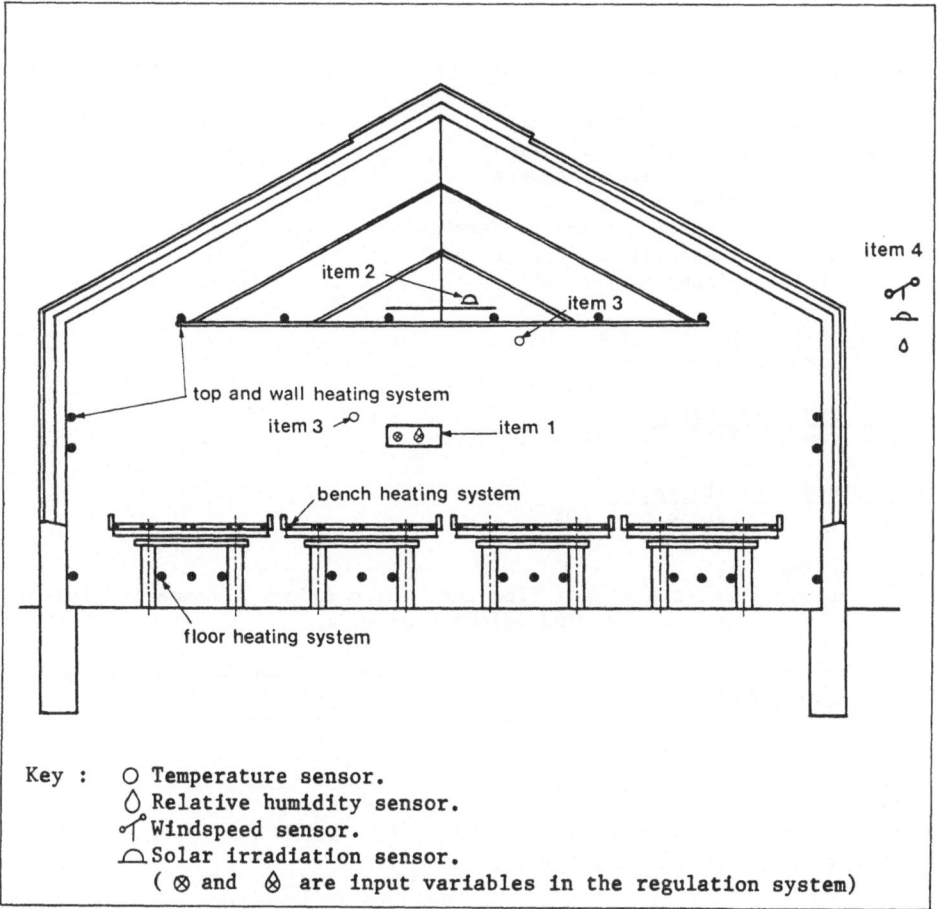

Key : O Temperature sensor.
       ◊ Relative humidity sensor.
       ⌐ Windspeed sensor.
       △ Solar irradiation sensor.
       (⊗ and ◊ are input variables in the regulation system)

Figure 2.5.3. : Scheme of the measurement points
                (For an explanation of the indicated items, see next
                page).

Item 1   In the centre of the greenhouse an aspirated box measures the air temperature and the air humidity.
Item 2   Above the top heating pipes the solar irradiation on horizontal plane is measured.
Item 3   By infrared measuring instruments the temperature of the roof inside and the temperature of the leaves are measured.
Item 4   Outdoor climate results are given by the meteorological station at the Research Centre.
Further measurements :   Inlet water temperature, outlet water temperature and circulating waterflow for bench, top and wall, floor and total system.

All measurements are collected and stored on line on disk and tape by the local computer of the Research Centre.

3. Products

3.1. Culture in greenhouse

1. Type of crop
   A selection of 7 different types of green foliage or flowering pot plants is used :
   Chrysanthemum 'Yellow Mandalay';
   Codiaeum variegatum 'Hollufiana';
   Diffenbachia maculata 'Compacta';
   Ficus benjamina;
   Hedera canariensis 'Gloire de Marengo';
   Kalanchoë blossfeldiana 'Anette' and
   Saintpaulia ionantha 'Ballet'.

3. Growing system
   Capillary benches.

4. Required temp. regime
   min. temp. : 17°C; max. temp. : 28°C.

5. Required humidity regime
   max. relative humidity : 90%

6. CO_2 regime
   CO_2 is supplied at a steady flow rate which allows a concentration of 1 200 ppm for an outside wind velocity of 4m/s.

4. Economic aspects

4.1. Economic environment

1. Energy price of oil	ECU/1	0.12 – 0.25
2. Energy price of electricity	ECU/kWh	0.063
3. Energy prices evolution prospects		
– of oil or gas	%/yr	10
– of electricity	%/yr	10

4.2. Investment expenses *(1)*

		HOUSE 1	HOUSE 2	HOUSE 3	HOUSE 4
1. Total investment budget	ECU	45 600	52 300	52 200	50 000
2. Cost of equipment :					
– cost of greenhouse per m^2	ECU	138	201	201	187
– cost of regulation equipment	ECU	5 500	5 800	5 800	5 800
Measurement – and registration equipment	ECU	12 100	12 100	12 100	12 100
3. Time of personnel for the conception, installation and start-up of equipment	man-months	9	9	9	9
Cost of personnel for the conception, installation and start-up of equipment	ECU	17 350	17 350	17 350	17 350
4. Lifetime of the equipment	Years	20	20	20	20

(1) Built in 1979   Monetary unit   year 1979

4.3. Operating expenses (over period 1/11/81-30/4/82)

		HOUSE 1	HOUSE 2	HOUSE 3	HOUSE 4
2. Maintenance cost (replacement of material or components)					
– spare parts (mostly plants)	ECU	1 550	1 550	1 550	1 550
– salaries (internal-external)	ECU	6 200	6 200	6 200	6 200
4. Other manpower costs	man-month	15	15	15	15
	ECU	31 000	31 000	31 000	31 000

5. Results

5.2. Energy

I. Results measured over the period of 1/11/81 – 30/4/82

		HOUSE 1	HOUSE 2	HOUSE 3	HOUSE 4
4. Heating energy (non-solar)	GJ	446	302	273	338
5. Energy for operation (pumps, fans, ...)	GJ	3	3	3	3

II. Calculated energy consumption per m^2 for a standard greenhouse of 20 x 50 m^2 based on a Danish reference year. (cf. figure 5.2)

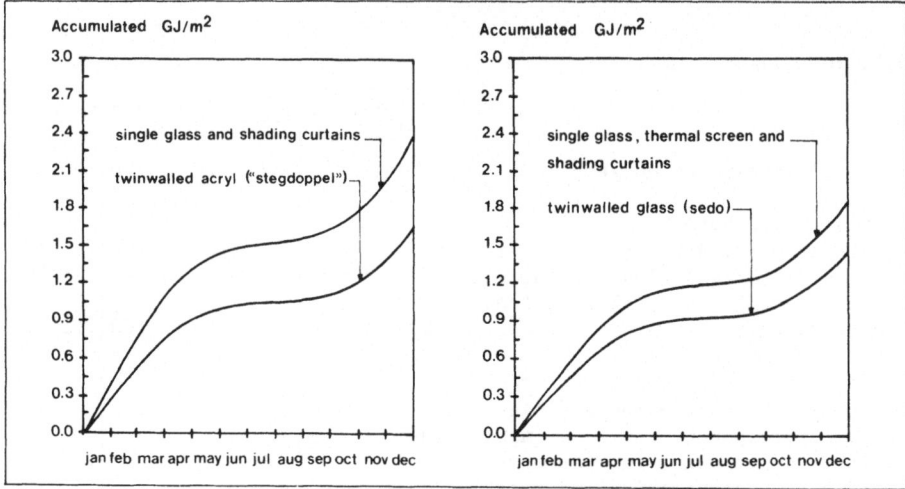

Fig. 5.2. : Calculated energy consumptions as a function of covering materials for a standard greenhouse (20x50 m²) in Denmark.

5.3. Inside climate (measuring period of 1/11/85 - 30/4/85)

		HOUSE 1	HOUSE 2	HOUSE 3	HOUSE 4
1. Temperature : min.	°C	15.9	16.1	16.4	15.5
average	°C	18.8	18.8	19.1	18.5
max.	°C	28.4	28.3	28.6	28.1
2. Humidity : min.	%	54	53	57	47
average		80	84	83	77
max.		93	95	94	93
3. Light : average	Wm^{-2}	107	109	105	105

5.4. Crop results

	HOUSE 1	HOUSE 2	HOUSE 3	HOUSE 4
1. Date of planting	1/10	1/10	1/10	1/10
2. . Harvesting date (production time in number of days)				
Chrysanthemum	62.3	66.7	64.8	62.5
Codiaeum	145.5	140.8	133.4	138.8
Diffenbachia	108.3	107.3	109.3	103.7
Ficus benjamina	143.6	139.7	129.6	132.5
Hedera	55.9	60.9	52.1	50.9
Kalanchoë	109.4	107.0	106.2	106.1
Saintpaulia	76.7	80.0	82.4	74.5
. mg. dry weight/ cm² leaf area				
Chrysanthemum	2.0	1.8	2.0	2.0
Codiaeum	65.2	64.0	72.0	67.8
Diffenbachia	29.6	31.4	29.7	35.0
Ficus benjamina	6.4	6.1	6.0	6.3
Hedera	3.3	3.1	3.2	3.3
Kalanchoë	6.9	6.5	6.2	6.4
Saintpaulia	5.6	5.3	6.1	4.9

	HOUSE 1	HOUSE 2	HOUSE 3	HOUSE 4
. Dry matter of saleable plants (g/plant)				
Chrysanthemum	3.8	3.4	4.2	3.7
Codiaeum	11.9	11.1	11.2	11.8
Diffenbachia	6.7	6.5	7.2	6.6
Ficus benjamina	8.8	8.0	8.2	7.8
Hedera	3.5	3.5	3.4	3.3
Kalanchoë	8.4	7.2	8.6	7.3
Saintpaulia	4.0	3.5	4.4	3.8
. Fresh weight of saleable plants (g/plant)				
Chrysanthemum	45.1	41.6	48.2	45.9
Codiaeum	62.1	56.6	60.1	60.3
Diffenbachia	82.4	77.2	89.7	83.9
Ficus benjamina	33.4	31.0	32.5	30.2
Hedera	22.8	22.2	21.9	21.4
Kalanchoë	170.9	152.2	179.3	149.3
Saintpaulia	89.2	76.4	94.6	95.6

It was expected that light reduction in the permanently insulated greenhouses (nos. 3 and 4) would have a severe influence on dry matter production and growth rate of the different types of potplants : this was only the case for a few plants.

5.6. Average Water temperature (°C)

	HOUSE 1	HOUSE 2	HOUSE 3	HOUSE 4
Inlet bench	38.9	32.9	30.4	36.1
Outlet bench	36.6	31.3	28.9	34.2
Inlet floor	60.3	53.3	49.9	53.5
Outlet floor	52.2	46.5	44.1	46.8
Inlet top/wall	43.2	35.1	35.2	30.7
Outlet top/wall	40.0	32.2	32.4	28.7
Inlet total	84.0	83.7	83.2	83.2
Outlet total	63.3	66.0	66.7	66.9

5.7. An example of some measured data (cf. figures 5.7.1 and 5.7.2)

The measurments are stored on disk and tape of the computer installation of the Research Station. The measurements are available for every hour and different computer programs allow a printout and a graphical scheme of the recorded values for each greenhouse. Below an example is given of the average leaf temperature over the period March 16-April 30 1982 in greenhouse 1 and greenhouse 2.

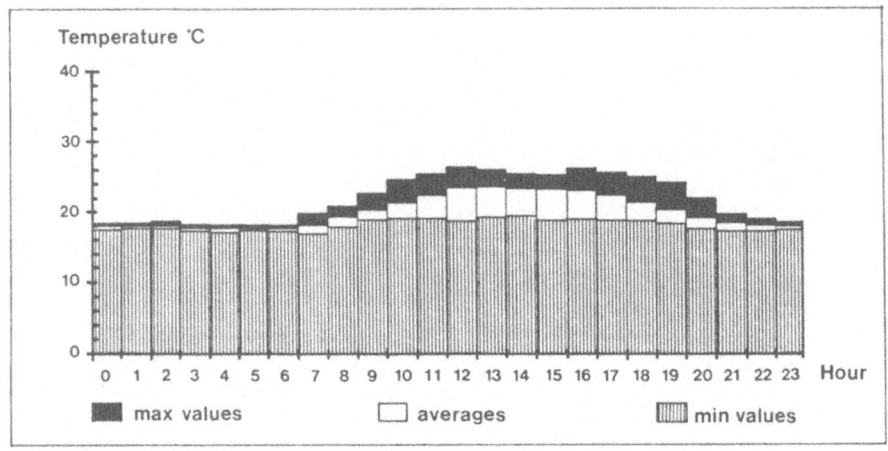

Fig. 5.7.1 Leaf temperature in the reference greenhouse, March 16–April 30 1982.

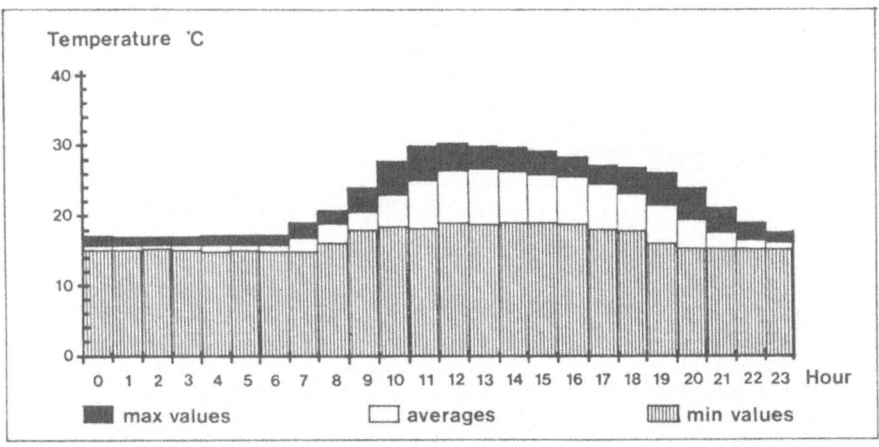

Fig. 5.7.2. Leaf temperature in the double glass greenhouse, March 16–April 30 1982.

3. Evaluation of double - covered polythene green-house for early tomato production

Project location

DUBLIN, IRELAND

Study Center :
- Project leader : Dr. T. O'FLAHERTY
- Institution : Kinsealy Research Centre
  Malahide Road
  Dublin 5
  IRELAND

Running since 1977

Project description

The project is based on a triple-span 30 m x 20 m double covered polythene greenhouse, with a piped hot water heating system.
The control of heating, ventilation and CO_2 supplementation in the greenhouse is incorporated in an integrated system involving a Nascom 2 microcomputer, with analogue back-up which takes over automatically in the event of computer failure. The input parameters to the control system are the temperature, relative humidity and CO_2 concentration in the greenhouse, and the temperature, relative humidity and light intensity outside.
The control strategy to be implemented through the microcomputer software

will have the following main objectives.

(1) to achieve desired values of air temperature in the greenhouse, in accordance with outside light intensity.
(2) to achieve desired levels of CO_2 concentration in the greenhouse, in accordance with outside light intensity (i.e. day or night).
(3) to moderate the rise of humidity levels in the greenhouse by the use of ventilation, within certain limits of energy consumption, as determined by prevailing levels of inside and outside temperature and humidity.

Scheme

Over a range of conditions the energy requirement of a single-clad polythene greenhouse and of a traditional glasshouse are closely similar. Therefore it is reasonable to consider that the heating energy data given here for the single-covered polythene reference solution (where no use is made of the studied climate regulation system) are applicable also to a classical single-glazed glasshouse fully equipped with conventional environmental control facilities.

The heating system in such a house is normally a piped hot-water system supplied from an oil-fired boiler. In a typical commercial unit of area 0.4 ha, the system is designed to supply a peak heating load in the glasshouse of about 900 kW. The temperature of the water in the heating pipes is controlled by a mixing valve, which in turn is controlled from a temperature sensor in the glasshouse.

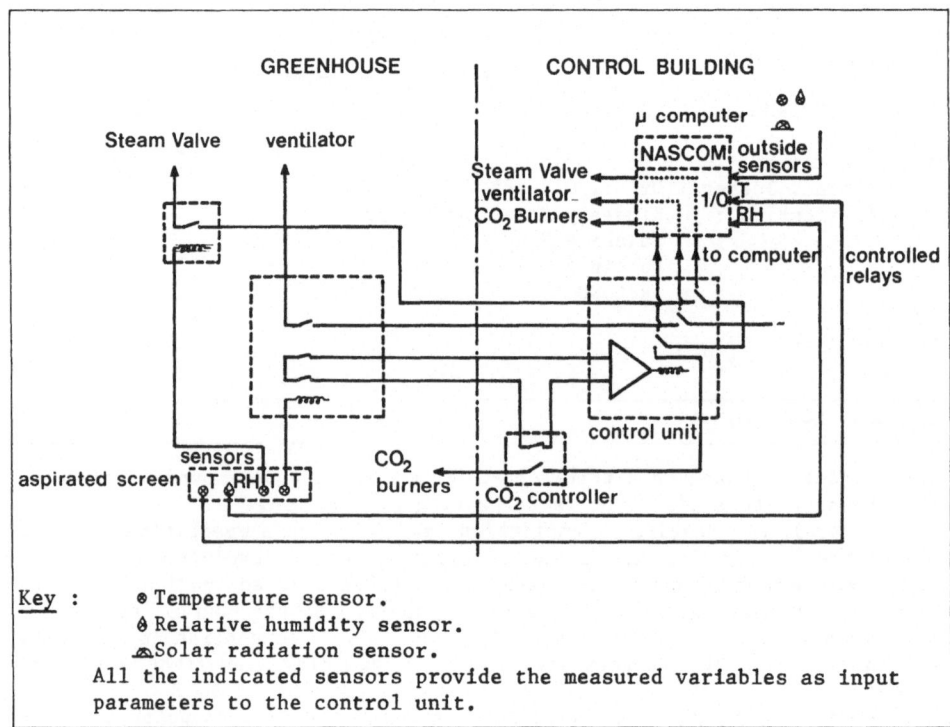

Key :   ⊛ Temperature sensor.
        ◊ Relative humidity sensor.
        ⚏ Solar radiation sensor.
    All the indicated sensors provide the measured variables as input parameters to the control unit.

Ventilation is provided by ridge ventilators in the roof of the glasshouse, giving a total ventilator opening area equal to approximately one sixth of the floor area of the house. The degree of ventilator opening under any given temperature conditions in the glasshouse is controlled by a thermostat, acting normally through a PI or PID controller.

No special provision is generally made for control of humidity in the glasshouse.

Carbon dioxide (CO_2) supplementation is provided by propane burners. In a 0.4 ha unit ten burners each with a consumption of 0.9 kg/h are sufficient to provide enrichment to a level of approximately 1 000 ppm.

Watering and plant feeding is provided through a trickle irrigation system whereby each tomato plant is supplied through an individual drip nozzle.

The growing system to be used is a ground NFT (nutrient film technique) system. This means that the floor of the house is graded to a slope of 1 in 50 and rolled and covered with fine gravel. The polythene gullies are laid directly on the floor, with polystyrene slabs under them for insulation. The system is split half way along the house so that the gully length is 14 m. Solution warming is provided by a 9 kW immersion heater in the reservoir.

Conclusions

The heating energy used by a triple-span double-clad polythene greenhouse over the period January 1 to November 9 was 1 624 MJ/m^2, while the amount used by the reference single-clad polythene house over the same period was 2 013 MJ/m^2.

The transmission of photosynthetically active radiation (PAR) by the triple-span double-clad house was 60.5%.

The yield of tomatoes by June 30 in the double-clad house was 14.3 kg/m^2, compared to 10.2 kg/m^2 in the reference house.

Double polythene greenhouses have wide potential application in the northern countries of the European community, because of the substantial reduction in energy requirements which they offer together with their low capital cost compared with traditional glasshouses. Their effective and economic application will be particularly aided by the implementation of lowcost microcomputer techniques for climate control.

The most important area where future work will be needed is likely to be in the further development and refinement of climate control techniques; this will relate both to the performance and reliability of sensing devices and control hardware, and also to the improvement of control strategy through further development of software. Work is also likely to be needed in the application of similar techniques to types of energy-conserving greenhouses other than double polythene houses.

Further references

O'FLAHERTY, T., Some methods of reducing fuel requirements for greenhouse heating. In "Technical and Physical aspects of Energy Saving in Greenhouses", CEC Publ. No. EUR 5679e, 67-71. (1977)
O'FLAHERTY, T. and MAHER, M.J., Fuel consumption and crop performance in double-covered polythene greenhouses. Acta Hort. 107, 81-85. (1981)
O'FLAHERTY, T. and GRANT, J., Use of double cladding techniques to reduce greenhouse heat loss. Acta Hort. (1983)

PROJECT ANALYSIS

1. Site and climate

1.1. Site

Latitude :	53°25' N
Longitude :	6°10' W
Altitude :	20 m
Nearest main city :	Dublin
Distance from main city :	11 km
Direction from main city :	North
Obstructions :	No significant obstructions

1.3. Annual long-term averages

			Project	Country in general
1. Prevailing wind direction	W		W-SW	SW
	S		W-SW	SW
2. Average wind speed	W	m/s	5.8	3.8-8.3
	S	m/s	4.6	3.8-6.7
3. Total precipitation		mm/yr	763	720-2 600
4. Absolute humidity	W	g/kg	4.5	
	S	g/kg	8.5	
5. Global irradiation on horizontal plane		MJ/(m² yr)	3 800	3 600
6. Diffuse proportion		%	60	60
7. Degree days (Base temp. = 4.4°C)	C days		2 004	1 770-2 270
8. Hours sunshine		hrs/yr	1 470	1 300-1 650
9. Ambient temperature		C	9.6	9.0-10.5
10. Average max. temp. (July/Aug.)		C	18.8	18.0-20.0
11. Average min. temp. (Jan.)		C	1.9	1.0-4.5
Note : W = Winter S = Summer				

12. Meteorological station :	Dublin Airport	
13. Latitude :	53°26' N	
14. Longitude :	6°15' W	
15. Altitude :	68 m	
16. Distance from site :	5 km	

2. Plant

2.1. Greenhouse

1. Dimensions – Floor area :   600 m²
              – Volume :   1 950 m³
              – Glazing area :   1 000 m²
2. Glazing : Double polythene (air-inflated)
3. Covers/Screens : None
4. Ventilation system : 2 x 1.2 m propellor fans.

2.5. Installation parts and measuring devices

2. Description of the sensors

- Temperature (T_{ai}, T_{ao}) : – Wirewound PRT
                                    – BS1904 Grade II
                                    – Precision : 0.1 C
                                    – Range : $\pm$100 ... + 100C

- Humidity (RH_{ai}, RH_{ao}) : – Vaisala : Humicap HMP 21U
                                    – Precision : $\pm$2%
                                    – Range : 0 ... 100%RH

- Solar radiation :   Kipp and Zonen Solarimeter CM 5/6
                                  – Accuracy : $\pm$0.1%

- PAR (R_o, R_i) :   – TFDL Selenium Flat, responsive to
                                    photosynthetically active radiation

- Flowrate (heating water)   – Kiercol SF 200 8
                                  – Accuracy : $\pm$ 1% FSD
                                  – Range : 200 litre/min

- Temperature   – PRT (as for Ti, To)
  (heating water)

- Heat consumption (H) :   – Computation from heating water flow-
                                    rate and flow and return temperature
                                    by Nascom 2 microcomputer

- CO_2-concentration (C) :   – Siemens CO_2 Analyser Type M52080 – A61
                                  – Accuracy : + 100 ppm at 0.1%
                                  – Range : 0 -3000 ppm

- Wind speed (v) :   – Vector Instruments photo-electric
                                    anemometer
                                  – A100M
                                  – Accuracy : $\pm$2% +-0.1 m/s

3. <u>Scheme of measuring points</u> (cf. figure 2.5.3.)

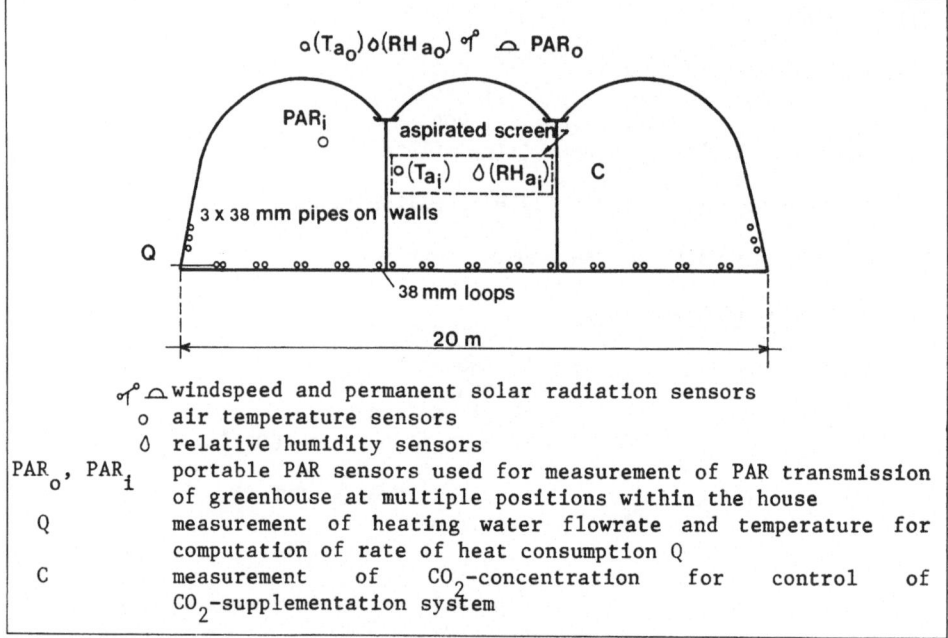

Figure 2.5.3. : Scheme of the measuring points

3. Products

3.1. Culture in greenhouse

1. Type of crop : tomatoes
2. Date of planting : February
   harvesting : April – October
3. Growing system : Nutrient film technique (NFT) with root
   zone warming
4. Required temperature regime : min. temp. : 16°C
   max. temp. : 25°C
6. Required CO_2 regime : min. CO_2 : 0.03%
   max. CO_2 : 0.1%
7. How sensitive is the crop to light ? Highly sensitive
8. Other (related), crops which require an analogous inside
   climate and luminosity :
   Cucumbers, peppers, aubergines.

4. Economic aspects

Incl. V.A.T. 1983

4.1. Economic environment

1. Energy price of oil   ECU/1   0.28
2. Energy price of electricity ECU/kWh   0.109
3. Energy prices evolution   Oil may remain stable
                                Electricity may increase by 5-10%/yr
                                (i.e. the same as inflation rate)
4. Manpower cost   261 ECU per man-day (professional)
                                180 ECU per man-day (technical)
5. Interest rate of loans for farm
   equipment (rate without subsidy)   17%/yr
6. Subsidies available for the solar
   and competitor farm equipment   25% (as capital grant)

4.2. Investment expenses

		double polythene	Reference
1. Total investment budget (materials only-600 m² floor area)	ECU	52 800	75 000
2. Cost of equipment :			
- expenses for buildings	ECU/m²	25	62
- cost of regulation equipment	ECU/m²	20	20
- other material costs	ECU/m²	43	43
3. - Time of personnel for the conception, installation and start-up of equipment	man-days	200	200
- Cost of personnel for the conception, installation and start-up of equipment	ECU	44 000	44 000
4. Lifetime of the equipment (evaluation)	years	15	20
5. Residual value of the investment	ECU	Nil	Nil

4.3. Operating expenses

		Double Polythene	Reference
1. Energy cost	ECU/year	4 500	7 000
2. Maintenance cost (replacement of materials or components)		450 (polythene)	
- spare parts		450	None
- salaries (internal-external)		750	750
3. Hire of surface for the installation		None	None
4. Other manpower cost	ECU/yr	15 000	15 000
6. Raw material cost	ECU/yr	11 200	11 200

4.4. Hidden cost and benefits

		Double polythene	Reference
1. Production obtained (kg tomatoes/ m^2 floorarea by June 30)		14.3	10.2
2. Average sales price per unit produced	ECU/kg	0.75	0.75
3. Total income obtained per greenhouse of 600 m^2	ECU	6 420	4 560

5. Results

5.2. Weather conditions and energy

1.	Dec	Jan	Feb	Mar	Apr	May	Jun	Jul	Aug	Sep	Oct	Nov	Month	PERIOD	
2.	21	31	28	31	30	31	30	31	31	30	31	9	Days	reference year	
3.	7.5	4.1	5.3	7.3	7.6	10.5	12.3	15.6	14.7	12.8	11.7	6.0	°C	average ambient TEMP	
4.	4.0	3.5	3.5	4.1	4.1	2.5	2.7	2.8	2.6	3.1	3.6	4.0	m/s	average WIND	
11.	169	322	334	305	154	124	41	26	54	86	132	46	MJ/m^2	AUXILIARY HEATING energy	double polythene
13.	211	402	417	381	193	155	52	33	68	108	165	57	MJ/m^2	OIL consumption $\eta=0.7$	
14.	0	0.08	0.14	0.16	0.27	1.15	1.7	2.5	2.2	1.15	0.92	0.11	MJ/m^2	ELECTRICITY consumption	
15.	–	479	443	401	224	149	42	14	38	64	110	48	MJ/m^2	HEATING energy	reference (GJ)
17.	–	599	554	501	280	186	52	19	48	80	138	60	MJ/m^2	OIL consumption	
18.	–	0	0.1	0.3	0.5	1.3	3.9	9.5	6.1	1.8	0.9	0	MJ/m^2	ELECTRICITY consumption	

5.3. Inside climate of greenhouse

1.	Dec	Jan	Feb	Mar	Apr	May	Jun	Jul	Aug	Sep	Oct	Nov	Month		
2.	21	31	28	31	30	31	30	31	31	30	31	9	Days		
3.	13.5	17.0	17.0	18.0	15.0	15.0	15.0	15.0	15.0	15.0	15.0	15.0	°C	min ⎱ TEMPERATURE	
5.	15.5	17.0	17.0	25.0	25.0	25.0	25.0	25.0	25.0	25.0	25.0	17.0	°C	max ⎰	
9.	28	44	62	124	233	365	327	327	275	153	116	69	W/m^2	Radiation	
12.	0.1	0.1	0.1	0.1	0.1	0.03	0.03	0.03	0.03	0.03	0.03	0.03	%	Average CO_2	
14.	17.0	17.0	17.0	17.0	17.0	17.0	17.0	17.0	17.0	17.0	17.0	17.0	°C	min ⎱ TEMPERATURE	reference
16.	17.0	17.0	17.0	25.0	25.0	25.0	25.0	25.0	25.0	25.0	25.0	17.0	°C'	max ⎰	
20.	36	55	78	155	293	458	410	410	345	197	145	86	W/m^2	Radiation	
23.	0.1	0.1	0.1	0.03	0.03	0.03	0.03	0.03	0.03	0.03	0.03	0.03	%	Average CO_2	

5.4. Crop results

	double polythene	reference
1. Date of planting :	20 February	28 February
2. Harvesting date :	5 April-12 Nov.	25 Apr.-30 June
3. Flowering date :		14 February
7. Production obtained : (kg/m^2)	14.3 (by 30 Jun) 30.3 (by 12 Nov)	10.2 (by 30 Jun) –

5.7. An example of some measured data (cf. figure 5.7)

The diagram (below) shows the percentage transmission of photosynthetically active radiation (PAR) at 56 points at 2 m height in the project greenhouse. Rows 3 and 5 represent points on a line 1 m directly below the north-south running gutters. Rows 2,4 and 6 correspond to the centre lines of each of the three bays, while rows 1 and 7 represent lines 1 m from the inner polythene "wall" on either side of the house.

The mean of the 56 values is 60.5%. The mean for the sixteen points under the gutters is 49.8%, while that of all the other points is 64.8%.

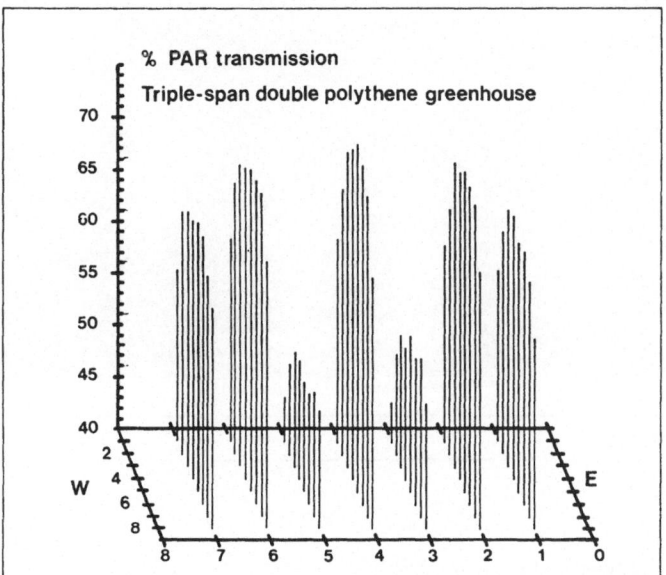

Fig. 5.7 : Distribution of the PAR radiation transmission throughout the project greenhouse.

4. Comparison of the light transmission of green-houses covered with twinwalled sheets of poly-carbonate and acrylic

Project location

BEDFORD, ENGLAND (1980)

Study Center:
- Project leader : Dr. B.J. Bailey
- Institution :   National Institute of Agricultural Engineering
                   Wrest Park.
                   Silsoe, Bedford.
                   England

Project description

The average light transmissions across the central spans of two three-span double clad greenhouses at gutter level are measured. In one house the double cover is made of polycarbonate, in the other acrylic is used.
The objective is to compare the light transmission of acrylic and polycarbonate, double clad greenhouses under diffuse light and sunlight conditions within the PAR and total radiation waveband ranges.

Scheme

The internal light detectors comprise ten PAR sensors, distributed uniformly across the central span, and two total radiation detectors (Kipp) at one-third and two-thirds of the way across the same central span. External light is monitored by one PAR and two Kipp detectors, K_1, K_2, one of which, K_2, is shaded against the sun. Transmissivity measurements are made between the hours of 10.00 a.m. and 2.00 p.m. under conditions of cloudy weather only (diffuse radiations), ($K_2 : K_1 > 0.9$) and bright sun only (diffuse and direct radiation), ($K_2 / K_1^2 < 0.25$).
For other conditions transmissivity measurements are not carried out, though total light integrals are logged on a daily basis. Records are made every 30 secs, under the control of an APPLE 48K mini-computer and Solartron data logging A/D converter.

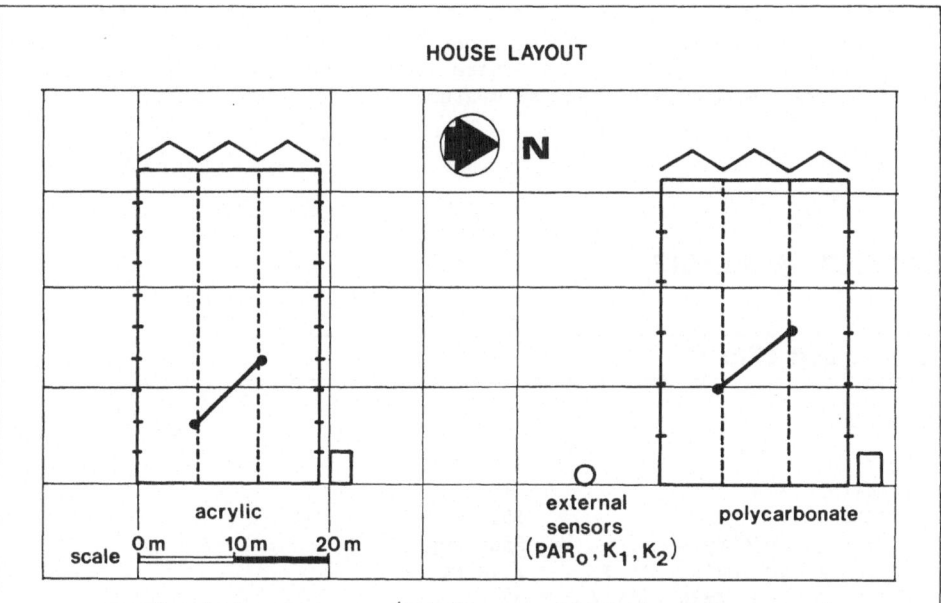

HOUSE LAYOUT

N

acrylic

external sensors (PAR_0, K_1, K_2)

polycarbonate

scale   0 m   10 m   20 m

●━━● location of internal sensors (10 PAR sensors , 2 total irradiation sensors)

Note : During early June, a hailstorm virtually destroyed the polycarbonate house, thereby terminating the current experiment.

Conclusions

The acrylic house transmits 5% more light under both cloudy and sunny weather conditions, for both PAR and total (Kipp) radiation.

	Acrylic House	Polycarbonate House	Research
Average diffuse radiation transmissivity % (PAR) (cloudy weather)	59	54	June 82–June 83
Average diffuse radiation transmissivity % (KIPP) (cloudy weather)	61	56	"
Average global radiation transmissivity % (PAR) (sunny weather)	59	54	June 82–Sept.82
Average global radiation transmissivity % (KIPP) (sunny weather)	65	61	"

Total radiation appears to have a higher transmissivity than PAR, but this result is not conclusive as there are only two total radiation measurements compared with ten PAR measurements across the span.

PROJECT ANALYSIS

1. Site and climate

1.1. Site

Latitude :	50°44' N
Longitude :	1°34' W
Altitude :	15.5 m
Nearest main city :	Southampton
Distance from main city :	21.4 km
Direction from main city :	SSW
Obstructions :	none

1.3. Annual long-term averages

		Project
1. Prevailing wind direction		SW
2. Average wind speed	m/s	2.5
3. Total precipitation	mm/yr	777.8
5. Global irradiation on horizontal plane	MJ/(m^2 yr)	3 936
8. Hours sunshine	hrs/yr	1 725
10. Average max. temp. (July/Aug) C		20.2
11. Average min. temp. (Jan) C		2.1
12. Meteorological station :	on site	

2. Plant

2.1. Greenhouse

		Poly.	acryl
1. Dimensions :	– Floor area : m^2	627	614
	– Volume : m^3	2 584	2 445
	– Glazing area : m^2	1 136	1 156
	– roof area : m^2	701	686
	– sides area : m^2	435	470

2. Glazing :   Bayer Makrolon polycarbonate (Thermoclear)
ICI Diakon PMMA (Ecocal)

3. Covers/screens : none

4. Ventilation system : Double sided ridge ventilators 75 m^2
ventilator area.

2.5. Installation parts and measuring devices

2. Description of the sensors :

Measurements were made of solar radiation and photosynthetically active
radiation (PAR).   Solar radiation was measured over the wavelength
range 0.3-2.5 µm using Kipp and Zonen solarimeters.   These black body
(total energy) detectors used a thermopile as the sensing element.   The
PAR was measured using quantum sensors which contained a silicon
photodiode as the detecting element.   These sensors were filtered to
respond to wavelengths of 400-700 µm and their response over this range
was dependent on photon energy, thus they measured photon flux density.

3. Scheme of measurement points

Ten PAR sensors were equally spaced along a beam installed diagonally
across one span near the middle of each greenhouse just above the crop
wires at a height of 3.2 m.   Two solarimeters were also fitted on each
beam at one-third and two-third of the way across the beam.   The
radiation outside the greenhouse was detected by two solarimeters and
one PAR sensor mounted on a mast at a height of approximately 7 m.   The
arrangement is indicated in the scheme on p. 116.   One solarimeter was
fitted with a shade ring so by comparing the signals from the two
solarimeters the quality of the light, i.e. whether it was direct
(bright sunlight), diffuse (overcast sky), or a mixture of both, could
be determined.   Signals from all sensors were measured automatically by
a computer control data logger at 30 second intervals whenever there
was a finite amount of light.   When the light was predominantly diffuse
or direct the ratios of the signals of the internal to external sensors
were also recorded.   At midnight information on the total amount of
solar radiation and PAR received externally and inside each greenhouse
was printed on a paper chart.   In addition, all data were transferred
to disc stores which were periodically sent to NIAE for detailed
analysis.

3. Products

3.1. Culture in greenhouse

1. Type of crop : Tomatoes (cv. Marathon)

2. Date of sowing/planting : 21 Nov. 1980 - 18 Feb 1981
   harvesting : 25 March 1981 - 12 October 1981

3. Growing system : peat bags and NFT

4. Required temp. regime : min. temp. : 16, 18, 20°C (depending on plant stage)
   max. temp. : Ventilation at 27°C up to end April, later ventilation 21°C.

6. Required CO_2 regime : min. CO_2 : % approx. 1 000 ppm December-April

7. How sensitive is the crop to light ?
   1% light equivalent to 1% crop value.

8. Other (related), crops which require an analogous inside climate and luminosity : Cucumbers, peppers

4. Economic aspects

4.2. Investment expenses

		Polycarbonate	Acrylic
1. Total investment budget	ECU	27 450	30 600

2. Cost of equipment :
  - expenses for buildings
  - cost of regulation equipment
	
	ECU
	ECU
		
		21 000
		6 450
			
			25 300
			5 300

4. Lifetime of the equipment (evaluation)	 (years)	 10	 10
5. Residual value of the investment	ECU	3 220	3 220

4.3. Operating expenses

		Polycarbonate	Acrylic
1. Energy cost	ECU/yr	21 000	19 350

4.4. Hidden cost and benefits

		Polycarbonate	Acrylic
1. Production obtained (cf. section 5.4)	 kg/m²	 29.9	 32.7
2. Average sales price per unit produced	 ECU/kg	 0.85	 0.88
3. Total income obtained	ECU/m²	25.4	29.0

5. Results

5.1. Method of calculation

The data stored after midnight on disc consists of the following time integrals taken over the immediately previous day :

(1) Total, diffuse and PAR radiation for the previous 24 hours.
(2) Transmissivities (R) over the period between 10 a.m. and 2 p.m.; summation of readings in each house, (Σ(R)).
(3) The square of the transmisssivities in (2), Σ(R)^2.

Subsequent processing consists of adding successive totals over each month to produce monthly totals for the light integrals, and average monthly transmissivities. The transmissivities are further space-averaged to produce mean house transmissivities.

5.4. Crop results (greenhouse) (growing season '80/81, cf. section 3.1.2)

1980/81

		Acrylic	Polycarbonate
7. Production obtained :	kg/m^2	32.7	29.9
8. Other differences : (in results, aspects, dimensions, ...) Quality % of Class I fruit	%	87	83

5.7. An example of some measured data (cf. figures 5.7.1 and 5.7.2)

Transmissivity measurements are made between 10 a.m. and 2 p.m., separately under diffuse light and sunlight. Records are available every 30 secs. Comparative results are given below for 1982/1983.

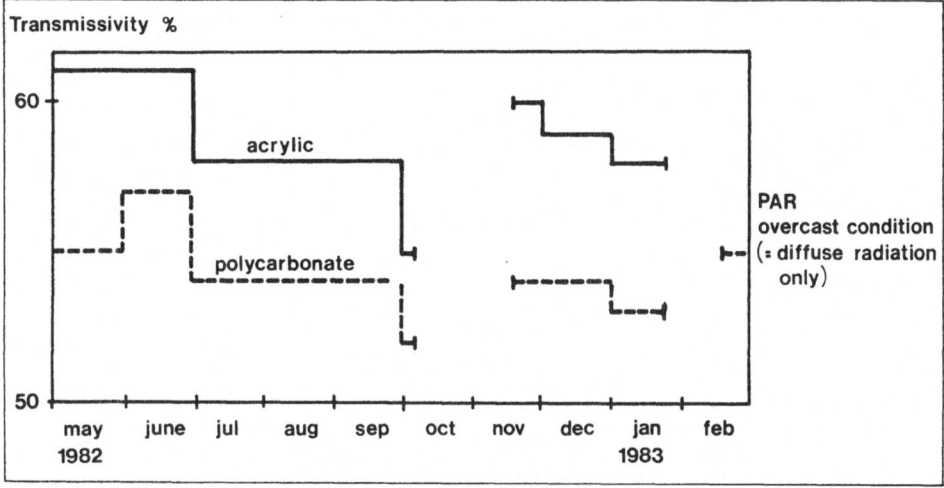

Fig. 5.7.1 : PAR transmissivity values for overcast conditions over time

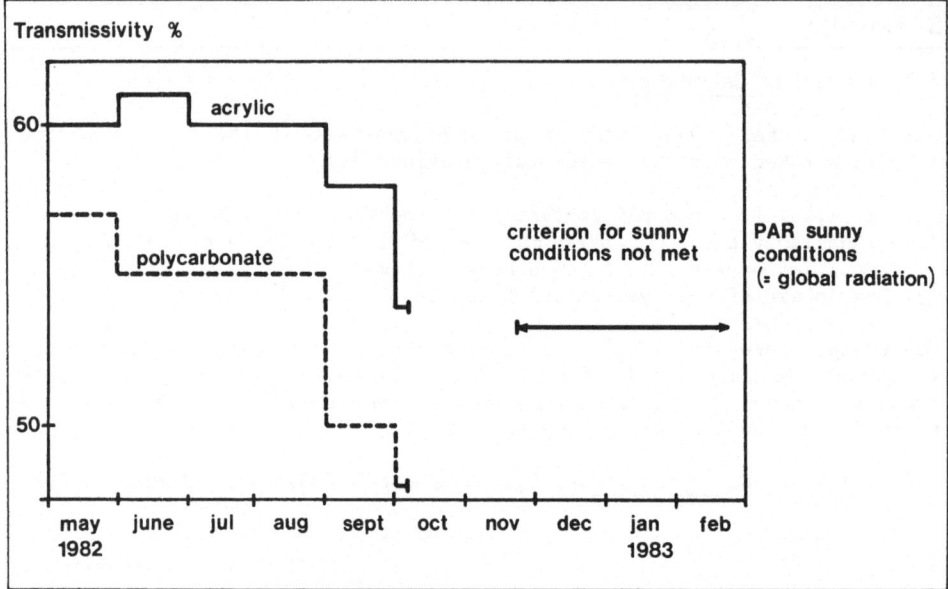

Fig. 5.7.2. : PAR transmissivity values for sunny conditions over time

5. Greenhouse climatization by means of "helio-blocs"

Project location

AUTEUIL (Paris), FRANCE

Study Center :
- Project leader : J.-P. Hauet and B. Letellier.
- Institution : Novelerg
  (subsidiary of the C.G.E. Group)
  Research and engineering company
  12 rue de la Baume
  75008 PARIS
  FRANCE

Project description

This study of greenhouses, totalling 4 800 m^2, is focused on energy conservation and presents a double interest :

- The old-fashioned conception (single glazing, non weather-proof, uncontrolled ventilation and humidity) leads to high energy consumption and puts further utilization in question.
- Tropical climatic conditions require a heating and regulating system that takes the living conditions of the plants (temperature, humidity, ...) very precisely into consideration.

The measures will enable the research team to :

- know the detailed performance of the system and to understand its functioning;
- understand the functioning of the installations such that remedies for the failures are expedited.

The measures will enable the perfection of different new systems (among other things of an active solar collecting system).

Scheme

The renovated test greenhouse has been weather proofed and equipped with :
- a cover of low emissivity glass or double paned glass.
- automated aeration, watering and shading systems (shades mounted on outside of greenhouse.
In a further step an active solar heating system will be built which functions on the basis of helioblocs. A site section of the greenhouses with helioblocs is given in figure S.1

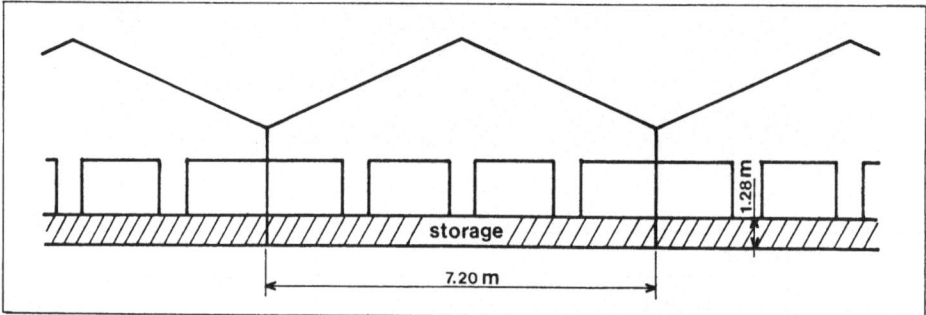

Fig. S.1 : Site section of greenhouses with helioblocs

The storage system of the active solar heating system would function as follows :
    I Excess solar energy (storage loading action) (cf.fig.S.2)

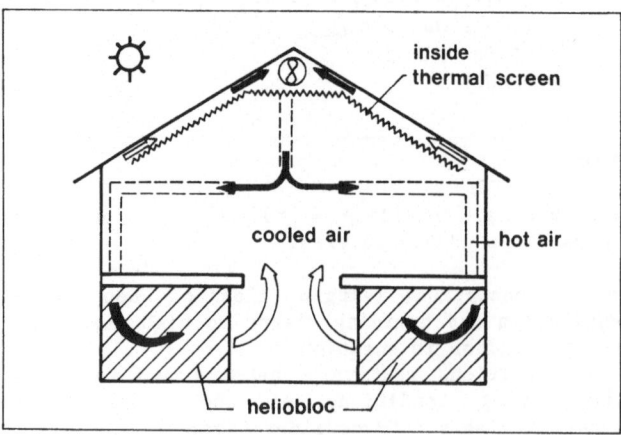

Fig.S.2 : Storage loading

The greenhouse and the variable screen collect solar energy and contribute to the elevation of the ambient air temperature.
When the programmed temperature is attained, the hot air is extracted along the slope of the roof and is directed to the storage area where it gives up its excess heat.
The fresh air coming from the storage area is then recirculated back into the greenhouse.

II Insufficient solar radiation (storage unloading action)

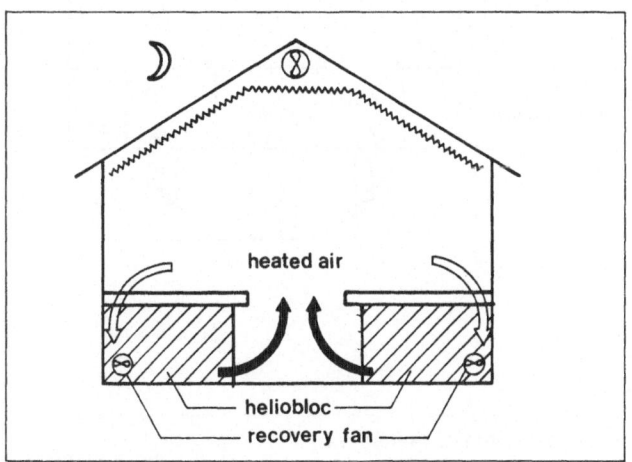

Fig. S.3 : Storage unloading

During the heating periods, the air of the greenhouse is heated in the storage area by forced circulation.
The traditional heating system using vented pipes is used as a back-up system.

Operation modes (For location of regulating sensors, cf. fig. S.4)

Storage and heat recovery will be monitored by a computer steered process :
    Storing:
    Storage is activated if $T1 > T$ desired $+ \Delta Tx$
                               $T2 > T5 + \Delta Ty$
                               $T4 < T3$
    Heat recovery :
    Heat recovery is activated if $T1 < T$ desired
                                     $T5 < T4$

In this report a comparison is made between the performance of a renovated test-greenhouse (still without active solar collection set-up) and that of a similar not renovated one (single glazing; manual ventilation).

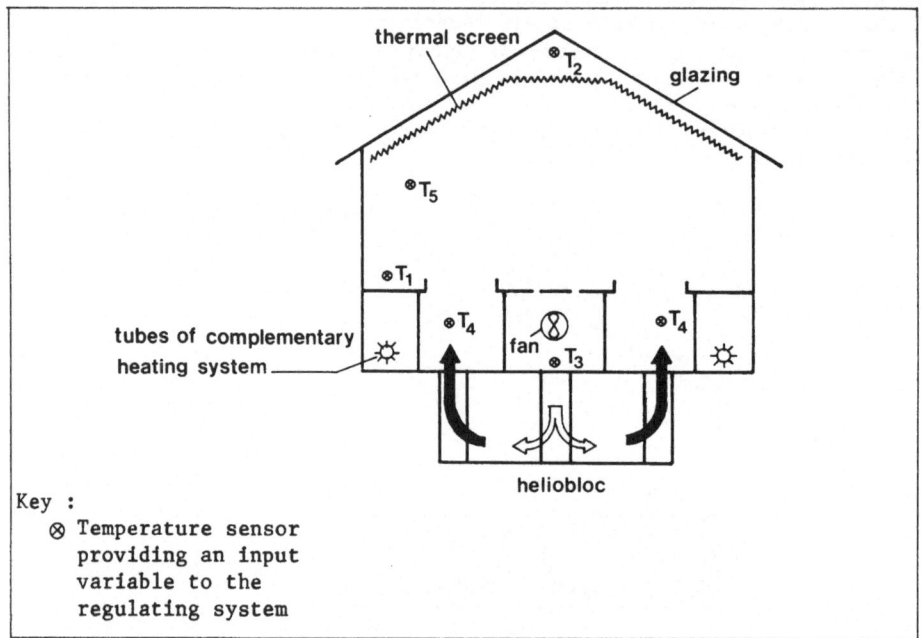

Fig. S.4 : Location of control sensors.

Conclusions

Collected data will enable to :

- understand thermal transfer between indoor climate, plants and outside climate.
- facilitate the design of energy saving systems, particularly by use of solar energy.
- verify the adequacy of the data processing simulation models used.
- check the performances of the already renovated greenhouses.

The preliminary results obtained indicate that :

- a 25% energy saving resulted from the use of double glazing.
- an additional 8% was realised due to the better air-tightness of the greenhouse.
- the relative humidity increased from 84% in a single glazed greenhouse to 89% in a double glazed greenhouse.
- the average temperature obtained increased from 22°C to 23°C.

PROJECT ANALYSIS

---

1. Site and climate

1.1. Site

Latitude :	48°49' N
Longitude :	2°20' E
Altitude :	75 m
Nearest main city :	Paris (located in)
Direction from center :	West

1.3. Annual long-term averages

			Project	Country in general
1. Prevailing wind direction	W		W NW	W NW
	S		W NW	W NW
2. Average wind speed	W	m/s	3.5	not available
	S	m/s	3.2	"
3. Total precipitation		mm/yr	643	"
4. Relative humidity	W	%	81	"
	S	%	70	"
5. Global irradiation on horizontal plane		MJ/(m^2 yr)	not available	3 960
6. Diffuse proportion		%	"	50
8. Hours sunshine		Hrs/yr	1 714	1 750 to 2 000
9. Ambient temperature		C	11.4	10.9
10. Average Max. temp. (Jul/Aug)		C	30.6	30
11. Average Min. temp. (Jan)		C	− 6.6	− 7

12. Meteorological station :   Montsouris
13. Latitude :   48°49' N
14. Longitude :   2°20' E
15. Altitude :   75 m
16. Distance from site :   5 km
Note : W = Winter : Jan., Feb., Nov., Dec.)
*   S = Summer : Jun., Jul., Aug., Sept.)*

---

2. Plant

2.1. Greenhouses (Data on the tested greenhouses)

1. Dimensions – Floor area :   175 m^2
               – Volume :   400 m^3
               – Roof glazing area :   200 m^2
               – Wall area :   55 m^2
2. Glazing : double glazing (k = 4.2 W/(m^2.K))
3. Covers/screens : inside the future greenhouses and outside the actual monitored greenhouses.

4. Ventilation system : natural ventilation for actual greenhouses and SOLPAC-system for future greenhouses.

Note : The municipal greenhouses in Paris consist of :
- Tropical greenhouses, totalling 1 300 m^2, in the process of being renovated, and dedicated to the culture of orchids and rare plants;
- Plain greenhouses, totalling 3 500 m^2, to be totally remodelled and dedicated to the culture of regular plants and flowering plants.

2.4. Heat storage

1. Medium of heat storage : HELIOBLOCS (beton blocks) provided with channels for air ventilation (NOVELERG patent).
2. Description, location in the system :
   The HELIOBLOCS are situated in the greenhouse soil and bedded in a caloritherm covering.
3. Insulation : Insulation by 10 cm of polystyrene.
4. Dimensions : - Volume of HELIOBLOC : 32 m^3
                 - Total dimensions, including channels :
                   19 m x 6.70 m x 0.70 m
5. Thermal capacity : 51.2 MJ/K
6. Overall heat loss coefficent : 112 W/K
7. Diagram : cf. fig. 2.4.7

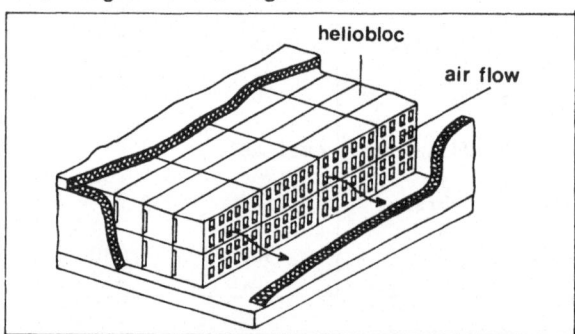

Fig. 2.4.7 : Diagram of heliobloc heat storage

2.5. Installation parts and measuring devices

2. Description of the sensors

   - Temperature (°C) (T1, T2, T3, Te, Tpo, Tpi; cf. sheme in section 2.5.3)
     Type : Platina sensor PT 100 (Louzot)
            Precision : + 0.3 °C
            Measuring range : - 70 °C
                              + 200 °C
            Variation : 0.39 Ω/°C
   - Relative humidity (%) (RHi, RHo)
     Type : Hygromer HTK (Rotronic)
            Precision : + 0.5%
            Measuring range : 5% - 100% RH

- Interior solar radiation intensity (W/m^2) (Ihi)
  Type : Photo-electric cell
         I.N.C. $\emptyset$ 80 (Photowatt)
- Exterior solar radiation intensity (W/m^2) (Iho)
  Type : CM 11 (Kipp and Zonen)
         Precision : + 3%
- Energy consumption (W/m^2) (Q) thermal energy meter
  Type : GWF WPM HK 125
  The consumption of individual greenhouses is measured by a
  microcomputer (System GERTEP).

3. Scheme of measurement points

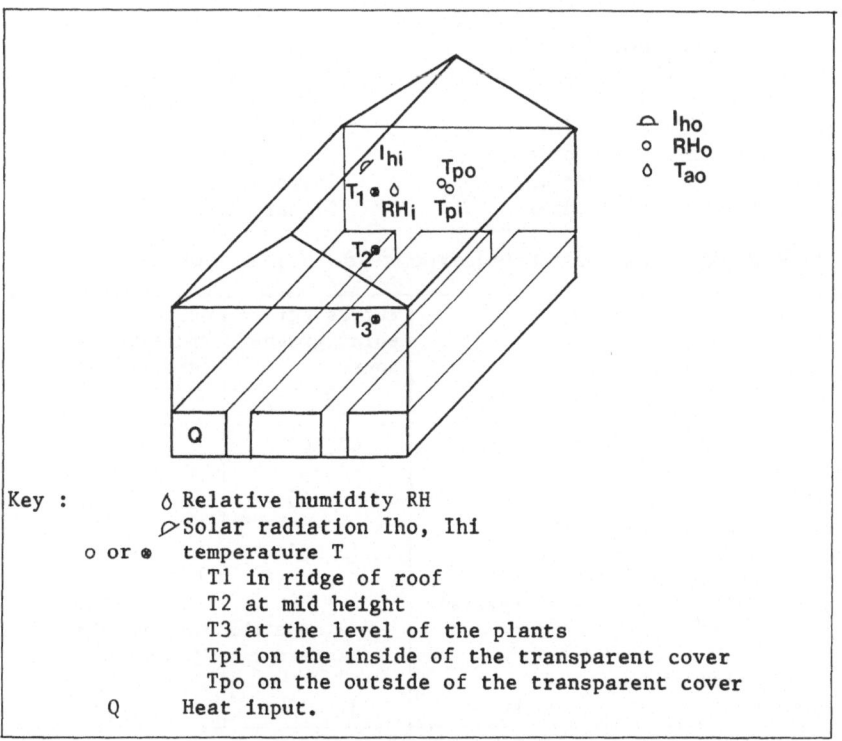

Key : δ Relative humidity RH
              $\triangleright$ Solar radiation Iho, Ihi
   o or ⊛   temperature T
              T1 in ridge of roof
              T2 at mid height
              T3 at the level of the plants
              Tpi on the inside of the transparent cover
              Tpo on the outside of the transparent cover
   Q   Heat input.

Fig. 2.5.3 : Positions of the sensors in the instrumental greenhouse

3. Products

3.1. Culture in greenhouse

1. Type of crop : potted plants; begonias, orchids...
2. Date of sowing/planting : permanent cultures (plant collection)
3. Growing system : conventional
4. Required temp. regime : min. temp. : 16°C
                            max. temp. : 22°C
5. Required humidity regime : min. Rel. humidity : 60%
                              max. Rel. humidity : 85%

7. How sensitive is the crop to light?
   Maximum 200W/m² *(1)*
   Therefore shading curtains are used.

(1) These are obviously shade plants. For such plants, long hours of shading will be required and hence equally long hours of active solar radiation collection will be possible in the future greenhouse set-up (N.O.T.E.)

4. Economic aspects

Base : 1982 prices

4.1. Economic environment

1. Energy price of gas	ECU/kWh	0.0216
2. Energy price of electricity	ECU/kWh	0.039
3. Energy prices evolution prospects		
– of oil or gas (real terms)	%/year	+ 3
– of electricity (real terms)	%/year	0

4.2. Investment expenses (estimated expense for 1 greenhouse)

		Future active solar green-house	Actual renovated "passive" solar greenhouse
2. Cost of equipment			
– expenses for buildings	ECU	42 000	27 400
– cost of regulation equipment	ECU	4 665	1 550
– other material (including positioning and start-up)		15 552	none
4. Lifetime of the equipment (evaluation)	(years)	+ 50	+ 50 except the fan and the screen : 5–10 years
5. Residual value of the investment	ECU	31 415	none

4.3. Operating expenses (estimated expense for 1 greenhouse)

		Future active solar greenhouse	Non-renovated reference sol.
1. Energy cost	ECU/yr	1 550	2 954
2. Maintenance cost	ECU/year	77.5	none

5. Results

5.1. Method of calculation

The energy consumption of each individual greenhouse is known from the global energy consumption, provided by the general calorimeter and the indexes, which are monitored by the GERTEP system for each greenhouse separately (cf. figure 5.1).
For comparison and extrapolation, the observed results are related to the number of degree-days. The number of degree days equals the numer of days heated, multiplied by the difference between the exterior temperature and the corrected interior temperature. For 1983 the total number of degree days in Paris-Montsouris was 1 492.

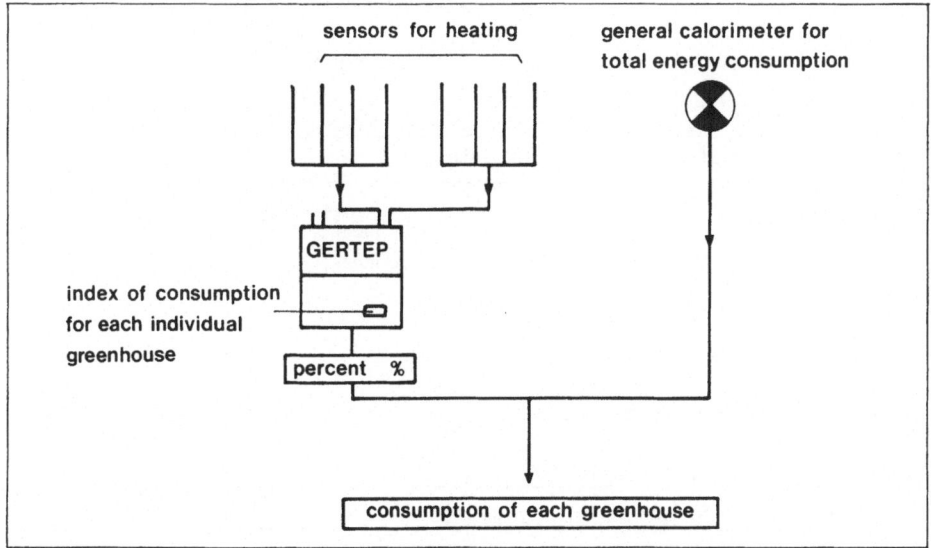

Fig. 5.1 : Scheme of the energy consumption monitoring system.

5.2. Weather conditions

1. 2.	Jan	Feb	Mar	Apr (1)	May	Jun	Jul	Aug	Sep (2)		from : January PERIOD until : September	
3.	6.8	3.1	8.1	9.8	12,2	18.8	26.5	22.3	16.6	°C	average ambient TEMP	
4.	4.0	4.2	3.7	4.0	4.1	N.A.	N.A.	N.A.	N.A.	m/s	average WIND	
5.	N.A.	N.A.	N.A.	N.A.	N.A.	N.A.	20.3	17.3	11.5	MJ/(m^2.day)	average SOLAR RAD	
5bis	82	75	77	78	78	N.A.	59.8	59.7	75.7	%	relative HUMIDITY	
11.	38	47	57	26	37	0	0	0	0	GJ/(greenhouse.yr)	HEATING energy (non-solar)	ref sol
15.	71	94	72	22	41	0	0	0	0	GJ/(greenhouse.yr)	HEATING energy	ref sol

(1) Defect from 16 to 30 April
(2) Data from 1 to 15 September

5.3. Inside climate of greenhouse

1. 2.	Jul	Aug	Sep	Unit	From: 28/07 PERIOD until: 15/09	
3.	20.6	17.1	19.2	°C	minimum	passive solar
4.	26.0	23.0	21.9	°C	average AIR TEMPERATURE	
5.	32.7	33.7	28.5	°C	maximum (at 1/2 height)	
6.	35.3	27.0	33.0	%	minimum	
7.	71.6	81.5	77.1	%	average RELATIVE HUMIDITY	
8.	100.0	100.0	100.0	%	maximum	
9.	3.8	3.5	3.1	MJ/(m^2.day)	average inside RADIATION	
14.	18.8	13.6	17.2	°C	minimum	reference
15.	26.4	22.1	21.9	°C	average AIR TEMPERATURE	
16.	35.9	33.4	32.3	°C	maximum (at 1/2 height)	
17.	N.A.	39.3	61.2	%	minimum	
18.	N.A.	82.0	88.8	%	average RELATIVE HUMIDITY	
19.	N.A.	100.0	100.0	%	maximum	
20.	4.2	2.3	3.3	MJ/(m^2.day)	Average inside RADIATION	

5.7. An example of some measured data

	Jul	Aug	Sep	Unit		
1.	26.7	23.2	20.9	°C	temperature of glass (interior side)	solar greenhouse
2.	26.1	22.0	18.0	°C	temperature of glass (exterior side)	
					opening of frame	
3.	8	36	8	%	LEFT CLOSED	
4.	27	26	74	%	50% OPEN	
5.	65	38	17	%	100% OPEN	
6.	69	90	96	%	RIGHT CLOSED	
7.	13	6	4	%	50% OPEN	
8.	18	3	0	%	100% OPEN	
9.	26.9	21.7	19	°C	temperature of glass	ref

The collected data for radiation, humidity and temperature are stored on disk. In figure 5.7 an example is given of temperature data on August 1st 1983.

On the basis of the characteristics of each greenhouse and the number of degree days (August-December) a comparison was made for the estimated energy use of the passive solar greenhouse and the reference greenhouse :

Energy consumption	Passive solar green- house GJ/(yr.green- house)	Tot. renovated reference greenhouse GJ/(yr.green- house)
Sep-Dec Jan-May	128 205	188 300
Total	333	488
Energy saving	31.7	

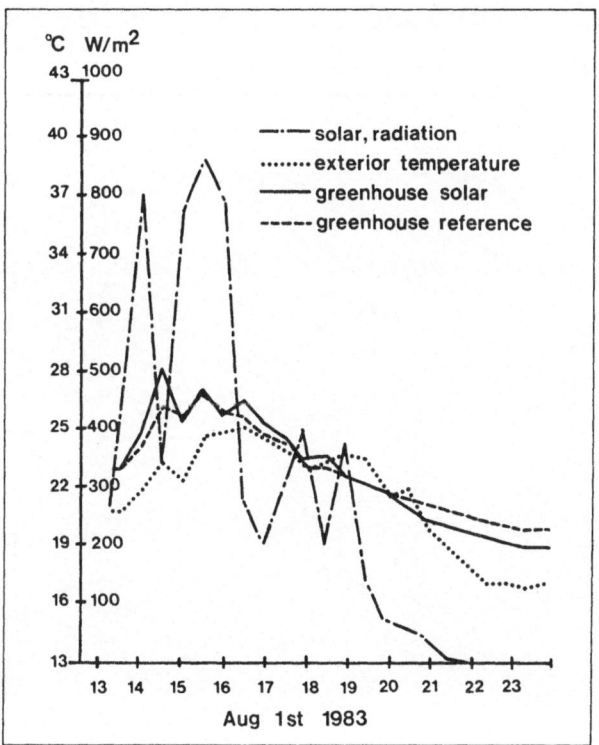

Fig. 5.7 : Some temperature measurement data.

6. Greenhouse climatization by an earth-air heat exchanger

Project location

ATHENS, GREECE

Study center :
- Project Leaders : Prof. S. Kyritsis
                    Dr. G. Mavrogianopoulos
- Institution :   Agricultural University of Athens
                    Iera odos 75
                    Votanikos Athens 11855
                    Greece

Running since 1982

Project description

In a 150 m² greenhouse an earth-air heat exchanger was built at a depth of 2 m. The heat exchanger consists of 20 aluminium pipes of 15 m length, 0.2 m diameter and 0.2 mm thickness.
The greenhouse air is blown through the pipes at a total rate of 15 000 m³/h whenever the greenhouse temperature falls below 12°C or exceeds 28°C.
The heat gained from the soil (Qso) was determined by measuring the energy consumption of an electrotherm placed in a completely similar greenhouse. In both greenhouses, the inside temperature was kept the same by means of

a differential thermostat, controlling both heating and ventilation.
The share of the solar energy input into storage which exceeds the energy
demand for daytime temperature stabilisation at a 28°C temperature level
was also calculated. (= Q_s')
This energy can be stored for night-time or longer range heating purposes.

The objectives of this study are :
- To determine the heat gain from the subsoil, during the period from
  November to May. (Q_{so})
- To observe the thermal environment inside the greenhouse.
- To determine the excess of greenhouse heat that could be stored in the
  subsoil over daytime. (Q_s')
- To determine the efficiency and the limitations of the earth-air heat
  exchanger.

Scheme

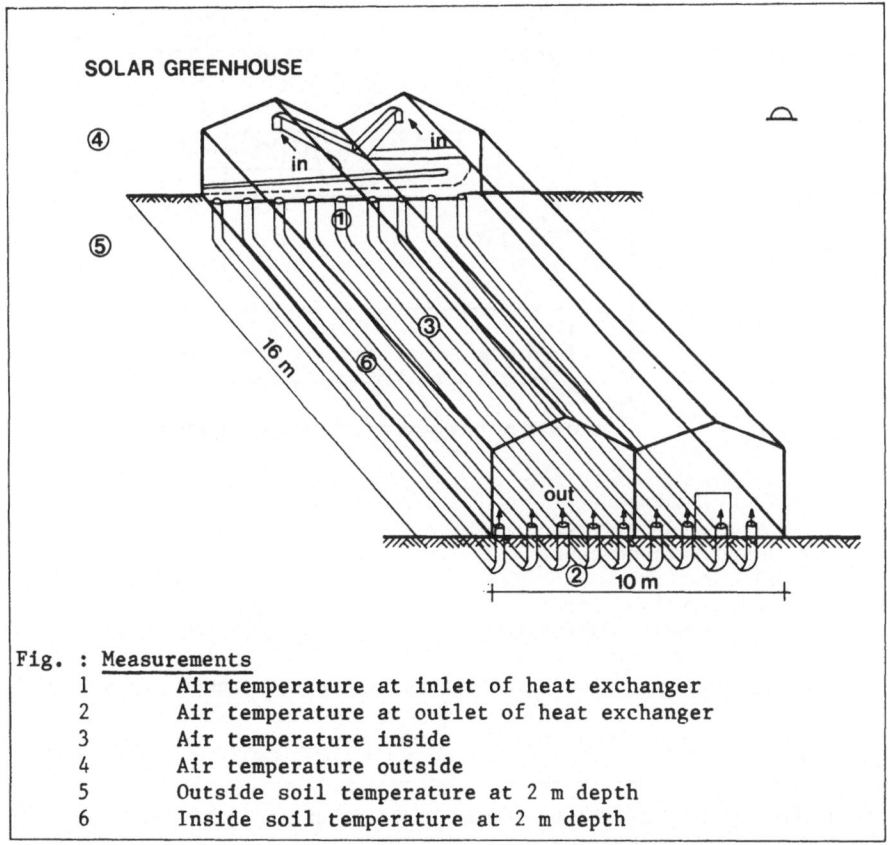

Fig. : Measurements
    1   Air temperature at inlet of heat exchanger
    2   Air temperature at outlet of heat exchanger
    3   Air temperature inside
    4   Air temperature outside
    5   Outside soil temperature at 2 m depth
    6   Inside soil temperature at 2 m depth

Also measured :
- Energy input for fan function in "solar" greenhouse
- Energy input for Electrotherm function in reference greenhouse
- The solar energy radiation penetrating inside the greenhouse

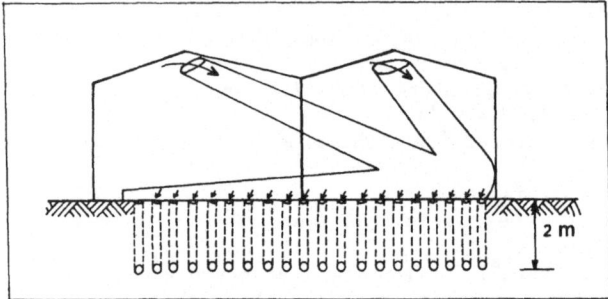

Fig. : Site section of the solar and underground heated greenhouse.

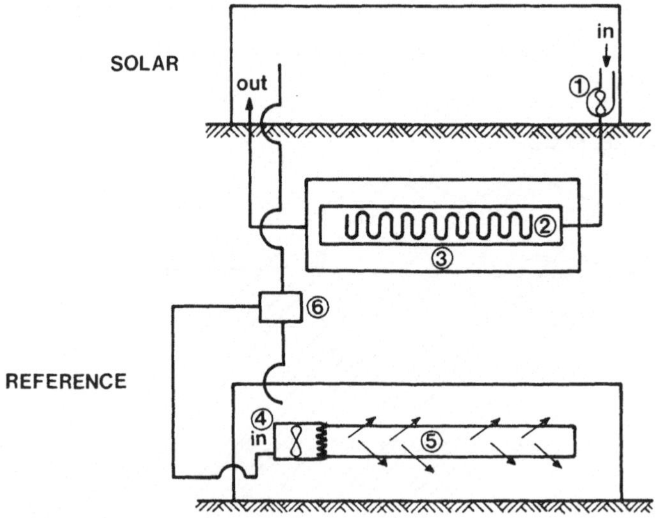

SOLAR

in

out

①

②

③

⑥

REFERENCE

④
in

⑤

1	Fan.
2	Heat exchanger.
3	Solar heat storage + underground heat.
4	Electrotherm.
5	Heat distribution.
6	Differential thermostat.

Fig. : Schematic representation of the solar versus reference greenhouse

Conclusions

- Measured heat output of the earth-air heat exchanger (Qso) was :
      4 410 MJ/month in November
     12 351 MJ/month in December   (for an explanation of the measuring
     15 904 MJ/month in January   method, cf. Project description,
     13 570 MJ/month in February   p. 122)
     11 300 MJ/month in March
      4 485 MJ/month in April

- Calculated energy storage (Qs') for night-time or longer range heating purposes : *(1)*

  4 730 MJ during November
  1 584 MJ during December
  2 001 MJ during January
  4 525 MJ during February
  11 805 MJ during March

- The efficiency of the earth-air heat exchanger during low night temperatures varied from 33.3% to 80%. The higher efficiency resulted at high excess heat storage. (The heat exchanger efficiency is defined in section 5.1.1)

- The lowest mean night temperature inside the greenhouse was 8.1°C, with an absolute minimum of 7°C from November to May, while the outside lowest mean temperature was - 0.8°C and the absolute minimum - 2°C for the same period.

- This method can offer very good results in energy savings in greenhouses and other spaces requiring heating or cooling.
  Thus in the Athenian conditions the method gave an energy saving for greenhouse heating of 70%.
  In more southern parts of the country it is expected even better results can be obtained..

(1) In December Qso > Qs'. This means the major part of the contribution of the underground storage consists in daytime temperature stabilisation.
In March Qso < Qs'. This means the daytime heat demand is more than covered, and heat storage over a longer time range occurs (N.O.T.E.).

PROJECT ANALYSIS

1. Site and climate

1.1. Site

Latitude :	37°58' N
Longitude :	23°45' E
Altitude :	50 m
Nearest main city :	Athens
Obstructions :	none

1.3. Annual long-term averages

1. Prevailing wind direction	W		NNE/NE
	S		NE/SSW/SW
2. Average wind speed	W	m/s	2.3
	S	m/s	1.9
3. Total precipitation		mm/yr	408.2
5. Global irradiation on horizontal plane		MJ/(m^2 yr)	(1)
8. Hours sunshine		hrs/yr	2 871
9. Ambient temperature		C	17.5
10. Average max. temp. (July)		C	27.0
11. Average min. temp. (Jan)		C	8.9

12. Meteorological station :   Meteorological Institute National observatory of Athens
13. Latitude :   37°58'3"
14. Longitude :   23°45' E
15. Altitude :   107 m
16. Distance from site :   1 km

> Note : W = Winter (Oct-Mar)
> S = Summer (Apr-Sep)

(1) The long term annual average according to the map on p. 5 is $\pm$ 6 200 MJ/(m^2.yr).

---

2. Plant

2.1. Greenhouse

1. Dimensions   – Floor area :   150   m^2
                 – Volume :   457.5 m^3
                 – Glazing area : 293.5 m^2 (roof = 151.5 m^2/sides 142 m^2)
2. Glazing : Polyethylene
3. Covers/screens : none
4. Ventilation system : Fan.

2.4. Collector and heat storage

1. Medium of heat storage : Subsoil, Sandy clay; 23.5% moisture content around the heat exchangers.
2. Location of the system : 2 m deep under the greenhouse soil
3. Insulation : none
4. Dimensions : not limited
6. Conductive heat loss coefficient : λ = 1.04 W/(mK)
7. Diagram : cf. Figure 2.4.7

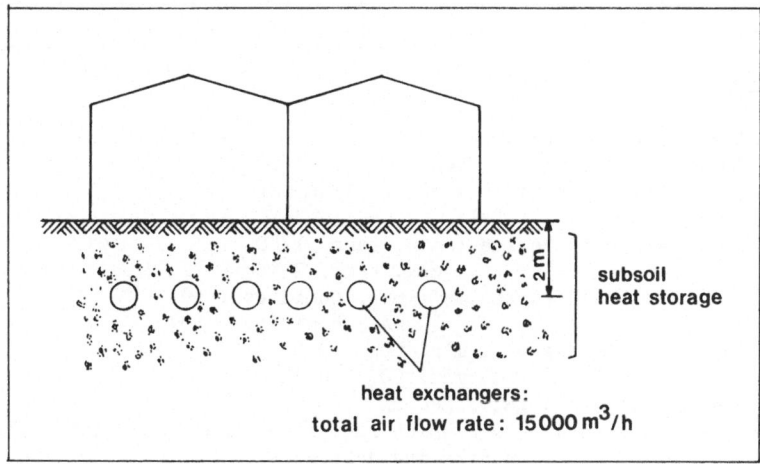

Fig. 2.4.7 : Diagram of "collecting system"

2.5. Installation parts and measuring devices

1. Technical description of the installation parts :

- Heat exchangers :
  - number : 20
  - length : 15.00 m
  - diameter : 0.20 m
  - wall thickness: 0.5 mm each

- Electrical fan :
  - capacity : 15000 m³/h
  - power : 5 kW
  - static pressure: 2 000 - 2 400 Pa

- Electrotherm 220 V electrical resistances
  - max Power : 90 kW

2. Description of the sensors.

- The air flow rate was determined by a wallas facility based on the principle of gas calorimetry.
- The electric energy consumption was measured by a kWh meter
- The indoor and outdoor air temperature were measured by thermographs and max-min thermometers.
- The solar energy in the greenhouse was determined by an Eppley type pyranometer and an AEG integrator.
- The air temperature on the control pipe of the Earth-Air heat exchanger at inlet and outlet, as well as the outside subsoil temperature were measured by thermistor sensors connected to an ULTRACUT 100 Ω temperature recorder.

3. Products

3.1. Culture in greenhouse

1. Type of crop : tomatoes
2. Date of planting : 1 Dec.
   harvesting : 10 March till 30 May
3. Growing system : N.F.T.
4. Required temp. regime : min. temp. 12°C
   max. temp. 28°C
5. Required humidity regime : min. humidity : 60%
   max. humidity : 80%
6. Required CO_2 regime : min. CO_2 : 0.03%
   max. CO_2 : 0.1%
7. Sensitivity to light : Medium
8. Other (related), crops which require an analogous inside climate and luminosity : peppers and aubergines.

4. Economic aspects

4.1. Economic environment

1. Energy price of oil	ECU/1	0.32
2. Energy price of electricity	ECU/kWh	0.064
3. Energy prices evolution prospects		
– of oil or gas	%/yr	15
– of electricity	%/yr	15
5. Interest rate of loans for farm equipment (rate without subsidy)	%/yr	14

4.2. Investment expenses

		Solar greenhouse	Reference greenhouse (with electrotherm)
1. Total investment budget (materials only)	ECU	15 300	5 350
2. Cost of equipment :			
– expenses for buildings	ECU	3 060	3 060
– cost of regulation equipment	ECU	9 180	1 530
– other material cost	ECU	3 060	760
3. Time of personnel for the conception, installation and start-up of equipment		1 man year	
Cost of personnel for the conception, installation and start-up of equipment	ECU	7 650	
4. Lifetime of the equipment	years	8 years	8 years
5. Residual value of the investment	ECU	none	none

4.3. Operating expenses

	Solar greenhouse	Reference greenhouse
1. Energy cost for the period of Dec-Febr (ECU/period)	150 for fan operation	740 for electrotherm operation

5. Results

5.1. Method of calculation

Air-Earth heat exchanger efficiency (η)

$$\eta = \frac{Tgi - Tgo}{Tgi - Tsoil}$$

Tgi : air temperature at the inlet
Tgo : air temperature at the outlet
Tsoil : the temperature of the outside soil at 2.00 m depth.

5.2. Weather conditions and energy

1.	Nov	Dec	Jan	Feb	Mar	Apr	Months	PERIOD	
3.	–	10	8.1	7.6	11.7	16.3	°C	average ambient TEMPERATURE	
6.	19 854	16 563	21 528	22 201	36 734	18 054	MJ/(month.greenh.)	input SOLAR energy [1]	solar
7.	4 730	1 584	2 001	4 525	11 793	–	MJ/(month.greenh.)	input energy STORAGE	
14.	1 360	2 671	3 074	2 740	2 250	1 792	MJ/(month.greenh.)	EXTRA ELECTRICITY consumption (fan operation)	
15.	4 410	12 351	15 904	13 370	4 485	–	MJ/(month.greenh.)	HEATING energy (electric)	ref. (GJ)

(1) Solar energy penetrating inside the greenhouse

5.3. Inside climate of greenhouse

1.	Dec	Jan	Feb	Mar	Apr			PERIOD	
3.	7.5	7	7	8.8	12		°C	min	both solar and reference greenhouse
4a.	11.5	11.0	11.0	13.9	14.3	night	°C	average TEMPERATURE	
4b.	19.2	20.4	20.3	23.7	24.4	day	°C	average TEMPERATURE	
5.	28	28	28	33	35		°C	max	
14.	0	-3	-3.6	-0.8	5.5		°C	min	greenhouse without heating
15a.	8.4	6	5.6	10	13.8	night	°C	average TEMPERATURE	
15b.	19.8	19.8	21	26.8	34.8	day	°C	average TEMPERATURE	
16.	35.5	30.0	36.8	45	47		°C	max	

5.4. Crop results

	Solar	Reference
1. Date of planting :	1 Dec 1982	
2. Harvesting date :	10 March till 30 May	2 March
7. Production obtained :	3 800 kg	4 200 kg
8. Other differences : (in results, aspects, dimensions, ...)	Normal	Normal
Variety	Sonetino	Sonetino

5.7. Some additional data

The effect of the energy storage on the minimum night and the maximum day
temperature is shown on figure 5.7 : here a comparison is made between the
solar greenhouse and a non-heated greenhouse.

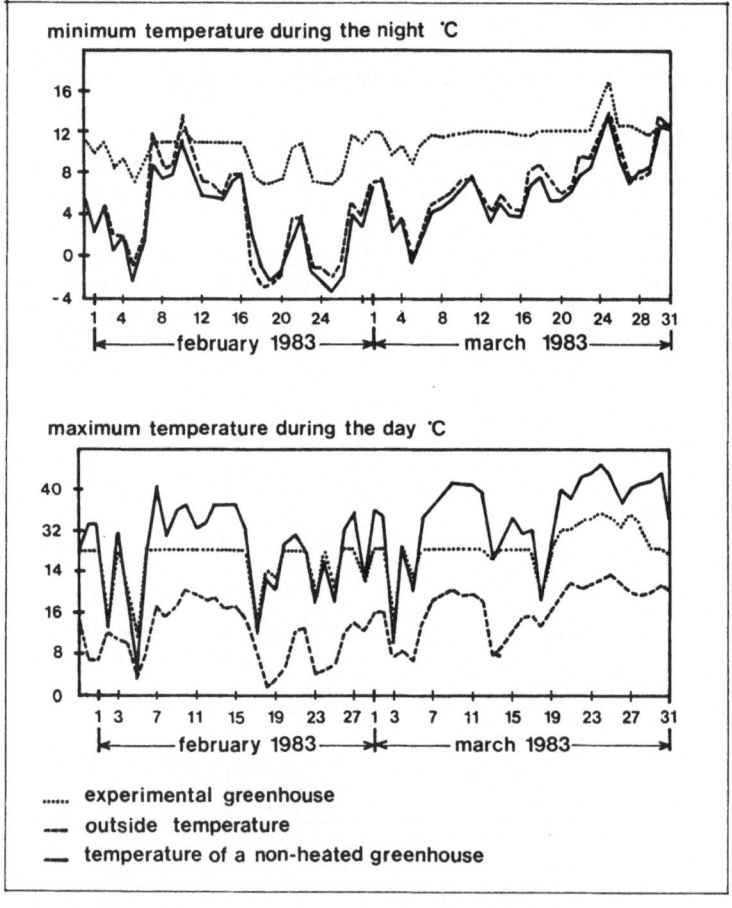

Fig.5.7 : Effect of energy storage on minimum night and maximum day
temperature

7. Development of a solar system for heating greenhouses

Project location

THESSALONIKI, GREECE

Study Center :
- Project Leader : Dr. Mavr. Grafiadellis
- Institution : Agricultural Research Center of Northern Greece
  Elliniki Georgiki Scoli - Thessaloniki
  Greece

Running since 1981

Project description

The "solar system" consists of an underground water reservoir, a fan, a water pump and two plastic tubes through which water and air are circulating simultaneously. The solar system itself is actually an air-water plastic heat exchanger. During the day, the water absorbs excess solar heat collected by the greenhouse; at night the previously solar heated water in turn heats the greenhouse air.
The goals are the determination of :

- the size of the heat exchanger, the air and water quantity needed and the size of the water reservoir for heating a greenhouse.
- the performance of the solar system.
- the effect of the solar system on temperature and light in greenhouses.

- the effect of the solar system on the earliness, quality and the yield of the most important vegetable crops.
- the cost of the solar system and the pay-back period.

Scheme

Two identical greenhouses of 150 m² are covered with polyethylene. The materials and the construction techniques utilised are currently available in the Greek greenhouse industry.

Site plan :

Orientation east-west. There are no obstructions to the sun. (cf. Figure S1)

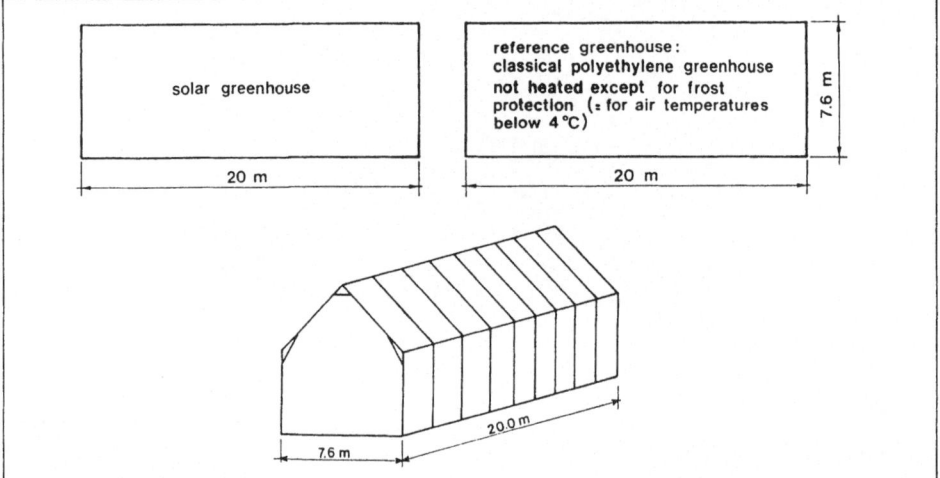

Fig. S1 : Ground-plan and perspective elevation of greenhouses

Operating modes

In the first two years (1981 and 1982) the water pump and the fan were connected to a time switch and were adjusted to operate in the warmer and colder hours of the day. (For example from 9 a.m. to 4 p.m. and from 9 p.m. to 7 a.m.). In 1983 the system was improved by using differential thermostats. When the temperature of air in the greenhouse is 5°C higher than the temperature of water in the reservoir, water pump and fan are switched on. Again at night time, when the air temperature in the solar greenhouse drops below 12°C, the water pump and the fan are activated.

Scheme of the "solar system" (cf. figure S2)

The different parts of the system are as follows :

- a water pump of 0.75 kW delivers 2.8 m³ of water per hour. The pump is connected to a time switch and to two thermostats.
- a fan is circulating 2 200 m³/h of air inside the solar greenhouse and is connected to the same controllers as the pump. (In 1981 and in 1982 a smaller fan of 1 000 m³/hr capacity was used.)

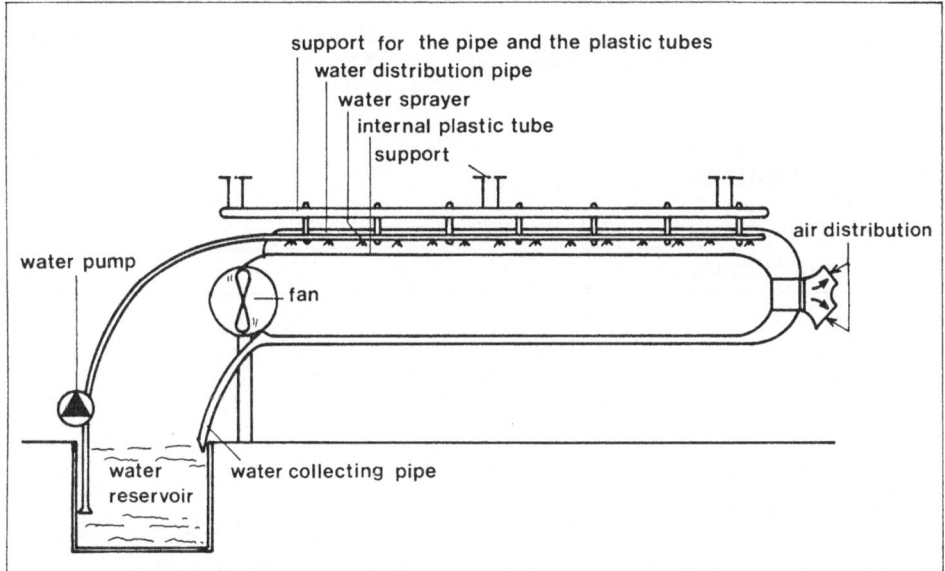

Fig. S2 : Scheme of the solar system

- two polyethylene tubes which serve as air-water heat exchangers. The external tube has a diameter of 76.4 cm and the internal black polyethylene tube has a diameter of 70.1 cm.
- an underground water reservoir of 11 m³.

Conclusions

The results obtained indicate :

- Solar heating increased the total yield by 10%.
- The solar heated greenhouse gained approximately 2-3 weeks earliness for the most important vegetable crops.
- For the data given in section 5.2 the solar system transferred into useful energy 19% of the solar radiation measured inside the greenhouse.
- When the air temperature in the non-heated greenhouse was 0°C, the solar heated greenhouse usually had an air temperature of 8°C. In warmer weather conditions the difference of temperature in solar heated and in non-heated greenhouses was smaller.
- At night, a considerable quantity of heat is also liberated from the soil through air circulation into the greenhouse air. (The soil actually also plays the role of some kind of a solar energy storage.)
- The cost of the solar system for heating, extrapolated to a greenhouse of 1 000 m², was estimated to be about 2 270 ECU (1).
- Based on the relatively low investment price and the significant yield improvements, the payback period is expected to be very short.
- The main problem of the project was the lack of a temperature recording instrument.

(1) Counting equipment costs only, and at 1981 price rates.

Further references :

- GRAFIADELLIS, M. and KYRITSIS, S. 1981 : Heating greenhouses with solar
  energy. Acta Horticulturae. nr. 115; 591-597.

PROJECT ANALYSIS

1. Site and climate

1.1. Site

Latitude :	40°41' N
Longitude :	23° E
Altitude :	30 m
Nearest main city :	Thessaloniki – Greece
Distance form main city :	13 km
Direction from main city :	East
Obstructions :	Solar collector is inside the greenhouse and it shades some plants near the North wall.

1.3. Annual long-term averages

			Project	Country in general
1. Prevailing wind direction	Winter		North	North
	Summer		South	South
2. Average wind speed				
(is noticed 8 % of days of	W	m/s	15	12
the year)	S	m/s	12	10
3. Total precipitation		mm/yr	456	360-2 289
4. Relative humidity	W	%	73	71
	S	%	56	55
5. Global irradiation on				
horizontal plane		MJ/(m^2 yr)	5 660	6 020
6. Diffuse proportion		%	37	35
8. Hours sunshine per year		hrs/yr	2 555	2 700
9. Ambient temp. mean/year		C	16	18
10. Average max. temp. (Jul/Aug)		C	34.2	35

2. Plant

2.1. Greenhouse

1. Dimensions :   - Floor area :   150 m^2
              - Volume :   450 m^3
              - Glazing area :   280 m^2 (roof = 160 m^2; sides 120 m^2)

2. Glazing : Single long-lasting polyethylene which transmits 70% of the visible light and has a durability of two years.

4. Ventilation system : The greenhouses have a side ventilation system. One side is opened manually and the other automatically. The surface of the openings is 22% of the covered area of the ground.

2.3. "Solar system"

1. Material : Polyethylene plastic film.

2. Dimensions :
   - length : 15 m
   - External diameter of tube : 0.764 m
   - Internal diameter of tube : 0.701 m

3. Flow rate : In 1981 and 1982 the air flow was 1 000 m^3/hr and the water flow was 2.8 m^3/hr. In 1983 the air flow increased to 2 200 m^3/hr and the water flow remained the same.

2.4. Heat storage

1. Medium of heat storage :
   An underground water reservoir of 11 m^3.

2. Description, location in the system :
   Underground, inside the solar greenhouse.

3. Insulation :
   Two sheets of black polyethylene film and a sheet of triplex film with air bubbles.

4. Dimensions :
   2 x 2.2 x 2.5 m

5. Thermal capacity : 46 MJ/K

2.5. Installation parts and measuring devices

2. Description of the sensors

   Water quantity is measured with a mechanical water meter. The air flow rate in the solar system is measured with an anemometer.
   Solar radiation is measured with a Kipp solarimeter.
   Relative humidity of air is measured manually with an electronic relative humidity indicator.
   Air temperature is measured with thermometers and with an electronic temperature indicator. For soil temperature measurements special mercury thermometers are used.

3. Scheme of measurement points (cf. figure 2.5.3)

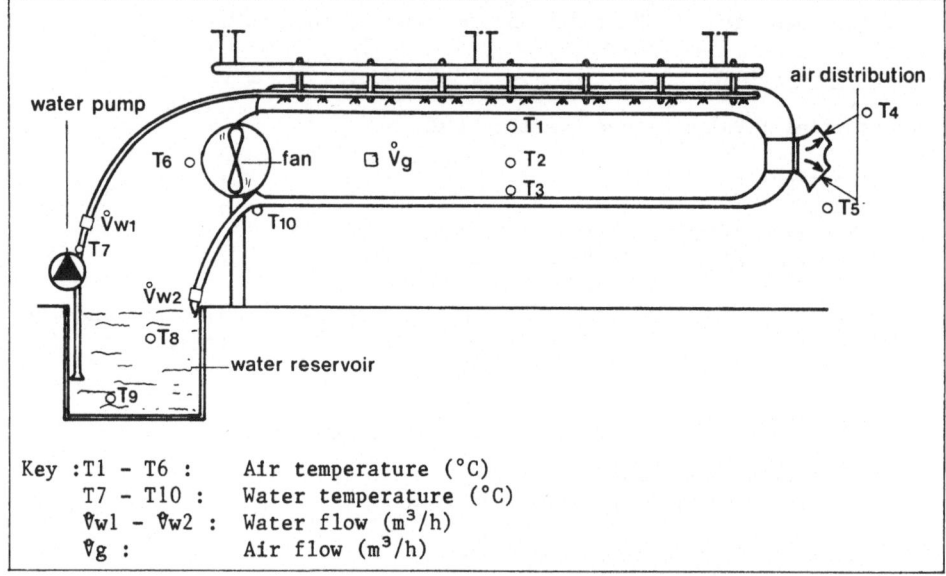

Key : T1 - T6 :   Air temperature (°C)
    T7 - T10 :   Water temperature (°C)
    $\mathring{V}$w1 - $\mathring{V}$w2 :   Water flow (m³/h)
    $\mathring{V}$g :   Air flow (m³/h)

Also measured :
- Air temperatures at different places inside and outside the greenhouse (°C).
- Different soil temperatures (°C).
- Relative humidity inside and outside (%).
- Solar radiation <u>inside</u> (W/m²).

Fig. 2.5.3 : Scheme of measurement points

---

3. Products

3.1. Culture in greenhouse

1. Type of crop : Tomatoes, peppers, egg plants and cucumbers.
2. Date of sowing/planting : 22 to 24 December / First week of March.
                 harvesting : Beginning of May till September.
3. Growing system : Conventional
4. Required temp. regime : min. temp. 8°C
                        max. temp. 35°C
5. Required humidity regime : min. rel. humidity : 60 %
                             max. rel. humidity : 80 %
6. Required CO_2 regime : min. CO_2 : 0.03 %
                      max. CO_2 : 0.18 %
7. How sensitive is the crop to light ?
   In summer time cucumbers need shading.
8. Other (related), crops which require an analogous inside climate and luminosity :
   Melons, beans.

4. Economic aspects

4.1. Economic environment

1. Energy price of oil	ECU/1	0.31
2. Energy price of electricity in agriculture	ECU/kWh	0.031
3. Energy prices evolution prospects		
– of oil or gas	%/yr	15
– of electricity	%/yr	20
5. Interest rate of loans for farm equipment (rate without subsidy)	%/yr	14

4.2. Investment expenses

		Solar greenhouse	Reference greenhouse (1)
2. Cost of equipment :			
– cost of regulation equipment	ECU	325	250
– other material cost	ECU	325	30
– additional costs of the collector per m^2 greenhouse surface	ECU	1.62	0
3. – Time of personnel for the conception, installation and start-up of equipment	(man-days)	20	2
– Cost of personnel for the conception, installation and start-up of equipment	ECU	490	N.A.
4. Lifetime of the equipment	(years)	8	8
5. Residual value of the investment	ECU	243	0

(1) Cost of frost protection (i.e. temporary use of gas oil heater).

4.3. Operating expenses

		Solar only	Reference solution
1. Energy cost	ECU/yr	31 (1)	32 (2)
2. Maintenance cost (replacement of material or components)	ECU/yr	64	51
4. Manpower cost	ECU/yr	190	180

(1) Electricity for fan and water pump operation.
(2) Oil for frost prevention.

4.4. Hidden cost and benefits

		Solar only	Reference solution
1. Production obtained	kg/yr	1 375	1 255
2. Average sales price per unit produced	ECU/kg	0.26	0.175
3. Total income obtained	ECU/yr	357	219

5. Results

5.2. Weather conditions and energy

1. 2.	7 Apr 1983 5 hr	8 Apr 1983 5 hr	11 Apr 1983 5 hr	21 Apr 1983 5 hr		PERIOD From 9 hr till 14 hr	
3.	12	10	13	17	°C	average ambient TEMPERATURE	
4.	1	0	2	2	m/s	average WIND	
5.	455	366	543	471	W/m^2	average SOLAR RAD on horizontal plane inside	
	1229	988	1467	1272	MJ/(5h.greenhouse)	the greenhouse	
7.	202	227	329	277	MJ/5h	input energy STORAGE	solar
8.	18	20	30	25	MJ/5h	STORAGE HEAT LOSSES	
9.	20	19	22	23	°C	average TEMPERATURE storage (1)	
10.	184	207	299	252	MJ/5h	USEFUL energy (1)	
14.	15	15	15	15	MJ/5h	ELECTRICITY consumption	

(1) Discharged later in the day.

5.3. Inside climate of greenhouse

1. 2.	7 Apr 1983 5 hr	8 Apr 1983 5 hr	11 Apr 1983 5 hr	21 Apr 1983 5 hr		PERIOD From 9 hr till 14 hr	
3.	6	7.5	10	13	°C	minimum	solar
4.	17	18	20	23	°C	average TEMPERATURE	
5.	30	27	30	33	°C	maximum	
6.	48	60	45	62	%	minimum	
7.	61	71	68	73	%	average RELATIVE HUMIDITY	
8.	75	82	70	85	%	maximum	
9.	455	366	344	472	W/m^2	radiation	
14.	1	2	4	8.5	°C	minimum	reference
15.	14	14	17	20	°C	average TEMPERATURE	
16.	30	26	30	32	°C	maximum	
17.	50	58	44	60	%	minimum	
18.	61	67	55	71	%	average RELATIVE HUMIDITY	
19.	72	76	66	82	%	maximum	
20.	455	366	544	472	W/m^2	radiation	

5.4. Crop results (greenhouse)

	Solar	Reference
1. Date of sowing	20 December	20 December
2. Beginning harvesting date :		
– Tomatoes	12 May	18 May
– Cucumbers	12 May	18 May
– Peppers	30 April	15 May
– Egg plants	20 May	30 May
5. Fruit number :		
– Tomatoes/plant	27.2	21.0
– Cucumbers/plant	34.9	30.9
– Peppers/plant	111.3	80.3
– Egg plants/plant	24.6	24.4
7. Production obtained :		
gr/plant		
– Tomatoes	7 829	7 160
– Cucumbers	12 359	10 706
– Peppers	3 815	3 376
– Egg plants	6 584	6 694

5.6. Heated water

	7 Apr 1983	8 Apr 1983	11 Apr 1983	21 Apr 1983		PERIOD
1.						
2.	5hr	5 hr	5 hr	5 hr		From 9hr till 14 hr
3.	17	18	18	20	°C	average input water temperature
4.	0.77	0.77	0.77	0.77	1/s	average flow rate
5.	23	23	23	26	°C	average heated water temperature

PART II

Applications of solar energy in drying processes

CHAPTER I :
Process requirements - technological options

1. DRYING, WHY ?

1.1. The problem : long-term conservation

Most agricultural crops within the EC are grown and harvested only once a year. The moment when these products are needed does not however usually correspond with the harvesting season.
Storage therefore plays a crucial part in the provision of agricultural products at the desired time.

Any stored biological product is subject to deterioration. This deterioration over time is a function of several variables, (cf. figure 1), influenced, among others, by the treatment and storage techniques :
- moisture content (m.c.) of the product/relative humidity (r.h.) of the environment
- temperature
- level of infection of the product by fungi, bacteria, insects
- contaminants (undesired plant materials, dust, etc. ...)
- physical damage to the product

Among these factors, the moisture content is a very important one.
Water is essential to all life forms. Due to its presence, the growth and propagation of organisms, feeding on the stored product, becomes possible. The respiration of these organisms, as well as of the agricultural product itself produces even more moisture and heat, both of which again stimulate further growth.

$$C_6H_{12}O_6 + 6\ O_2 \xrightarrow{\text{respiration}} 6\ CO_2 + 6H_2O \text{ (= additional moisture) (E.q. 1)}$$
$$+ 2\ 830\ J \text{ (= additional heat)}$$

Storage fungi, like Aspergillus and Penicillium with optimal growth rates around 35°C and 80% relative humidity, pose the main problems for grains, kept in similar circumstances. Wheat of 18% m.c. can be infected within 2-3 weeks with 88×10^6 colonies of fungi/gram (11).

The rapid propagation of these fungi, as well as the metabolic reactions of the product itself, result in :
- a lower nutritional value of the product
- raised temperatures, favourable for additional mite and insect infestation. The temperature can rise to such a high level that it results in loss of germination capacity, for grains for example, or even further to the point where spontaneous combustion occurs.
- health hazards :
  1. feed hazards : Mycotoxins of the fungi can decrease fertility, cause abortions, ulcerations, general ill-health ...
  2. handling hazards : Micropolyspora faeni and Thermoactinomyces

vulgaris in hay and grain can cause the notorious Farmers' lung disease.

1.2. Drying, a possible solution

Several treatment and storage techniques exist, which influence the above-described parameters in such a way that the deterioration over time is kept to a minimum. The most commonly used traditional technique has been drying and cooling of the product to such a degree that rates of metabolism are strongly inhibited, even without using chemical additives. Figure 2 gives an idea of the tremendous influence moisture content and temperature can have on life potential of fungi.

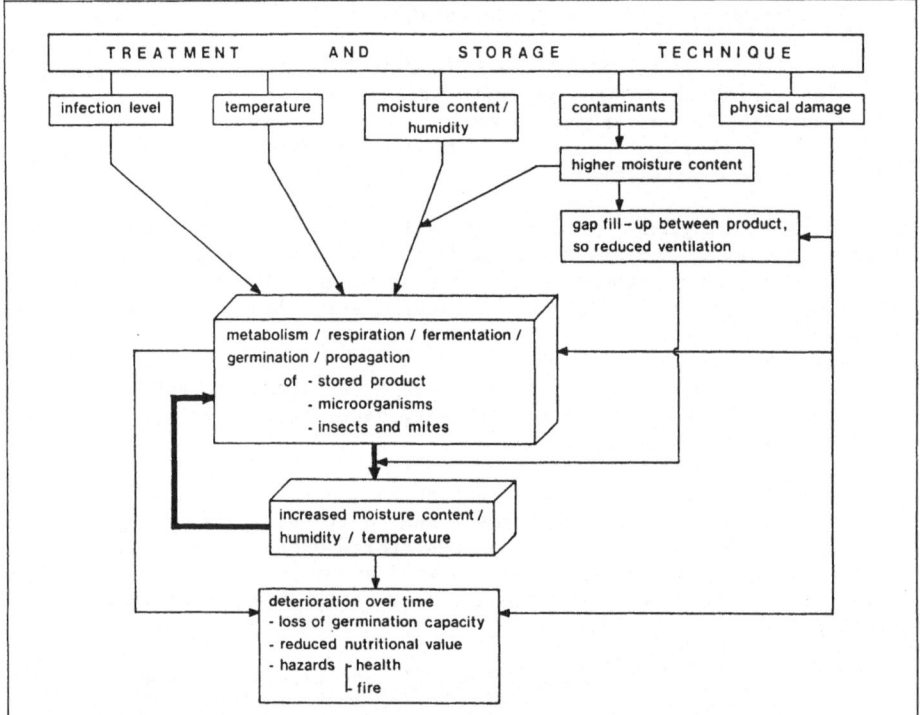

Fig. 1. Schematic view of the main factors that influence the conservation of agricultural products.

Depending on product characteristics, <u>artificial drying can be done in many different ways</u>. In certain cases, mechanical techniques are used (e.g. centrifuge, press, decanter ...), or reverse osmosis, cryogenation, etc. ... Most of the drying of agricultural products is however done by means of forced air ventilation.

In this drying method, the water is extracted from the product by providing energy – contained in the air – for the water's transition from liquid to gaseous phase. This vapor is then carried along with the ventilation air stream, and disposed of in the atmosphere outside the drier.

The basic elements in an air convection drying set-up are shown schematically in fig. 3.

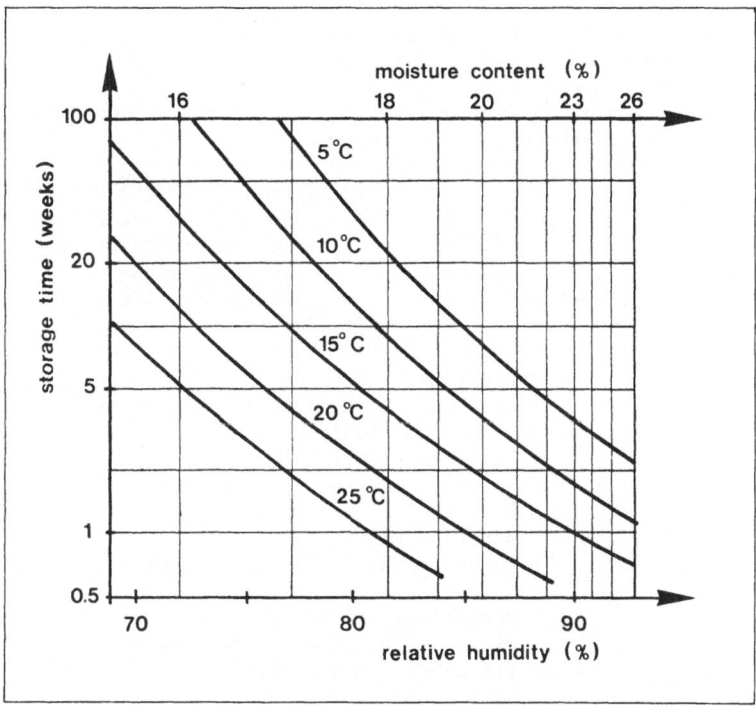

Fig. 2. Storage period of wheat – as a function of temperature and moisture content – before moulding becomes visible (Kreyger, 1972) (9).

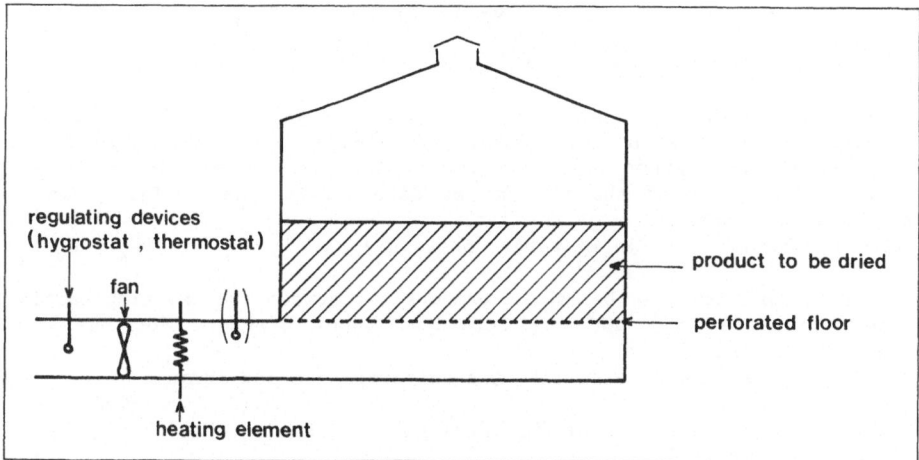

Fig. 3. Schematic representation of a simple drying set-up.

2. SOME ASPECTS OF DRYING PROCESS ENGINEERING

The aim, when developing an adapted drying process for a particular product (with its specific characteristics of heat transmissivity, water diffusivity, desorption isotherms, etc. ...), is to find an optimal combination of the drying parameters (air characteristics, drying set-up) so as to reach the desired moisture content in the desired time (before excessive deterioration occurs), and in such a way that the required properties of the product (germination capacity, vitamin content, ...) are not affected. (cf. figure 4)

In this context a more quantitative insight into the interaction of the different parameters is obviously desirable.

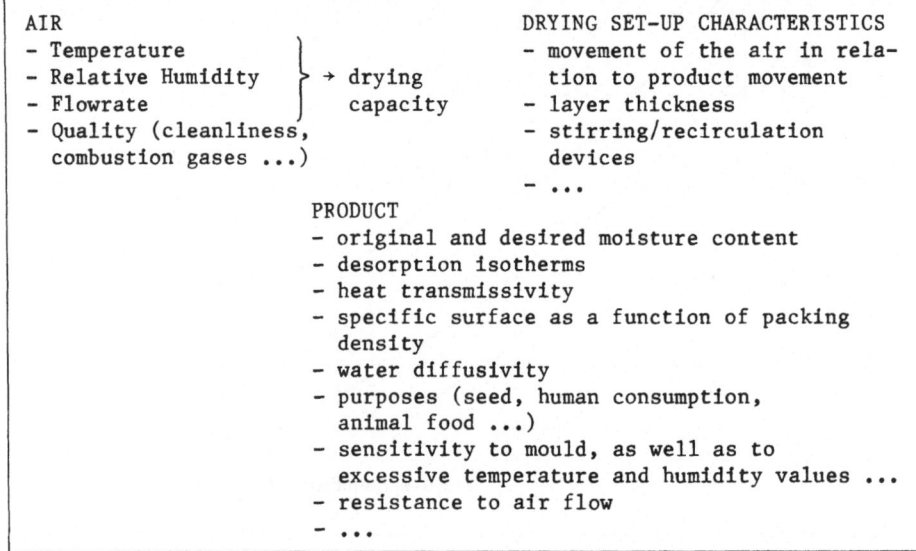

Fig. 4. Schematic overview of the different aspects of the drying parameters that need to be adjusted to best fit the characteristics and purposes of the product that is to be dried.

2.1. Characteristics of the drying air

Air mainly consists of "dry air" ($+$ 1/5 O_2; $+$ 4/5 N_2) and vapor. Its condition is characterised by several variables (e.g. air temperature, air humidity, energy content ...). If two of those variables are known, the energetic condition of the air can be determined. Air humidity can be expressed in 2 ways :

- absolute humidity x (in kg water vapor (G_V) per kg dry air (G_A); G_V/G_A) *(1)*
- relative humidity ψ (ψ being the proportion between the existing partial vapor pressure (p_V) and the saturated vapor pressure for the same temperature (p_V') *(1)*

If G_A or G_V are incorporated into the universal gas law formula $\dfrac{pV}{(T+273)}$ $=nR$

where p = partial pressure of the ideal gas in question (Pa)
  V = volume (m^3)
  T = temperature (°C)
  R = gas constant ($\simeq$ 8.31 J/mol K)
  n = weight (molar)

(1) Sometimes x is expressed in g/kg and/or ψ percentagewise. In those cases the derived equations will obviously need to be adapted to match with these different dimensions.

an equation can be derived, which links weight factors (G_A, G_V, x) to partial vapor pressures (p_V, p_V', ψ) :

$$x = 0.622 \; \frac{\psi \; P_V'}{P_{Tot} - \psi \; P_V'} \qquad \text{(Eq. 2)}$$

P_{Tot} = total air pressure (Pa), corresponding to the sum of the partial pressure values of dry air and vapor.

Note : The value of P_V' = for different air temperatures can be found in steam diagrams.

A very important variable for drying engineering is the energy content of the air. The "enthalpy" (j) describes the energy level of a specific volume consisting together with other gaseous components (mainly vapour) of 1 kg of dry air. The energy level of dry air and liquid water at 0°C are assumed to be zero. Thus defined, j can be found as a function of temperature and moisture content. Indeed, both vapour and dry air increase in energy content almost linearly as the temperature increases.

$$j = c_A T + x \; (r_0 + c_V T) \qquad \text{(Eq. 3)}$$

with j = air enthalpy (J/kg dry air)
      c_A = specific heat of dry air (1 006 J/(kg.K))
      T = temperature in °C
      x = absolute humidity (g/kg or kg/kg)
      r_0 = evaporation heat of water at 0°C (2.5 x 10^6 J/kg)
      c_V = specific heat of vapour (1 860 J/(kg.K))

For different conditions of the air, the value of the main variables of the air at a specific atmospheric pressure are given in Mollier diagrams or in so called "psychrometric" diagrams.

In those tables the variables are expressed per kg dry air, while data on ventilator and flow rate measurements in the drying set-up are available per m^3 of humid air. Here too application of the universal gas law makes a transition possible from one expression to the other via equation 4.

$$\rho_A = \frac{28.96 \times 10^{-3} \; (P_{Tot} - \psi P_V')}{R_A \; (T + 273)} \qquad \text{(Eq. 4)}$$

with ρ_A = the specific mass of dry air (kg dry air/m^3 humid air)
      R_A = adapted gas constant to compensate for the deviation of dry air from the behaviour of an ideal gas (R_A = 8.3362 J/(mol K))

2.2. The product

2.2.1. Product moisture content (m.c.)

Two quantitative definitions are used.
- m.c. on a "wet base" is the ratio of the water mass present in the product over the total mass of the product (water + dry matter) in the given condition. (Expressed decimally or in %)
- m.c. on a "dry base" is the ratio of the water mass present in the product over the mass of the dry matter only of the product (decimal or %)

For practical purposes the water mass present in the product is considered to be equal to the amount of water that can be removed given vapour pressure of 0 bar and under such circumstances that potentially disturbing reactions are avoided (Definition of Guilbot).

2.2.2. Sorption and desorption isotherms

Over time each system strives to reach a thermodynamic balance, with the free enthalpy equally distributed over the whole system. For this purpose water will be exchanged between product and air.

In figure 5 an example is given of situations of equilibrium given different air characteristics and previous treatment for potato starch. Those curves of equilibrium situations are called desorption or sorption isotherms depending on whether the equilibrium is reached after drying or wetting of the product.

The water contained in the product in zone A in the figure is strongly bound in a monomolecular layer to the dry matter. The water in zone B (zone with quasi linear relationship between m.c. and r.h.) is bound by "Van der Waals" - chemical bonds. Only the extra water in zone C, at most osmotically attracted to the product, is available for enzymatic and microbial activities and this at an increasing rate as the m.c. increases. Thus the desorption isotherms provide an indication of the conditions of safe storage after drying (e.g. in this case the r.h. must be below 70% for safe storage of previously dried starch).

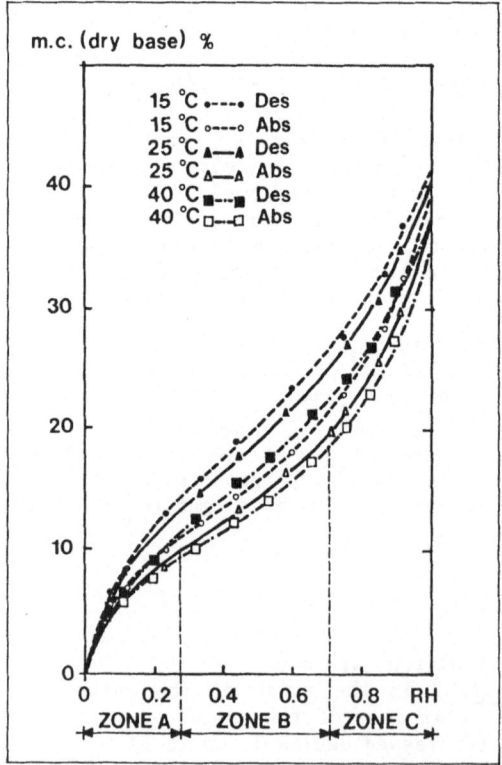

Figure 5 : Sorption (Abs) and desorption (Des) isotherms of potato starch at different temperatures (9).

2.2.3. The 2 main stages in product drying (cf. figure 6)

Stage 1

In figure 5 the equilibrium air relative humidity for a moisture content greater than 40% (dry base) is 100%. In these high moisture content zones the speed of water evaporation due to heat convection of air to the product appears to be significantly smaller than the speed of diffusion of the water through the product to the surface.

Hence the starch behaves as a free water surface and the relative humidity of the air closely in touch with the product is 100 %. After a first adaptation period the product takes on the wet bulb temperature of the air, at which point any convective heat addition is completely utilized for evaporation. This situation will last to the point where all the so called "free water" is evaporated (till a dry base moisture content of about 40 %).

In this first stage the change in moisture content can be calculated as a function of time since the heat convection is a measure of the evaporation rate :

$$\underbrace{ks\ (T_a - T_p)dt}_{\text{heat convection}} = \underbrace{r\ \rho_{dp}\ d(mc)}_{\text{evaporation}} \qquad \text{(Eq. 5)}$$

with k   heat transfer coefficient of the product ($W/(m^2\ K)$)

  s   specific surface of the product (m^2/m^3 storage volume)

  T_a air temperature (°C)

  T_p product temperature (°C)

  r evaporation heat of water at a specific temperature (J/kg water)

  ρ_{dp} specific mass of dry product (kg dry product/m^3 total product)

  time in seconds/temperature difference in Kelvin/moisture content on a dry base

The amount of water that is being removed per m^2 product surface and per second (D) equals :

$$D = -\ \frac{\rho_{dp}}{s}\ \frac{d(mc)}{dt} \qquad \text{(Eq. 6)}$$

Assuming no product shrinkage occurs, integration yields :

$$\int_{to}^{t} dt = -\ \frac{\rho_{dp}}{s}\ \int_{mc_o}^{mc}\ \frac{d(mc)}{D} \qquad \text{(Eq. 7)}$$

From equations 5 and 6 follows :

$$D = -\ \frac{k\ (T_a - T_p)}{r} \qquad \text{(Eq. 8)}$$

Filled into equation 7 and integrated this yields the following **approximate** equation.

$$t - t_o = \frac{\rho_{dp}\ r}{ks(T_a - T_p)}\ (mc - mc_o) \qquad \text{(Eq. 9)}$$

In practice minimization of drying time in this stage is done in 2 main ways.

- Increasing k by applying high air speeds
- Increasing T_a :
  The corresponding increase of T_p to the wet bulb air temperature is smaller than the air temperature increase, and due to the curved shape of the saturation line (cf. Mollier diagrams) the effect of an increased air temperature on the size of $(T_a - T_p)$ becomes greater as the absolute

air temperature increases. In cases where the natural drying capacity of the air is negligible, the specific thermal energy consumption *(1)* will thus decrease with increasing air temperature.

The allowable maximum temperature level is however limited by the temperature sensitivity of the product, which is especially high at high moisture content levels. Furthermore, as air temperature increases, the problem for different drying set-ups of non-uniformity of the dried product becomes more acute.
- Reducing the absolute humidity of the air would lower the wet bulb and hence product temperature, thus increasing (T_a-T_p). This possible option, which also applies to stage 2, is however almost never chosen in agricultural drying practice.

Stage 2
As all the "free water" is evaporated, the diffusion through the product of the osmotically or even more strongly bound water becomes the limiting factor in the drying process, and this increasingly so. The drying speed is no longer constant but decreasing, and the product moisture content asymptotically reaches the equilibrium moisture content.

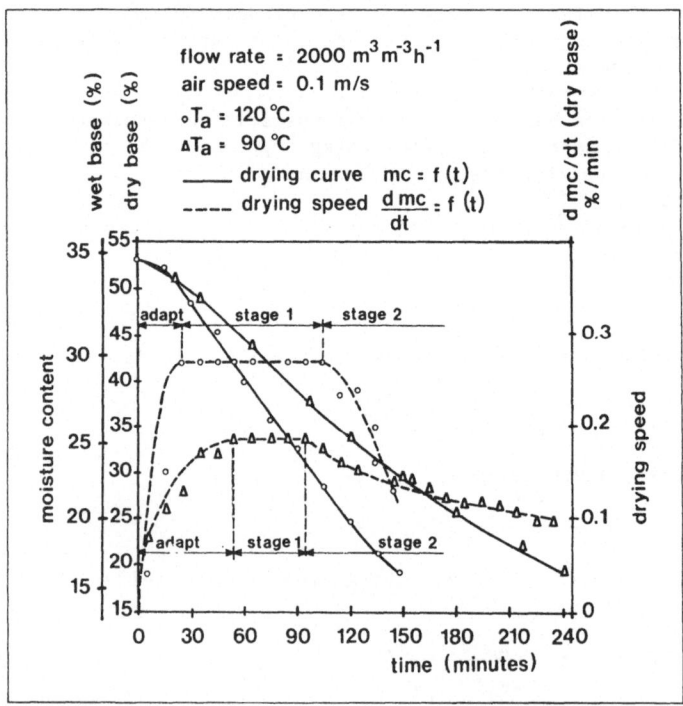

Fig. 6 : stages in the drying process (9)

Next to energy for evaporation, in this stage extra energy is required to overcome the osmotic forces and the resistance against diffusion. The addition of this energy is expressed in a temperature rise of the product

(1) "Specific thermal energy consumption" = Ratio of the thermal energy input over the amount of water evaporation.

above the wet bulb temperature of the air.

Note : In some cases the drying speed decreases quasi linearly with
moisture content. Inserting this linear relationship into equation
7 would yield after integration :

$$t - t_c = - \frac{\rho_d \, (mc_c - mc_E)}{sD_c} \ln \left(\frac{(mc - mc_E)}{(mc_c - mc_E)} \right)$$

with index c referring to the critical point of transition between stage 1
and 2

E referring to the equilibrium point between product and air.

Here too in practice the drying time is shortened by increasing the air
temperature. However in this stage the specific thermal energy
consumption increases with air temperature :
- The fraction of the energy input into actual evaporation over the input
  of sensible heat into the product itself decreases. This energy input
  into the product, which is expressed in a product temperature rise,
  helps overcome the resistance to diffusion (This resistance factor
  itself increases with drying speed). It does not however in any way
  contribute to the final absolute amount of water evaporation. Even
  worse, it has to be removed by costly cooling devices in order to
  establish safe conservation conditions.
- The hotter the drying air in this stage the greater the unutilized
  drying capacity of this air when leaving the drier. (This problem is
  inherent to certain drier types only).

High air temperatues can result in excessively high product temperatures,
leading to deterioration by e.g. maillard reactions. Furthermore, sharp
temperature and moisture content gradients are brought about which create
internal stresses, leading to microscopically small tears causing
fragility and susceptibility to deterioration.

In this second stage, the drying air speed has less of an influence on
drying performance, since the limiting-process is located within, rather
than at the surface of the product.

The optimization of the above mentioned and other variables in function of
product and purposes is a complex task. Hence the great variety in chosen
technical solutions.

2.3. The drying set-up

2.3.1. Simulation (2) (15)

Simulation of a drying installation gives quantitative insights into the
influence of the different system parameters. It enables one to predict
optimal component capacities and performance without building expensive
experimental systems.

For simulation purposes, a batch-grain drier could possibly be considered
as consisting of thin horizontal slices (thickness dl) (see fig. 7).

Known are :
- characteristics T_a and rh of the incoming air at time t_1
- characteristics T_p and mc of the product in layer de at time t_1

Unknown are :
- T_a' and rh' of the air after flowing through layer dl

- T_p' and mc' of the product at time t_2

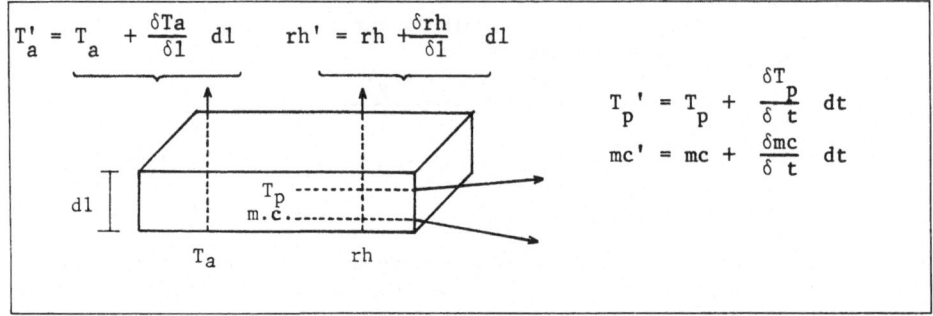

$$T_a' = T_a + \frac{\delta Ta}{\delta l} \, dl \qquad rh' = rh + \frac{\delta rh}{\delta l} \, dl$$

$$T_p' = T_p + \frac{\delta T_p}{\delta t} \, dt$$

$$mc' = mc + \frac{\delta mc}{\delta t} \, dt$$

Fig. 7. Change of air and product parameters over a thin horizontal layer of product to be dried.

If these 4 unknowns can be calculated, then T_a' and rh' can be used as input parameters for the slice higher up, while T_p' and mc' will function as the input values for slice dl in the next time-step. Thus the drying evolution can be calculated both as a function of space and time.

The four unknowns can be solved by calculating the energy and humidity/moisture balance of the air and the product (4 equations).

1. Energy balance of the air

Schematically, the flow of energy through the product can be depicted as shown in fig. 8.

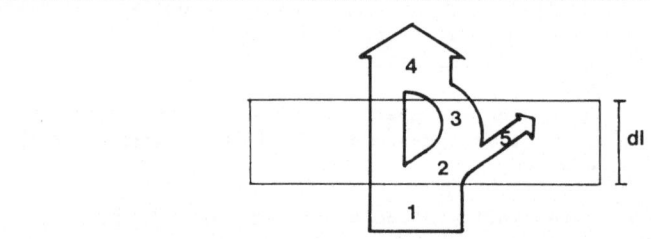

(1)   the energy contained in the incoming air ; can be calculated by combining flow rate with enthalpy of the air.

(2)   the energy transferred to the product by convection ; can be calculated with the formula for heat convection.

(3)   the energy input into evaporation of the water (this energy will be latently present in the outgoing air) ; can be calculated using the enthalpy formula applied to the evaporating water.

(4)   energy contained in the outflowing air ; (same mathematical expression as for the incoming air).

(5)   the perceptible heat input into the grain.

Fig. 8. Energy flow through layer dl with

Combining all these energy contents in an energy balance, namely $(1) - (4) = (2) - (3) = (5)$ would yield, after some rearrangement :

$$\frac{\delta T_a}{\delta l} = \frac{- ks (T_a - T_p) + \Phi_A \frac{\delta rh}{\delta l} (r - c_v T_p - r_o)}{\Phi_A (c_A + c_v rh + c_v \frac{\delta rh}{\delta l} \, dl)} \qquad \text{(Eq. 10)}$$

where $\dot{V}_A$ = flow rate of the dry air, expressed in $kg/(m^2.s)$

k = heat transfer coefficient between product and air $(W/(m^2K))$

s = specific surface of the product (m^2/m^3)

c_A = specific heat of dry air $(J/kg\ K)$

c_v = specific heat of vapour $(J/kg\ K)$

r_o = evaporation heat of water at $0°C$ (J/kg)

r = evaporation heat of water at the given product temperature (J/kg)

Following a similar strategy, the 3 other formulas are derived :

2. <u>The energy balance of the product</u> results in :

$$\frac{\delta T_p}{\delta t} = \frac{k\ s\ (T_a - T_p) - \dot{V}_A\ (r + c_v\ (T_a - T_p) - c_w T_p)\ \delta rh/\delta l}{\rho_p\ c_p + \rho_p\ c_w\ (mc) - c_w\ \dot{V}_A\ \frac{\delta rh}{\delta l}\ dt}\qquad (Eq.\ 11)$$

where c_w = specific heat of liquid water in the product $(J/kg\ K)$

c_p = specific heat of the dry product $(J/kg\ K)$

ρ_p = specific mass of the dry product (kg/m^3)

3. <u>The air humidity balance</u>

$$\frac{\delta rh}{\delta l} = -\frac{\rho_p}{\dot{V}_A}\ \frac{\delta mc}{\delta t}\qquad (Eq.\ 12)$$

4. <u>The moisture content balance</u>
It is difficult to determine this balance accurately by theoretical formulas (cf. section 2.3). Most often, empirical equations are developed.

The foregoing are the 4 basic formulas for drying simulation. Based on these formulas, and using small time and space steps, a computer can calculate the evolution of the variables in function of space and time.

Note : (Some additional subroutines will obviously have to be provided in order to deal with special cases, for instance when rh = 100 % or when the mc is in equilibrium with the drying air humidity - cf. desorption isotherms, section 2.2.2).

2.3.2. Characteristics of the drying set-up

1. <u>Drier types</u>
All driers can be classified in 4 different categories
   1. Static driers : the air moves through a static product layer.
   2. Cross flow driers : the air moves at an angle of 90° with the direction of the product movement.
   3. Counter flow driers : air movement parallel but in the opposite direction to the product movement.
   4. Concurrent flow driers : air movement parallel and in same direction as product movement.
Based on simulation, the performances representative for grain driers belonging to the different categories can be predicted and compared with one another.

Cross flow driers (1)

Data on the performance of cross flow driers are given in figures 9 and 10.

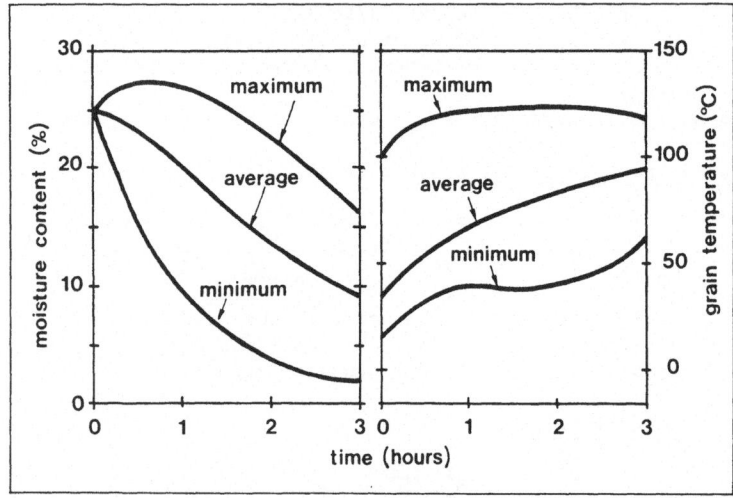

Figure 9 : Moisture content (w.b.) and temperature gradients in a cross-flow grain drier; simulated with T_{air} = 120°C and air flow 0.3 m³/(m².s) (2).

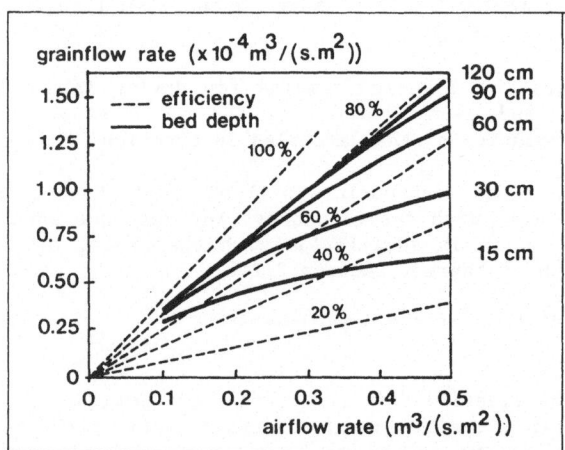

Figure 10 : Efficiency (2) of a cross-flow grain drier; simulated with T_{air} = 120°C; m.c. = 25 % – 15 % (w.b.) (2).

(1) Static driers have comparable performance characteristics.

(2) The efficiency of the drier is defined as the ratio of the simulated drying speed over the theoretically maximum possible speed obtainable if the drying capacity of the air were fully used for water evaporation (i.e. air flowing out of drier is saturated ; the net perceptible heat input into the product is zero).

This technically simple method is often used for grain drying. A major disadvantage is however the non uniformity of the moisture content throughout the product layer. In order to keep this problem within acceptable limits layer thickness is reduced and air flow rate increased. (Hence, a significant part of the air drying potential is not utilized). Efficiency values of 60% with layer thickness of 30 cm and air flow rates of $\pm$ 0.5 m³/(s.m²) can commonly be found.

Counter flow driers

Data on the performance of counter flow driers are given in figures 11 and 12.

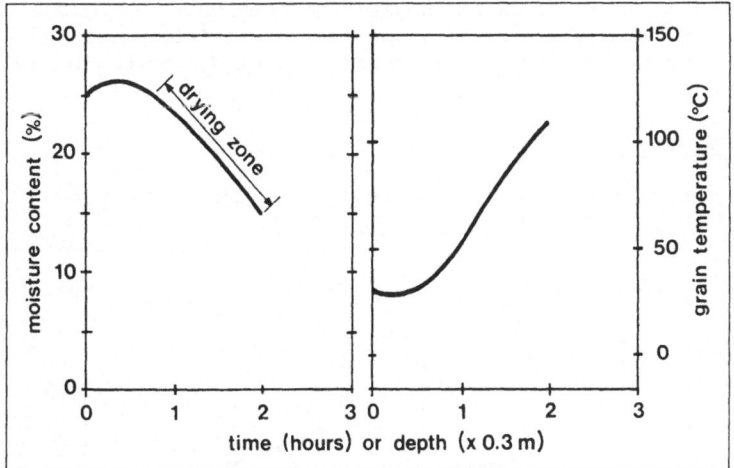

Fig. 11 : Moisture content (w.b.) and temperature gradients in a counterflow grain drier; simulated with T_{air} = 120°C; airflow = 0.3 m³/(s.m²) (2).

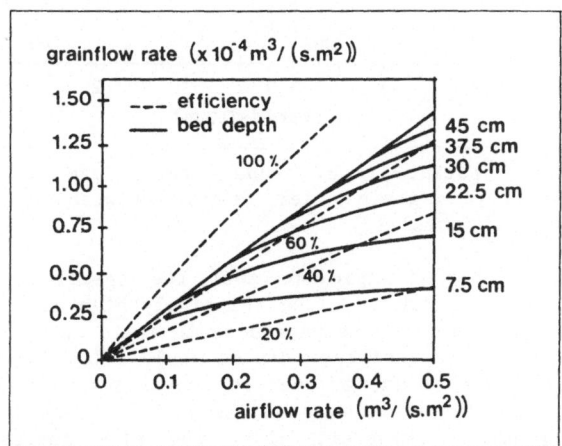

Fig. 12 : Efficiency *(1)* of a counterflow grain drier; simulated with T_{air} = 120°C; m.c. : 25 % – 15 % (w.b.) (2).

(1) See footnote 2 on p. 154.

In this and the following case the problem of non-uniformity throughout
the product layer is theoretically speaking inexistent.
Counterflow drying allows the drying capacity of the air to work on the
product with a maximum intensity. Indeed, as the product dries, (and as
the air must be drier to exert the same drying effect on the product), it
comes across ever fresher drying air.
Due to this intensive interaction a relatively high drying speed is
reached, and the required passage length of the air through the products
is short.
The maximal efficiency, obtained for exhaust air near or past the
condensation point is 70%. The relatively strong heating of the product
accounts for the other 30%.
This category of driers is often used for temperature sensitive products.
The product temperature in the moist -extra sensitive- stage is low, and
for relatively low air temperatures reasonably quick drying remains
possible.

Concurrent flow driers

Data on the performance of concurrent flow driers are given in figures
13 and 14.

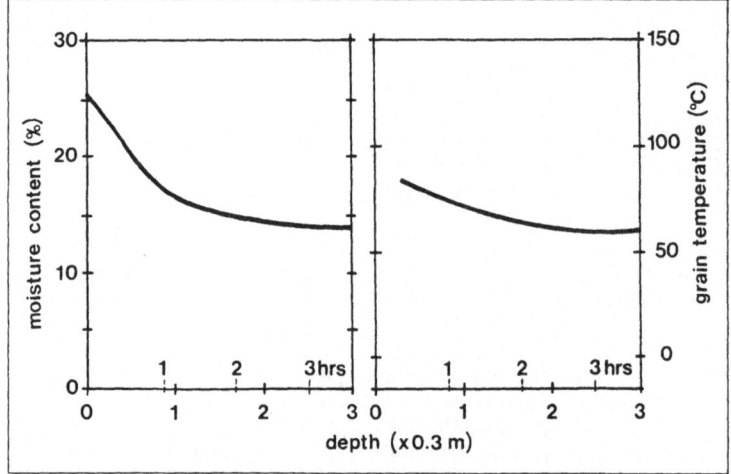

Fig. 13 : Moisture content (w.b.) and temperature gradients in a
concurrent-flow grain drier; simulated with T_{air} = 120°C; air
flow 0.3 m³/(m².s) (2)

In section 2.2.3. it was explained that the specific thermal energy
consumption can be minimized by using relatively **high** air temperatures in
stage 1 and relatively **low** air temperatures in stage 2.
Therefore in theory concurrent flow drying would allow for minimal energy
consumption. The ratio of energy used for water evaporation over product
heating is large and so efficiency values of 90% could be reached, and
cooling needs reduced. However, given the slow interaction between
product and air in the late stages of drying, the required layer thickness
to allow for full use of the air drying capacity is impractically large.
Layer thicknesses of 1 metre and air speeds yielding efficiency values of
60% are commonly found.

Given the high initial product temperature only less temperature sensitive products are dried following this technique.

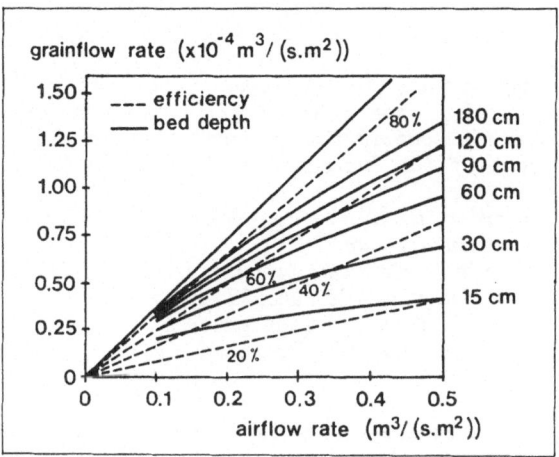

Fig. 14 : Efficiency *(1)* of a concurrent-flow grain drier; simulated with T_{air} = 120°C; m.c. : 25 % - 15 % (w.b.) *(2)*.

2. Instrumentation
Ventilator and heater are the two most fundamental elements in convection air driers.

The ventilator

The choice of the ventilator will depend both on the desired air flow rate and on the resistances against the air flow that must be overcome.
The Bernouilli equation can be used to determine the fan pressure required. The generated pressure must be at least as great as the sum of the continuous and accidental pressure drops (η) incurred by the air when blown through the drying system.

$$(p_{s1} + \frac{\rho_{a1} v_1^2}{2}) = \eta + (p_{s2} + \frac{\rho_{a2} v_2^2}{2}) + (z_2 \rho_{a2} - z_1 \rho_{a1}) g \qquad \text{(Eq. 13)}$$

where p_s = static air pressure (N/m^2)
v = air speed (m/s)
ρ_a = specific mass of air (kg/m^3)
g = gravity acceleration (m/s^2)
z = height above a certain reference point (m)
index 1 refers to the air just past the ventilator
index 2 refers to the exhaust air at the exit of the drying system.

The pressure that is to be generated $(p_{s1} + \rho_{a1} v_1^2/2)$ equals the pressure losses η (continuous and accidental) incurred, plus the remaining pressure at exit (zero, if atmospheric pressure), added to the height difference between inlet and outlet (usually negligible).

(1) See footnote 2 on p. 154.

By means of example, in figure 15, representative values are given of pressure drops incurred by the air on its way through a one metre thick product layer.

Notes on figure 15 :
a. *The figure gives values for a loose fill (not packed) of clean, relatively dry grain. For a loose fill of clean grain having high moisture content (in equilibrium with relative humidities exceeding 85 %), use only 80 % of the indicated pressure drop for a given rate of airflow.*
b. *Packing of the grain in a bin may cause 50 % higher resistance to airflow than the values shown.*
c. *When foreign material is mixed with grain, no specific correction can be recommended. However, it should be noted that resistance to airflow is increased if the foreign material is finer than the grain, and resistance to airflow is decreased if the foreign material is coarser than the grain.*

The values of the required air flow and pressure generation are fundamental data in the choice of an adequate ventilator. A great variety of other variables is also involved, so that the selection of fans is most often delegated to specialists.

Centrifugal ventilators are most often used because of the relatively high pressure which they can generate. Their characteristic pressure airflow-rate curve is also rather steep, so that differences in pressure drops have a relatively small impact on the flow rate. Characteristic curves for a specific r.p.m. rate of a centrifugal ventilator are given in figure 16.

The heater

The capacity of the heater is relatively easy to calculate, based on consideration of the air properties.
- The net maximum necessary increase in enthalpy of the air before entering the drier can be calculated, when the properties of the ambient air and the intake air of the drier are defined for the operating condition that is most demanding of the heater (cf. equation 3).

- The transition from enthalpy input (J/kg dry air) to energy input (J/per m^3 ambient air) can be done by making use of equation 4.

- This net heat demand value needs to be increased in several instances
  - to account for chimney losses and other heater efficiency reducing factors.
  - to account for heat exchanger losses in the case of indirect air heating. (Indirect air heating prevents contact between the undesirable components of combustion gases and product.)
  - where direct air heating is applied, to account for water production as part of the combustion process (e.g. when using natural gas).

Note : In order to allow both the prevention of condensation and the minimization of unused heat, the operation of fan and heater are controlled by regulation devices, based on temperature, humidity and possibly flow measurements of the air, as well as by regular checking of the product moisture content.

159

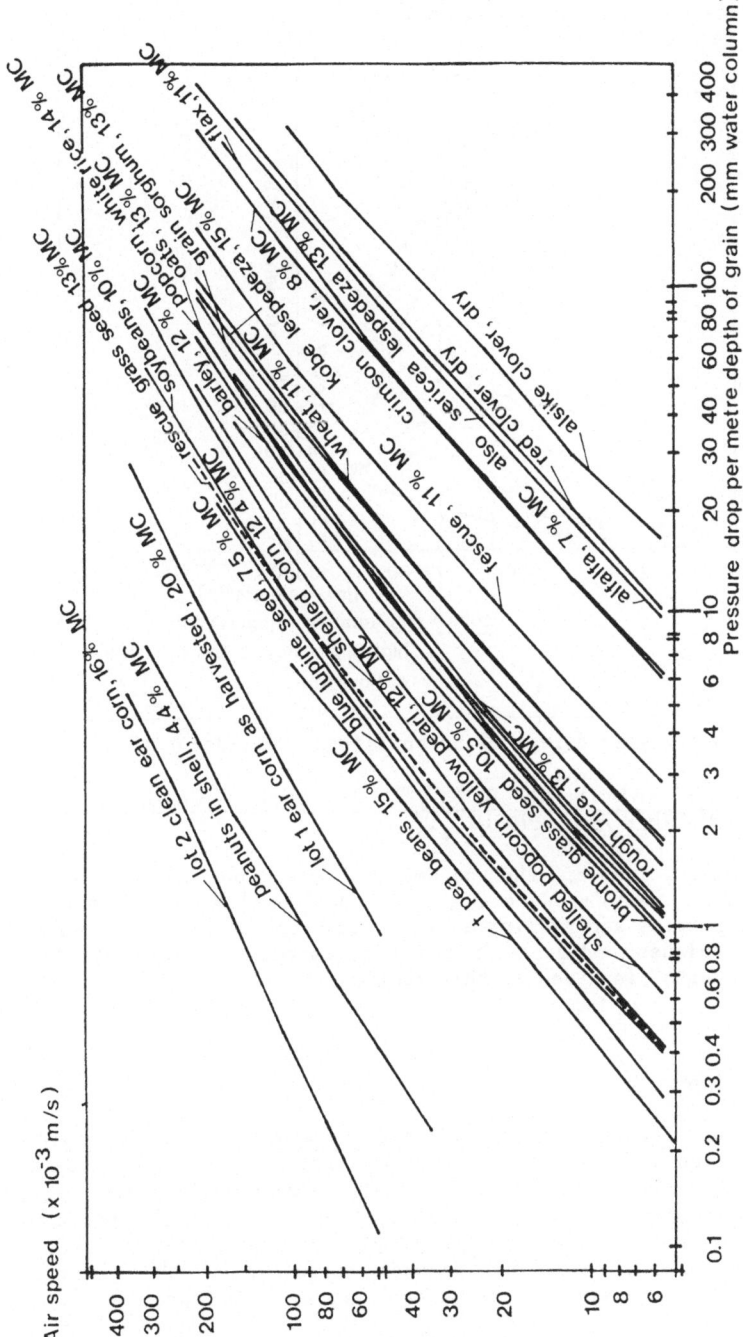

From Shedd
(1953)

Fig. 15 : Resistance of grains and seeds to airflow. (2)
        (For explanatory note, cf. p. 158)

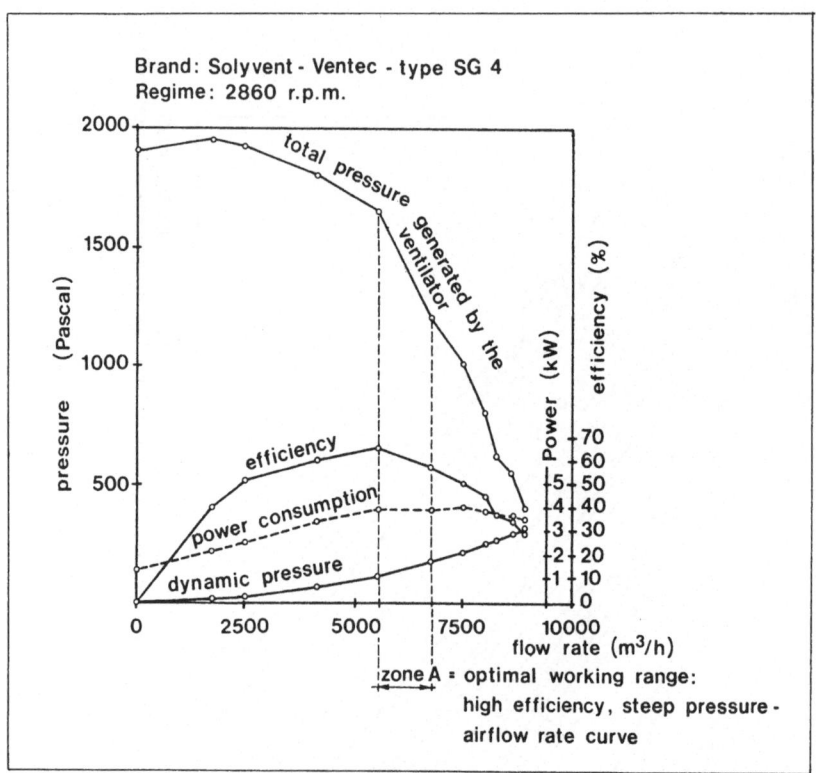

Figure 16 : Characteristic curves of a centrifugal ventilator (9)

3. REQUIREMENTS AND SOME ADAPTED DRYING TECHNIQUES FOR SPECIFIC AGRICULTURAL PRODUCTS

The optimal combination of the parameters (air and drying set-up characteristics) is very product specific, due to the wide range of product characteristics and requirements. A description of the currently used drying techniques, adapted to the requirements of some specific agricultural products, is given in this section.

3.1. Grain

3.1.1. Introduction

The recent revolutionary changes in the harvesting and processing techniques have led to a regular need for artificial drying. Indeed, mechanization has had a tremendous impact. Practically everywhere, combines are now being used, leaving behind the traditional pattern of picking, bundling into sheaves, further drying on the field to 13-15 % m.c., threshing in the shed, and gradual spreading of the grain on storage attics, where possible extra natural drying took place.

In order to allow for reasonable pay-back periods (1) of these combines,

(1) The payback period of an instrument is defined as the ratio of the total costs of the instrument (investment and working costs) over the annual earnings yielded by the instrument.

their utilization obviously needs to be spread as much as possible over time. This means that, during less favorable weather conditions, part of the grain needs to be harvested while it is not yet sufficiently dry. In regions with quite variable weather conditions, the moisture content of the mechanically harvested grain usually varies between 13 to 25 %, while, in some extreme cases, levels of 28-30 % m.c. can occur! (13)

Hence, each year, the need for drying will differ. In some cases practically none will be necessary, while in other cases, a good bit will have to be dried. For Belgium, the average amount that needs artificial drying is 15-20 % of the harvest. (13)

This grain, harvested and threshed practically all at once needs to be dried in the shortest possible time. Hence the need for rather sophisticated installations like static or dynamic driers in combination with horizontal and vertical silo's. While the smaller farms had to delegate this activity to cooperatives or grain merchants, the larger ones have kept the drying process on the farm, so as to be able to sell the grain at the best possible time (price of grain varies with time and season) as well as at the moisture content rate for which maximal earnings are obtained.

Figures on the importance and location of grain production in the E.C. are given in table 1.

Table 1 : Importance and location of grain production in the E.C. (1983) (3)

	Country	Soft Wheat	Hard Wheat	Barley	Maize	Oats and mixed grains	Rye and mixed grains	Other	Total (except rice)
production values given in x 1000 tons	F	24 436	399	8 814	9 659	1 922	311	317	45 840
	D	8 998	-	8 944	934	2 489	1 646	-	23 010
	UK	10 883	-	10 084	-	498	24	-	21 489
	I	5 613	2 901	1 174	6 669	307	28	104	16 796
	Dk	1 548	-	4 423	-	93	315	-	6 379
	Gr	1 465	561	572	1 622	53	9	-	4 285
	B	1 043	-	670	39	99	25	-	1 877
	Eir	389	-	1 459	-	100	-	-	1 909
	Nl	1 043	-	177	2	61	26	-	1 308
	Lux	19	-	35	-	11	3	-	68
	Tot. EC	55 437	3 861	36 352	18 924	5 634	2 388	421	122 973
	Cultivated area x 1000 ha	11 012	2 164	8 863	2 957	1 755	677	101	27 528

3.1.2. Requirements in grain drying

- For safe cool storage over one year for most grains the moisture content must be reduced to around 14-15 % (w.b.), and this well before moulding occurs (= function of moisture content, product temperature and drying time).
  It is an asset to be able to control the final moisture content between tight limits : on the one hand, exceeding the maximum allowable moisture content level for safe storage would result in grain

deterioration over time.
The cooperatives and merchants will obviously pay less for such moist grain, since they need to dry it themselves, on the other hand however, in the E.C. no bonus is paid for overdried grain, which costs the farmer more in drying energy, while yielding him a lower final weight (measure of income) than the product would have had at the desirable moisture content.

- Since most, if not all grain drying is situated in stage two of the drying process, the commonly applied air flow rates are relatively low; about 170 m^3/(h.ton). (1)

- The allowable maximum drying air temperature depends on several factors including grain-type, desired qualities corresponding with purpose of use, relative humidity of the air, resulting maximum product temperature (function of drying set-up; cf. section 2.3.2.1), moisture content and drying time.
An overview of recommendations applied in different countries for maximum drying air, or final grain temperature is given in table 2.

These recommendations fall short in comparison to what is currently known about the interaction of the different above mentioned factors. E.g. : in many cases the maximum air temperature is given as the sole criterion, while in reality the maximum product temperature is more of a determining factor. As was demonstrated in section 2.3.2.1 the relationship between product and air temperature can differ significantly depending on the category of drier.

- Excessive product temperatures may cause loss of germination capacity, a requirement for seed and malting grains; loss of baking quality (requirement for milling grains); destruction of vitamins, proteins, and carbohydrates (constitutive elements in human and animal feeds); discoloration; cracking; etc...

3.1.3. Description of commonly used grain drying set-ups

1. Stationary driers
The scheme shown in figure 3 (cf. p. 145) is typical of a stationary drier.

It consists of a fan, often controlled by a hygrostat, a thermostat and possibly other regulating devices; a heating element; a perforated floor.
- In the simplest form of stationary drier (the full bin drier, also called a storage drier) the grain is loaded in the drying bin, possibly all at once. If the capacity of the ventilator and (possibly) the heat source are large enough, the whole batch can be dried before moulding occurs. Drying is followed by cooling, aeration (crucial -among other factors- to prevent problems in locally wetter spots), and storage, all taking place in the same bin.

- This method poses certain problems:
 + The bottom layer of grain is overdried, while moulding often occurs on top. (Problem of non-uniform drying; cf. figure 9.) Stirrers and recirculators are sometimes introduced to help acquire a greater uniformity of moisture content of the product.

Table 2 : Some national recommendations for maximum drying air, or final grain temperature. (12)

Grain	Purpose	Initial moisture content % w.b.	Inlet Temperature of drying air °C		Final grain temperature
United Kingdom			Continuous drying	Batch drying	
All grain	Stock feed		82-104 [a]	82-104 [a]	
Wheat	Milling	< 25 / > 25	66 / 60	66 / 60	–
Barley	Malting	< 24 / > 24	49 / 43	49 / 43	–
All grain	Seed				
Oilseeds	Oil extraction	–	46	–	–
Sweden					
All grain	All purposes	20 / 25 / 30	65 / 60 / 55	60 / 55 / 50	–
Wheat and rye	Milling	> 22	–	–	–
France					
All grain	Seed	–	–	–	
Barley	Malting	–	} 40-42	40-42	30
Oilseeds	Oil extraction	–	–	–	
Wheat	Pasta & Milling	–	80	60	45
	Industrial processing (dry and wet milling)	– / –	80-90 [b] / 100-110 [c]	– / –	} 60
	Animal feed	–	100-110	–	65-70
West Germany [e]					
Wheat	Seed or milling	< 18	–	–	45
Barley	Seed or malting	18-20	60-68	–	40
Oilseeds	Seed or oil extraction	> 20	–	–	36
Grain	Feed	–	–	–	66-75 [d]
Maize	Feed	< 35	90-99	–	–

(a) Recommendation contains the rider : – 'Feeding properties are not harmed by temperatures up to 104°C during drying; but the higher the temperature used the more difficult it is to cool the grain effectively before storage.'
(b) Standard drying
(c) 'Dryeration' and drying in 2 passes
(d) These temperatures are defined as 'maximum heat in material being dried' i.e. measured at the hottest zone of the drier. Normally this will be the same as the final temperature prior to cooling
(e) Directly heated hotair driers not allowed because of danger of combustion gas residues.

+ With limited ventilation and heating capacity, a thick layer of grain often cannot be dried within the allowable time. Layer drying is a possible solution. With layer drying small moist layers of grain are gradually added to the drying bin as the drying front moves upward through the grain.

- In order to use the more sophisticated driers more than once in a harvest time, the grains are transferred to a storage bin after drying and cooling. (If the drying technique is the same as described above, this is called <u>batch-in-bin drying</u>, or simply <u>batch drying</u>)

Note : 1) The general term "<u>in-bin-drier</u>, or simply "<u>bin drier</u>" is used to designate both the full bin drier and the batch-in-bin drier.

     2) <u>Column batch</u> driers are a type of stationary drier in which the thickness of the drying layer and hence, the problem of non-uniform drying is reduced. (The air is blown across relatively thin stationary grain columns.)

2. <u>Continuous driers</u>

Most of the more sophisticated driers belong to this category. Their design often allows for automation as well as quick quality drying.

- The simplest form of continuous drier is a bin drier in which the grain layer under the drying front is gradually removed. (This is a <u>counterflow-type of drier</u>, since the air and grain movement are in opposite directions. Uniform drying and intensive interaction product-drying air throughout the drying process are typical characteristics of these drier-types. Cf. section 2.3.2.2.)

- Some schemes of other possible types of continuous driers are given in Figures 17 and 18.

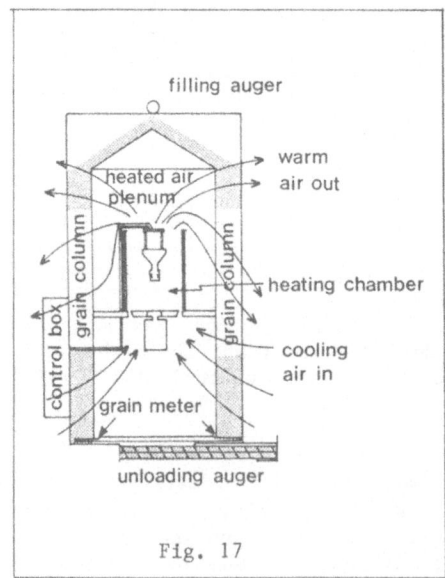
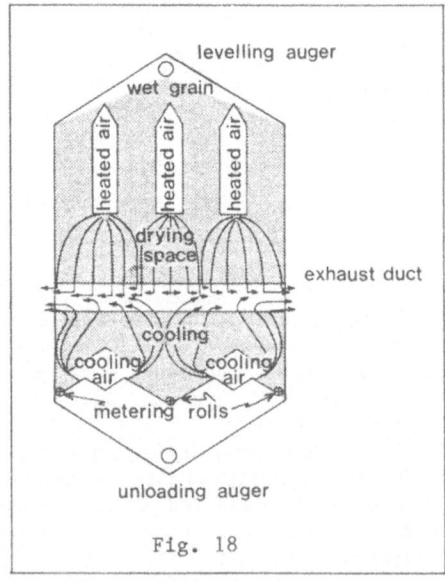

Fig. 17. Cross-flow drier with forced-air drying and reverse-flow cooling (2). Typical for this drying set-up is the conversion of undesirable sensible heat of the dried grain into useful drying capacity of the drying air. (Cf. section 2.2.3).

Fig. 18. Scheme of a concurrent-flow drier (2). Main characteristics of this drying set-up : intense drying interaction air-product in early drying stage. Tempering effect in later stages. (Cf. section 2.3.2.2).

3. "Dryeration"

Dryeration is an addition to the common dry and cool techniques. In this process, not yet fully dried grain is first "tempered" for 6-10 hours (i.e.: it is left in a separate bin, so that the moisture content distribution becomes more uniform). After tempering, the grain is cooled (cooling also implies some extra drying), and then transferred to a storage bin. This technique can lead to a spectacular reduction of the occurrence of internal stress-cracking (cf. section 2.2.3), especially in the case of larger (e.g. maize), or extra sensitive grains (e.g. rice).

3.2. Hay *(1)*

3.2.1. Introduction

The quality of the results of any drying process is determined by the extent to which optimal drying conditions can be established.
When hay is dried in the open air, it is the changeable weather which determines these drying conditions. This results in sometimes extreme variations of hay production quality and quantity over time. An illustration from the UK is given in figure 19.

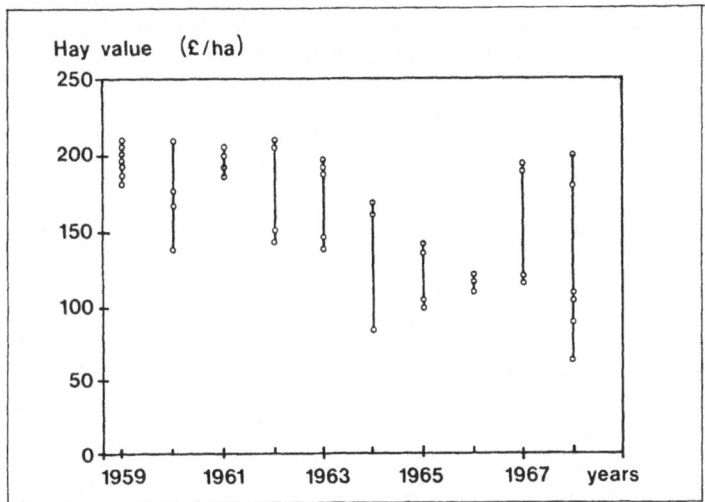

Fig. 19. Variation in the value of different batches of hay within a
season over a 10 year period, 1959 - 1968 at Plymouth (14).

Indeed it can take up to a week of good weather to dry long mowed grass. The occurrence of rain and dew during that period can significantly prolong this time and rewetting of the grass goes together with extra leaching losses of nutrients, extra respiration of the hay, and even with moulding.

As well as its implications for labour savings, mechanization has led to an accentuation of these problems. Indeed, the previously common hay stacks which allowed for significant further drying under more protected

(1) Although this section focuses exclusively on hay-drying, several of the discussed elements apply also to other green forages like alfalfa for instance.

conditions have been replaced in many cases by more compact and more easily handled hay bales, the moisture content of which must be lower than usual in order to prevent moulding. Hence, longer drying times are required.

Loosely chopped hay, allowing for easy pneumatic transportation is furthermore extra sensitive to leaching losses from the relatively large amount of cut and bruised areas.

To summarise, in the case of hay making in the open air, dry matter losses can vary between 10 and 60 % depending on weather conditions. (60 % dry matter loss corresponding to 80 % loss of nutrient value!) (11) (13)

In order to obtain a more reliable production, and to further reduce labour costs in highly mechanized regions with rainy climates, a transition is being made towards silage of green or pre-dried grass and of other forages. (13)

An estimation of the distribution of hay and silage-making is given in table 3 for several European countries.

Table 3 : Production of hay and silage in Europe : estimates for countries of the EC and Scandinavia (Figures expressed in Mtonnes of dry matter) (1978/1979) (14).

Country	Hay	Silage (grass and other forages) (Mt of dry matter)	Total
Belgium	1.7	1.9	3.6
Denmark	0.4	1.1	1.5
Finland	1.7	0.9	2.6
France	22.3	16.0	38.3
Germany, Federal Republic	9.1	10.5	19.6
Ireland	3.6	2.4	6.0
Italy	9.3	5.9	15.2
Luxembourg	0.2	0.2	0.4
Netherlands	1.3	4.1	5.4
Norway	0.7	1.1	1.8
Sweden	4.6	1.0	5.6
United Kingdom	6.9	5.5	12.4 *(1)*
Total	61.8	50.6	112.4

The question of the desirability of the silage making option from the point of view of ration-making, pathology, microbial environment in the cowshed, manipulation, nutrient quality, ingestion, etc... is the subject of heated debates between advocates and opponents of the silage technique. For instance, in certain regions where a significant amount

(1) Statistics collected later on at the same Animal and Grassland Research Institute in Berkshire for the U.K. indicate a strong further evolution in that country of the trend towards silage making : by 1984 the share of haymaking in the total hay and silage production was no more than 35 %.

of milk is used for hard pasta cheese production, application of the silage technique is prohibited to the farmers, since its application is said to have an undesirable influence on the cheese quality.

In order to obtain 1 ton of dry hay, starting from grass at 85 % m.c., about 5 000 kg of water needs to be evacuated. In view of the relatively low commercial value of hay, controlled drying under shelter right from the fresh grass can hardly be economically feasible.
However the advantages of controlled drying are especially significant in the post-drying stage of the grass :
- In the early stages drying takes place very swiftly and large absolute quantities of water can be removed in a relatively short time. (1) (Stage 1 in the drying process. - cf. section 2.3.2). Since short time good weather spells are far more frequent and predictable than long ones, the risks of poor drying are reduced.

- Nutrient losses due to leaf shatter caused by turning of the swaths (requirement in open air drying only) are especially significant in the lower moisture content ranges.

Therefore, in the regions with rainy weather conditions, where hay drying is preferred to silage making, examples of hay driers, post-drying hay from 35 to 45 % m.c. down to 16 to 18 % m.c. by means of forced air convection are not uncommon.

Some characteristic data of controlled post-hay drying are given in table 4.

Table 4. Controlled post-hay drying(characteristic data)

- Original moisture content : 75-85 % m.c.
- Required moisture content for safe cool storage over one year : 16-18 %.
- Maximum allowable initial moisture content for safe barn drying : 35-45 % depending on the permeability of the hay for air (function of storage condition-loose, chopped, baled) as well as on drier characteristics (heating or not; uniformity of air stream ...).
- Optimal airflow rates : 840 - 3 360 m^3/(h ton). These high rates and corresponding pressure drops are necessary to ensure sufficient air penetration in denser parts of the hay-stack.

3.2.2. Description of a few commonly found hay drying set-ups

Several of the artificial hay driers are in principle comparable to the in-bin driers for grain. (Stationary driers; characteristics, cf. section 2.3.2.2.)
For example the common English "Storage Barn Hay Drier" consists of a storage area with airtight side walls in which hay bales are stacked tight and in a cross-bonded way, so no air-leaks occur, up to a height of 5.5 m and on top of a perforated floor. A strong fan produces a flow of unheated or only slightly heated air (for instance 2 500 m^3/h/ton of hay) and this usually dries the hay from 30 % down to 17 % m.c. within 2-3

(1) Pre-drying of grass from an original moisture content of 85 % down to a moisture content of 40 % involves the removal of nearly 90 % of the absolute amount of water originally contained in the product.

weeks. In the case of hay, storage driers are often preferred to batch driers in order to avoid extra loading and unloading expenses.

Instead of a perforated floor, duct systems are often used to blow air through the hay. Usually secondary ducts are branched off from the central ducts in order to provide sufficient air flow rates throughout the haystack.
In a simple set-up of this type, bales are tightly packed around a horizontal duct, which runs lengthwise through an open sided barn (cf. figure 20).
A more sophisticated version is to be found in the Dutch driers. They consist of an open sided barn with a suspended "bung" (1) that can be moved up or down. The fan is in some cases built into the top of the 'bung'. Several ducts are made to branch off radially from the main shaft.
Drying is started as soon as some hay is stacked. This technique is used for chopped, bailed as well as loose long hay, for which it was originally designed (cf. figure 21).

Fig. 20 : Bales dried by the "tunnel" method. (4)

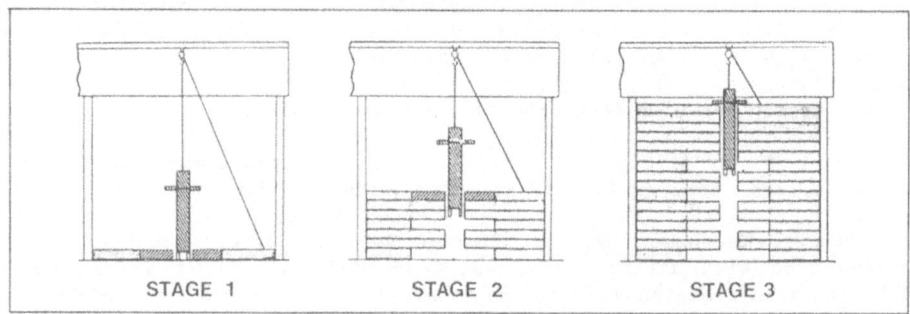

Fig. 21 : Diagram showing stages of loading a Dutch hay dryer. (4)

(1) A "bung" is a cylinder which forms a shaft in the hay that is being loaded onto the stack.

Some of the very sophisticated hay driers work with short hay (cut at an early stage of growth, or chopped).

For example, in the "German drier" the hay is blown into the drier and evenly distributed by a rotating "grass-spreader". As the stack moves upward, a "bung" moves along, both to form a shaft and to prevent the air from escaping before it flows radially through the hay itself.
To discharge the hay, the bung is removed; hayrakes rake the top hay toward the central shaft, at the bottom of which a conveyor belt removes the hay from the drier. (cf. fig. 22)

Note: In some places a modified version of the "old brown hay process" is again being advocated. In this method the moist hay is allowed to heat up to a certain temperature, at which point the fan is operated. The self-generated heat causes the water to evaporate, and the hay is dried without artificial heat application. This system is not only wasteful of digestible nutrients, but includes all the potential dangers attached to fungal growth.

3.3. Fruit (5) (6) (16)

Cooling and controlled atmosphere storage are commonly used techniques for stretching the fruit supply over time after the harvest season.
However, in remote places for instance, where the fruits cannot possibly remain cooled throughout the marketing chain, the drying technique is sometimes applied. Indeed, dried fruits can easily be conserved for a long time without large and expensive cooling installations. An additional benefit of drying is that it significantly reduces the volume and weight of the fruit. Therefore, fruit is often dried where long transportation distances have to be covered, with limited means. In this context grapes, figs, dates, plums, apricots, peaches, apples, pears, etc... are sometimes subject to drying.
Some characteristics related to fruit drying are given in table 5.

Table 5. Some characteristics related to fruit drying

m.c. at harvest		m.c. dry	requirements	
Apples (slices)	85 – 87 %	15 %	$T_{air\ max.}$	60–70°C
Pears (slices)	80 – 86 %	25 %	$T_{air\ max.}$	70–80°C
Grapes	79 %	24 % (10–14 %)	$T_{air\ max.}$	$\pm$ 80°C
Apricots + peaches	85 %	30 %	$T_{air\ max.}$	$\pm$ 65°C
Plumbs/chevines	77 – 81 %	20 %	$T_{air\ max.}$	40–60°C

The characteristics and requirements of different fruits vary greatly. Hence specialized drying methods and instruments are needed. Drying is often part of a whole series of steps, including washing, grading, cutting, sulphuring (for some fruits to prevent discoloration), drying, tempering, processing.

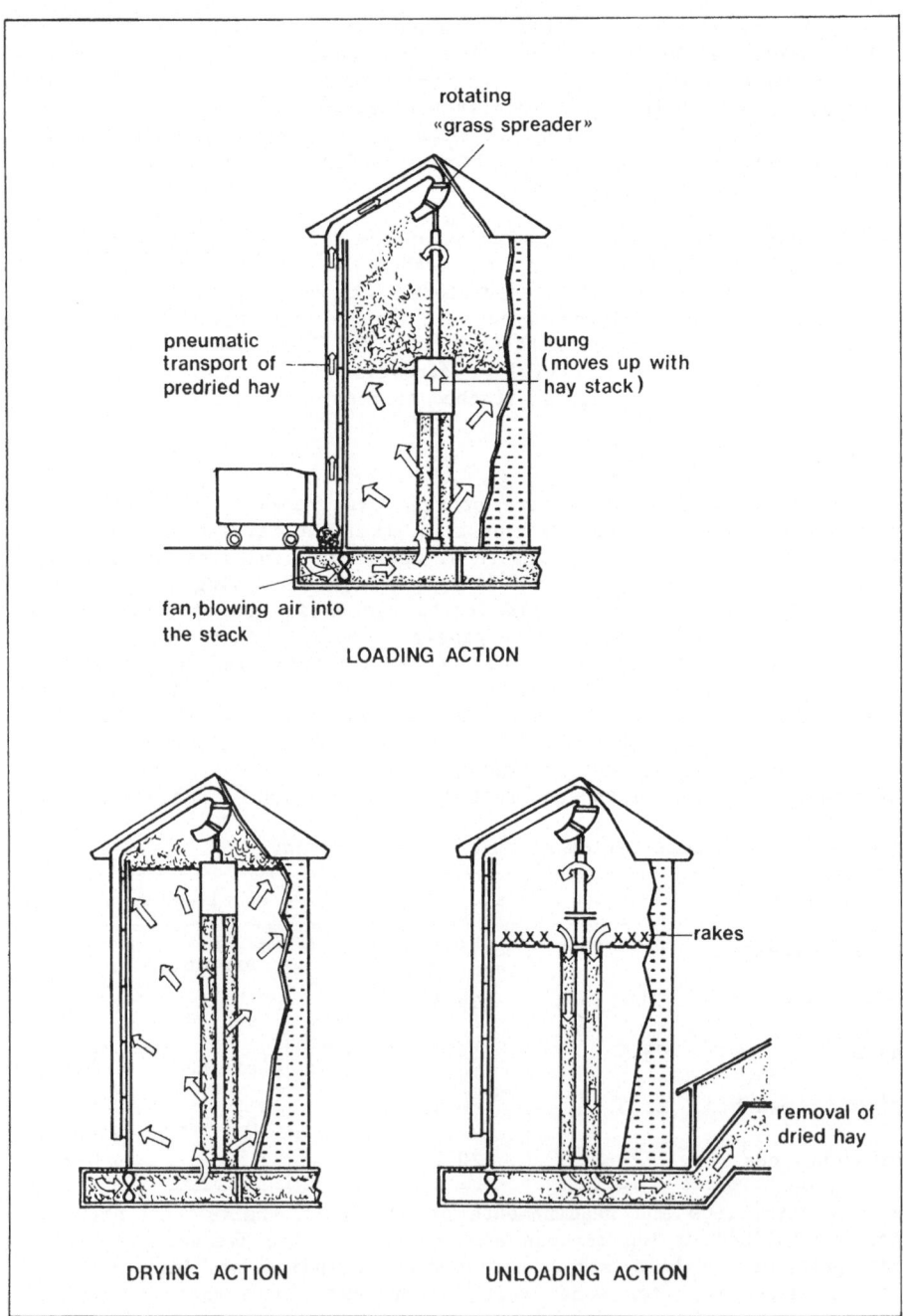

Fig. 22 : Schematic representation of a "German drier".

Different pretreatments are often applied. For instance, grapes are sometimes dipped in caustic solutions of NaOH or Na_2CO_3 to remove the waxy bloom and check the skin, which significantly speeds up dehydration.

The variety of drying devices reflects the difference in characteristics of the different fruits. They can however be categorised as
- kiln driers (basically the same as in-bin-driers for grains)
- tray driers, in which different trays of fruits are superposed in the drying device.
- continuous driers, most commonly used, which apply the different flow possibilities: cross-flow, concurrent flow and counterflow. (cf. section 2.3.2.1 on p. 153)

4. EVOLUTION AND ALTERNATIVE DRYING TECHNIQUES (9) (16)

The motives behind the current evolution towards alternative techniques are very different in nature. The two main ones, in order of importance to agriculture are:
- production cost considerations (mainly energy costs, but also installation costs)
- ecological considerations

4.1. Production cost considerations

4.1.1. Energy cost

In the area of production costs, the energy cost is a major factor.
The amount of energy needed to remove 1 kg of water from hay or grain, using traditional methods is in the neighbourhood of 5 000 kJ. (The heat required for evaporation of 1 kg of H_2O is 2 500 kJ - but the efficiency of the drying process is only about 50 % (8).)
The evolution of energy prices has led to different adaptations :
+ Evolution towards more energy-efficient drying set-ups.
- More efficient use of the traditional driers (regular maintenance, adjustment of burners, regular controls and good management) can lead to an average of 10 % lower energy needs. (9)
- A more economical use of the drying capacity of the air :
In cross-flow type driers the drying and cooling air blown through the already partially dried grain near the exit of the drier can be recycled partially or totally (in the case of total recycling after renewed heating) and used in the part of the drier where the grain is still moist. (1)
This application can economise the energy need by 10 - 20 %, as well as in the case of total recycling raise the capacity of the drying system. (For this purpose, good regulation devices are a must).
- Maximum use of the natural drying capacity of the ambient air : the Low temperature drying" technique. (cf. section 5)
- Also dryeration (cf. section 3.1.3), with its extra drying effect during cooling, is an adaptation in the direction of more energy efficient drying installations (up to 20 % energy savings).
+ Evolution towards the use of alternative energy sources.
- extracting the waste heat from the drying air at its exit from the system, and preheating the incoming air with it by means of heat

(1) In the case of renewed heating special precautions need to be taken against ignition of the dusty recycled air.

exchangers.
 - using the secondary products of the grains that are to be dried as a
   source of energy.
   E.g. : - Straw produces 13 800 kJ/kg if burned. This is about 1/3 of
            the heat produced by burning the same weight of oil.
            In France, this way of producing heat can be 3.5 times less
            costly than traditional fuel heating.
          - Maizecobs produce about 11 500 kJ/kg if burned. Burning the
            cobs of the maize that is to be dried should normally
            suffice as energy input for the whole drying operation.
 - using solar energy. (cf. section 5)

4.1.2. Investment costs

In order to minimize the importance of the investment costs into the
drier per unit of dried product, applications are sought which allow
multiple use of the drier spread as much as possible over time.
. In the case of grass, the cutting frequency can be increased up to
  7 to 10 times per year. The reduced volume of the regularly added
  loads calls for a drier of relatively small capacity, applied at a high
  utilization rate.
. Aerated (possibly cooled) wet storage facilities, which make this same
  drying time spread possible, are gaining in popularity. (For a grain
  drying plant which dries about 35 000 m^3 of grain per year an optimal
  combination would be a drier with drying capacity of 70 m^3/hour and a
  temporary wet grain storage of about 1 800 m^3).
. The use of one and the same drier to dry several product types or even
  for other purposes at different times is another option. In this
  context, different users could possibly benefit from the same drier.
. During the off-drying season the drier could in some cases also be used
  for other applications (e.g. as a storage area).

4.2. Ecological considerations

In certain countries, regulations against air-pollution might impose
restrictions on drying techniques which use large amounts of air
(typically air convection drying). Additional exhaust control and
processing equipment, necessary to help meet these requirements, would
make this type of drying economically unfeasible. In this context,
experiments are being done in the area of conduction heating and low
volume aeration.

5. INCORPORATION OF SOLAR ENERGY INTO THE DRYING PROCESS

The application of solar energy is especially attractive in the context
of the "low temperature drying technique" when practised on products
which are dried during periods of high solar radiation and requiring
large quantities of drying air (e.g. hay, certain grains ...) (cf.
sections 3.1 and 3.2).
The low temperature drying technique aims at obtaining a dried product of
reasonable quality, making maximal use of the freely available natural
drying capacity of the air.
Since this natural drying capacity in itself often yields insufficient
drying speeds for quality drying, the air is heated by 5 to 10 K only.
On the one hand this temperature rise significantly affects the drying
capacity of the air, and hence drying speed (cf. section 2.1). On the

other hand, especially in colder climates, the allowable drying time before the occurrence of mould is still reasonably long (cf. fig. 2).

Losses associated with this technique are of a very different nature than for high temperature drying. The main weak point here is the occurrence of deterioration due to excessively long drying times. (Hence the limitations on product types and initial moisture contents for successful application of this technique.)
However, in low-temperature drying, brittleness, thermal destruction of e.g. vitamins, proteins and carbohydrates, undesirable input of huge quantities of sensible heat into the product, loss factors typical for high temperature drying, are practically non-existent (cf. section 2.2.3).

Thanks to the small temperature gradients, simple stationary driers with reasonably thick drying product layers can be used, without the occurrence of excessively non-uniform drying. In contrast, these problems are rather acute for simple high temperature driers (stationary or cross-flow driers), where often a small layer thickness needs to be chosen, combined with high air flow rates, resulting in high specific thermal energy consumption rates (cf. section 2.3.2.2).

Given the very small air temperature increase requirements in low temperature drying, only very simple, low cost and easy to manage solar systems are required, which yield reasonable efficiency values, and hence can produce high quantities of low grade energy for reasonable costs. A great variety of simple solar collector types exists. Where storage and drying occur in one and the same place, these simple collectors are usually integrated into the barn-roof or silo-walls. In other cases plastic greenhouses, black plastic tubes and other elements constitute the solar collector.

The situation is different for products requiring well defined and high drying temperatures. In this context, complex systems, involving among other factors concentrating water collectors, storage elements, etc... have also been studied.

Also in solar drying, major research efforts are made to increase the annual number of hours of operation of the solar driers, so as to effectively spread the investment cost over time. (cf. also section 4.1.2.)

A typical application in this context is called the "desiccant system", used on certain grain farms in the USA to enable drying at a time of high solar supply, even if the drying season doesn't fall at that time. (10)
This system involves keeping a stock of grain over the period of one whole year. This grain is then overdried during summertime (up to 8 - 10 % mc).
The new grain, harvested later on, is mixed with the overdried old grain so as to reach an average moisture content of 20 %. This mixture is then further solar dried.

A great variety of different solar drying applications is presented and discussed in chapter 2.

References

1. A.S.A.E. "Moisture relationships of grains", ASAE data: ASAE D 245.2, Agricultural Engineers' handbook 1977 pp. 396.

2. Brooker D.B., Bakker-Arkema F.W., Hall C.W. "Drying Cereal Grains" Second printing (1975); Westport, Connecticut, USA; The AVI Publishing Company, 1974 pp. 265.

3. C.E.C. "The Agricultural Situation in the Community", 1984 Report, Brussels - Luxembourg 1985.

4. E.D.A. "Greencrop drying" Farm electrification handbook n° 15, A guide to practical design of installations. E.D.A., London, 12/1967.

5. Eissen W. "Trocknung von Trauben mit Solarenergie", Doctoral dissertation at the University of Hohenheim W.G., Promotor Prof. Dr. Ir. H.-D. Kutzbach, 1983, pp. 58

6. Höhn E. "Das Trocknen (Obst, Gemüse, Gras)", E. Höhn publisher, 1943 pp. 155.

7. Lasseran J. Cl.; Station expérimentale de L'I.T.C.F., France; personal communication, 1985.

8. Maltry W., Pötke E., Schneider B. "Landwirtschaftliche Trocknungs-technik", 2nd edition, Berlin, VEB Verlag Technik Berlin, 1975 pp. 524.

9. Multon J.L. "Conservation et stockage des grains et graines et produits dérivés" (vol. 1 & 2) 1st edition, Paris France, 1982, Technique et documentation Lavoisier, APRIA pp. 1155.

10. MWPS - 22 "Low Temperature and Solar Grain Drying Handbook", Ames (Iowa) Iowa State University press 1980, pp. 86.

11. Nash M.J. "Crop Conservation and Storage in Cool Temperature Climates" First edition, Oxford, UK, Pergamon Press, 1978, pp. 393.

12. Nellist M.E.; "Safe temperatures for drying grain", A report to the home-grown cereals authority; N.I.A.E. Bedford, UK; Report N° 29, 1978; pp. 90.

13. Pletinckx A. "Utilisation de collecteurs solaires-air pour le séchage des fourrages et des céréales. CEC action 3 report (cf. detailed description p. 239).

14. Thomas C. "Forage conservation in the 80's, Occasional symposium N° 11 of the British grassland Society held in Brighton UK, 11/79, 1980, pp. 474.

15. Tuerlinckx G. (Former assistant at the Laboratory of Agricultural Building Research at the Catholic University in Leuven, Belgium), Unpublished documents.

16. Van Arsdel W.B., Copley M.J. Morgan A.I., "Food dehydration" Second
    edition. Westport, Connecticut, USA. The Air Publishing Company,
    1973 pp. 347 & 529 (vol. I & II).

CHAPTER II
Present research within the EC on the application of solar energy in the drying sector

The overview of research presented in this chapter is based on the materials received from CEC action 2 and 3 participants. This material is not complete : it neither cover all the research within the EC, nor do the given basic descriptions always contain all the elements, essential in the given context.

However, based on the received materials, which obviously leave much to be desired, an overview is presented, which will give the reader a realistic picture of the state of the art of research in this area, the direction in which it is moving, and the potential of the applications in question.

The main current research tracks in the area of solar drying are presented first. The major part of this chapter follows, with an overview of research projects and centres. (The format of the basic descriptions of the projects is explained in the "Addendum", cf. p. 467)

The general experience gained over the whole range of the technical projects studied is summed up and discussed in the conclusion section of this chapter (cf. p. 218).
The conclusions are supplemented by a "Prospects" section, in which future research activities of interest are suggested.

The "conclusions" and "Prospects" sections sum up the core findings and message of this book as far as solar drying is concerned

1. THE MAIN RESEARCH TRACKS

In trying to cope with the fundamental criteria (cf. p. 1),researchers have chosen to follow certain tracks systematically, as explained below.

1.1. The "physical feasibility" track

In the first track, SOLAR SYSTEM EFFICIENCY, as defined in the fundamental criteria is analysed in experimental testing set-ups.

Main questions:
Is it physically possible to succesfully combine the drying process with a solar set-up?
- Can the desired operating temperatures be reached given the climate?
- What is the quality of the dried product?
- Compatibility problems?
- How do the characteristics and dimensions of the different components (collectors with different degrees of sophistication, measuring and regulating devices, possible storage and auxiliary heating, ...) influence the parameters of the process (drying speed, product quality, energy consumption, ...)

1.2. The "economic feasibility" track

When an insight into the above questions has been obtained, the next logical step consists in the assessment and optimization of the "SOLAR SYSTEM COST".

Main question:
How can the economic feasibility be maximized?
Or, translated into technical terms:
How can a maximum ratio ($\eta*$) be obtained of

       Economic benefit of the solar intervention
over   Cost of the solar system (mainly investment costs; operating costs are often negligible)

To reach this objective, several options of parameter manipulation were chosen:

1.2.1. Choice of collector type

Certain products (e.g.: some fruits) are rather demanding in their drying process characteristics and in order for the solar system to be effectively integrated in this process, a high degree of sophistication is required. This means the results of the process as well as the value of $\eta*$, will only become acceptable at high levels of investment cost. A good part of the research on <u>sophisticated solar drying systems</u> can be found in this context.

Certain other products (e.g.: hay, grain ...) are not quite so demanding. Since in this case, good drying results can be obtained at low levels of installation sophistication, $\eta*_{max}$ will obviously correspond to lower investment costs.

A good part of the research described is concerned with just this question:
How cheap can we make a system, while still maintaining acceptable levels of system efficiency?
In this context are situated the research projects treating the development of cheap collectors with low price components and simple, easy to construct design. (do it yourself collectors).

In this same category can also be classified the research in the area of integration of collectors as part of an existing roof or wall. The decrease in collector efficiency caused by the use of cheaper materials, as well as the frequently occurring limitations in orientation and tilt of the collector, is in those cases expected to be less significant than the decrease in investment costs.

1.2.2. Choice of utilization factor

The ratio η^* can also be optimized by making maximum use over the years of the investment considered. Since a lot of products to be dried have a limited drying season, a significant part of the research is aimed at creating the possibility of using the same installation for the same or different products at different times or even for other applications : Multi-purpose use.

1.2.3. Choice of component dimensions and regulation systems

Based on the knowledge obtained in the "physical feasibility track" concerning the influence of different system components and operating modes on the drying process parameters, research is being done in the area of component dimension and regulation system optimization in light of the objective of maximum economic feasibility.

2. OVERVIEW OF SOLAR DRYING PROJECTS

2.1. Classification of the different research projects

In classification table 1 each of the basic descriptions that can be found in this chapter is listed together with an indication of the critical elements which it treats.
Following the table is a list of research centres other than the ones already mentioned, likewise with an indication of their assumed research activities, based on the limited available information (see table 2).
Key : ● element applicable to the given study (centre)
      ○ element planned for future consideration
    (●) side-aspect in the given study (centre)

Table 1 : Classification of the research projects described briefly in this chapter. (The "badges" at the heads of columns are printed with each description, as appropriate, to provide a visual index.)

B.D.	page	Country	hay/forage	cereals	fruits	spices/tobacco/vegetables	wood	other products	general approach	significant	secondary (operational set-ups)	physical feasibility	economic feasibility	more sophisticated types	simple collector types (not integrated in roof)	simple collector types (to be integrated in roof structure)	Multi-purpose use	Component dimension & regulation system optimization
1	185	B	●	●						●		●	○			●	●	○
2	186	B				●				●		●						
3	187	D	●	●			●				●	●	●				●	
4	189	D	(●)	●						●		●	●			●		●
5	191	D				●					●	●	●			●		
6	193	F	●							●		●			●			
7	194	F	●								●	●			●		●	
8	195	F	●								●	●				●		
9	195	F	●								●	●	●		●		●	●
10	198	F	●	(●)							●	●	●			●		
11	199	F		●						●		●	●		●	●	●	●
12	200	F			●					●		●	●		●			●
13	202	F			●	●				●		●	●				●	●
14	203	F					●			●		●				●		●
15	204	F						●		●		●	○		●		○	●
16	205	Gr			●					●		●						
17	206	Gr			●					●		●	○				○	●
18	207	Gr			●					●		●		●				●
19	209	Gr			●					●		●	●		●			
20	211	I	●							●		●	●		●		●	
21	212	I	●							●	●	●			●			
22	214	I	●	●						●		●					●	●
23	214	Nl						●		●		●					●	○
24	216	UK	●							●		●	●			●	○	○

Table 2 : Supplementary list of research centres, with indications of activities.

Research center	product							scient. focus		assess. type		collector types				
	hay/forage	cereals	fruits	spices/tobacco/vegetables	wood	other products	general approach	significant	secondary (operational set-ups)	physical feasibility	economic feasibility	more sophisticated types	simple collector types (not integrated in roof)	simple collector types (to be integrated in roof structure)	Multi-purpose use	Component dimension & regulation system optimization
Institute for Plant cultivation Hohenheim University Fruwirthstrasse 340 7000 Stuttgart-70 West Germany Representative : H. Jacob; M. Elsässer	●							●		●						
State Research Institute for greens and fodder cultures Lehmgrubenweg 5 7960 Aulendorf West Germany Representative : Schöllhorn	●							●		●	●		●			
Institute for Thermodynamics and Heat Technology University of Stuttgart Pfaffenwaldring 6 7000 Stuttgart 80 West Germany Representative : A. Haug; N. Fisch Y. Tanes; et al.	●	●							●	●			●		●	
MEUVO Department of Solar Energy and Solar Drying Sackgasse 6 8050 Freising West Germany Representative : S. Vogt; K. Meuren	●	●		●				●	●	●	●		●	●	●	○
Willibard Grammer Department of Solar-Climate-Technique 8450 Amberg West Germany Representative : H. Barthel				●				●		●			●	●		○
Institute for Wood Research University of München Representative : A. Schneider; F. Engelhardt L. Wagner					●			●		●						
Chamber of Agriculture of the department of Vienne, France 14, Rue Scheurer-Kestner-B.P. 129 86004 Poitiers Cedex France Representative : Mr. Marechal	●							●		●	●	●	●			●
E.N.S.A.T. (National School for Tropical Agronomy) 145 Avenue de Muret 31076 Toulouse Cedex France Representative : A. Peyre; S. Toekasap; C. St-Joly, et al.	●	●						●		●	●		●		●	●

(Table 2 continued)

Research center	hay/forage	cereals	fruits	spices/tobacco/vegetables	wood	other products	general approach	significant	secondary (operational set-ups)	physical feasibility	economic feasibility	more sophisticated types	simple collector types (not integrated in roof)	simple collector types (to be integrated in roof structure)	Multi-purpose use	Component dimension & regulation system optimization
A.S.D.E.R. (Association for the development of renewable energies in Savoie, France) 299 Rue du Granier 73230 Saint Alban Leysse France Representative : G. Savatier			●					●		●			●		●	●
CEMAGREF (National center for mechanisation in Agriculture, for agricultural engineering, and for water and forest management) Route d'Arles 30001 Nîmes France			●					●		●	●	●			●	●
C.E.E.M.A.T. (Study and experimental center for the mechanization of tropical Agriculture) Parc de Tourvoie 92160 Antony France Representative : A. Themelin						●		●								
Laboratory of Thermodynamics and Energetics University of Perpignan Avenue de Villeneuve 66025 Perpignan Cedex France Repesentative : M. Daguenet							●	●	●	●	●		●	●	●	●
Dept. of Agricultural Engineering School of Agriculture Aristotelian University Thessaloniki Greece Representative : G. Martzopoulos C.B. Akridis		●					●	●		●				●	○	
Institute of Agricultural Mechanics University of Sassari Italy Representative : P. Piccarolo F. Paschino	●							●		●		●	●			●
Zootechnical Institute University of Bologna Italy Representative : U. Chiappini	●							●		●			●	●		

(Table 2 continued)

Research center	product							scient. impact		assess. type		collector types				Multi-purpose use	Component dimension & regulation system optimization
	hay/forage	cereals	fruits	spices/tobacco/vegetables	wood	other products	general approach	significant	secondary (operational set-ups)	physical feasibility	economic feasibility	more sophisticated types	simple collector types (not integrated in roof)	simple collector types (to be integrated in roof structure)			
C.N.R. Experimental Institute for Mechanization in Agriculture Rome Italy Representative : G. Colzani A. Scarpini				●				●		●		●					
Research Center for Animal Production Reggio Emilia Italy Representative : Bussi L. Valli														●			
The Scottish Institute of Agricultural Engineering Bush Estate Penicuik Midlothian Scotland, U.K.	●	●						●		●				●		●	
Ministry of Agriculture, Fisheries and Food London Rd. Slough laboratories - Slough United Kingdom Representative : N. Burrell		●						●		●				●			

2.2. Location of the projects (see fig. 1)

Fig. 1 : Location of the projects described.

(Note : The circled numbers refer to projects described in more detail in the appendix to part II.)

2.3. Basic project descriptions

---

1. Research concerning the use of solar air collectors for forage and cereal drying

+ Project in Gembloux, Belgium (since 1982)

+ Description :
  5 flat plate solar air collectors were built, using different cheap
  conventional construction materials (see fig. 1.1).
  Intentions being, to establish the relationship : materials used
  – collector efficiency – cost for solar drying applications.
  Also other parameters that influence the collector efficiency were
  studied.

+ Conclusions :
  - The collector efficiency values ranged between 56 and 73 % with air
    flow rates of 150 m^3/(h.m^2 coll. surface), and temperature rises of
    3.5 – 13 K. Even better results were reached at the optimal air flow
    rate of 190 m^3/(h.m^2 coll. surface).
  - Several of the collectors with good performance unfortunately showed
    signs of deterioration after 4 months.
  - Based on efficiency and durability criteria the best collector was
    selected. It consists of a polycarbonate cover (10 mm), a flat
    asbestos cement plate painted black as absorber (5 mm), expanded
    polystyrene (20 mm) and chipboard (12 mm) as insulation and
    protection. (material costs 31 ECU/m^2.)
  - In a basic economic estimation the cost per unit of feeding value of
    hay dried with the help of similar cheap solar collectors was gauged
    to be in the same range as that of oil heated drying and predried
    ensilaging.
  - Future research will cope with the practical aspects of the
    integration of the developed collectors in an experimental grain
    drying set-up.

+ Study Center : A. Pletinckx
  Center for agricultural engineering.
  Chaussée de Namur 146
  5800 Gembloux
  Belgium

+ For detailed description of this project, cf. p. 239.

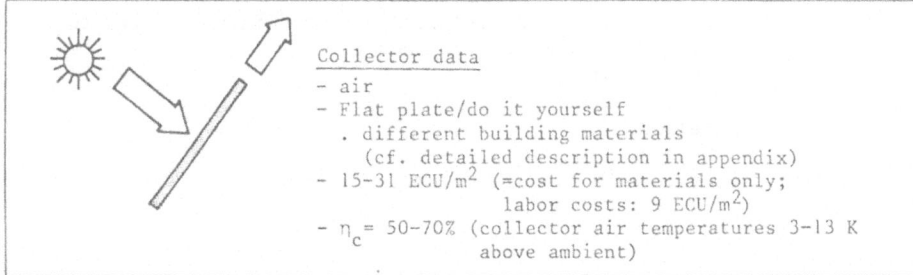

Collector data
- air
- Flat plate/do it yourself
  . different building materials
    (cf. detailed description in appendix)
- 15-31 ECU/m^2 (=cost for materials only;
                   labor costs: 9 ECU/m^2)
- η_c= 50-70% (collector air temperatures 3-13 K
                   above ambient)

Fig. 1.1 : Collector data.

Note : At the same institute, the performance of a solar grain drier
was simulated, and the results of this performance analysed to
establish its economic feasibility. The conclusion (made in
1980) was that unless the (1980) prices of the considered solar
collectors (= 75 ECU/m^2) were diminished by 50 %, and the oil
price (0.25 ECU/1) doubled, solar grain drying could not be
economically competitive with classical drying. The
development of cheap self-made collectors is a logical next
step after the simulation and economic analysis in question.

---

2. Solar tobacco curing

+ Project in Malawi, Africa (1983 - 1984)

+ Description (see fig. 2.1) :
  Quality tobacco curing requires a considerable amount of wood for
  heating purposes.
  Over the 7-9 day drying time - during more than half of which the
  temperature of the drying air is at a level of 50 to 70°C - an average
  of 1-2 kg of wood is being burned per kg of fresh tobacco in the
  traditional barn driers.
  In order to assess the potential performance of a solar system in this
  application, a pilot plant was designed, and monitored over the
  3 month drying season (January-March).
  The installation, which was meant to dry a batch of 6 600 kg of fresh
  tobacco leaves to about 950 kg of dry product, included 75 m^2 of
  air-collectors, as well as 50 m^2 of a non-covered water solar
  collector ("energy roof") with heat storage (1 m^3), together with a
  possibility of heat recovery or recycling of the exhaust air.

+ Conclusions :
  - The measured data are still being processed. However, based on the
    acquired experience, a potential energy saving of ± 20 % is being
    projected for a well designed system.
    This value is low, due to the fact that the drying period
    corresponds with the rainy season of the year (low solar energy
    supply).
  - The expensive high efficiency collectors can only be used 3 months
    per year. Given the lack of political and economic stability of the
    country in question, the resulting high payback period is expected
    to be prohibitive for this application.

+ Study Center : W. Dutré/P. Cromphout
                 Department of Applied Mechanics and Energy Conversion
                 Catholic university of Leuven
                 Celestijnenlaan 300 A
                 B-3030 Heverlee
                 Belgium

+ Industrial corporator Tractionel, Brussels Belgium.

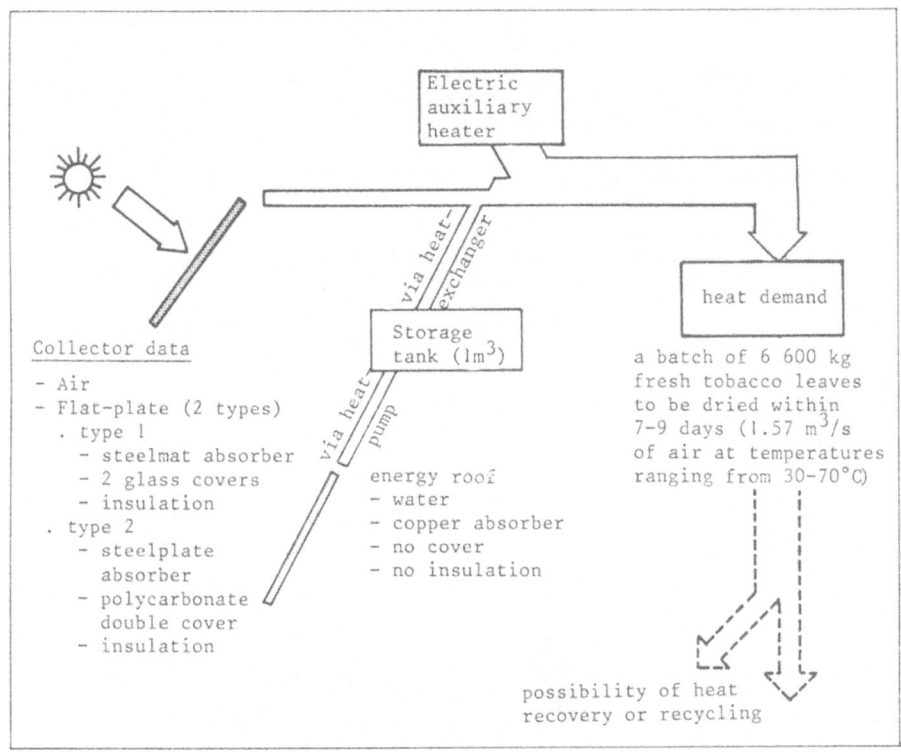

Fig. 2.1 : Data on the tobacco curing plant.

Note : In the study center in question a lot of simulation work is also taking place in the area of solar system performance, among other things on solar maize drying.

3. Solar hay and grain drying

+ Project in Beuerberg, W. Germany. (since 1983)

+ Description (see fig. 3.1) :
  116 m² of commercial flat plate air collectors were installed to heat up air with which to dry 24 round pre-dried bales of hay (1.30 m x 1.30 m).
  The installation would also allow for grain drying (up to 30-40 tons). In out of season periods for grain and hay, even wood chips can be dried, in order to reach a higher combustion value.
  Goals are : higher hay quality compared to natural drying.

+ Conclusions :
  - at air-flow rates of 4.7 - 6.7 m³/s, collector efficiency ranged between 65-75 %. Temperature rises of up to 14 K were observed. Over 2 sunny days, a decrease in the moisture content of the hay from 35-16 % m.c. was measured.
  - Estimations, based on some measured values would predict that the collector cost should drop to nearly 1/3 of its current value, in

order for the solar system to become competitive with a conventional oil fired air heating installation.
- High pressure drops were recorded, which means higher fan capacity demands, and higher air-leakage potential. Future design should take this into consideration.

+ Study Center : K. Meuren
Bayerische Landesanstalt für Landtechnik
Vöttingerstrasse 36
D-8050 Freising, Weihenstephan
W. Germany.

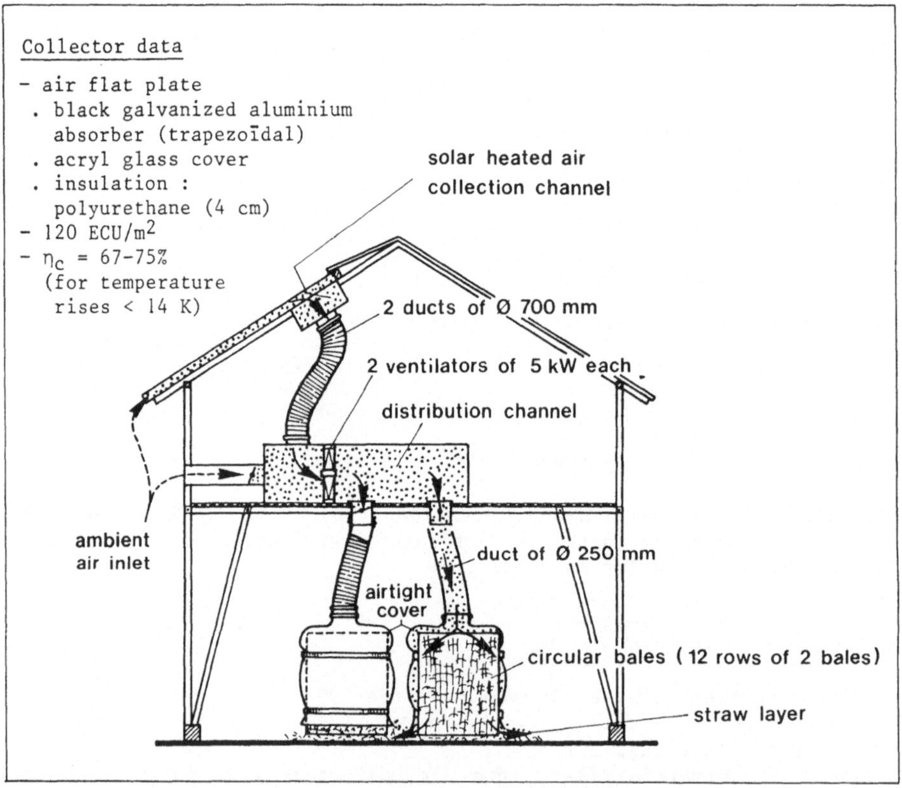

Collector data

- air flat plate
  . black galvanized aluminium
    absorber (trapezoïdal)
  . acryl glass cover
  . insulation :
    polyurethane (4 cm)
- 120 ECU/m^2
- η_c = 67-75%
  (for temperature
  rises < 14 K)

solar heated air
collection channel

2 ducts of Ø 700 mm

2 ventilators of 5 kW each

distribution channel

ambient
air inlet

duct of Ø 250 mm

airtight
cover

circular bales (12 rows of 2 bales)

straw layer

Fig. 3.1. : Scheme of the installation in Beuerberg.

Note : in this same institute, new solar collector types are also developed. In one instance, a non- perforated black draintube of light pvc plastic (φ 5-10 cm) was fastened on top of a roof. The air to be heated was sucked through the tube.

4. Solar drying of agricultural products

+ Project in Göttingen, W. Germany (1977-1983)

+ Description :
Research has been done in the area of low temperature batch-in-bin drying of grain (also applicable for hay).
The planned research was divided in 3 steps :
Step   1 : Experimentation with different collector types.
In a lab set-up, the performance of expensive collectors, designed for high temperature operation, was compared with that of cheaper flat-plate collectors, and very cheap-easy to make-roof collectors.
Three types of cheap easy to make roof-collectors were designed :
* An existing roof of corrugated asbestos-cement plates, painted black and
covered with flat acrylglass (type 1)

or covered with corrugated acrylglass (type 2)

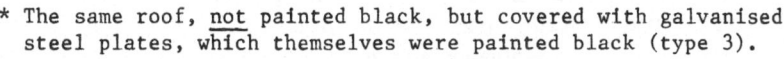

* The same roof, not painted black, but covered with galvanised steel plates, which themselves were painted black (type 3).
Step   2 : Drying experiments
The drying experiments were performed in 3 batch-in-bin driers, containing 4.5 m^3 of wheat or barley (stacked 1 m high, and with initial m.c < 30 %).
In one set-up, no heating of the air occurred (electric ventilator).
In another set-up, solar heating was done (electric ventilator).
In a 3rd set-up, the air was heated by the waste heat of a diesel driven fan.
For each of the driers, the decrease in moisture content of the product was measured as a function of ventilation air flow rate.
Step   3 : The conception, based on steps 1 & 2, of a model intended to help design optimized drying set-ups.

+ Conclusions :
- In the case of forced ventilation of unheated air, the drying can be speeded up (and uniformity of the m.c. over the drying layer improved) by increasing the air flow rate.
This increase not only influences transfer coefficients of heat and vapor, but also causes a higher air temperature, due to increased heat generation by the electric fan.
($\Delta 1$ K per $\Delta 360$ m^3 air/(h.m^3 product).)
It is clear that the above mentioned temperature rise goes along with a rise in the energy that is being spent per kg of evaporated water. (every increase in air-flow rate of 1 m^3/(hr.m^3 product) corresponds to an increase of mechanical energy input requirement of about 3 kJ/kg evaporated water).
- The use of solar collectors significantly speeds up the drying process as compared to a non-heated set-up. The mechanical ventilation energy savings for equal ventilation rates by a collector raising the air temperature an average of 6 K are about 250 kJ/kg evaporated water (see fig. 4.1).
The uniformity of the dried layer is however not as good as for a non-heated set-up. Increasing the air flow rate has little effect on this problem.

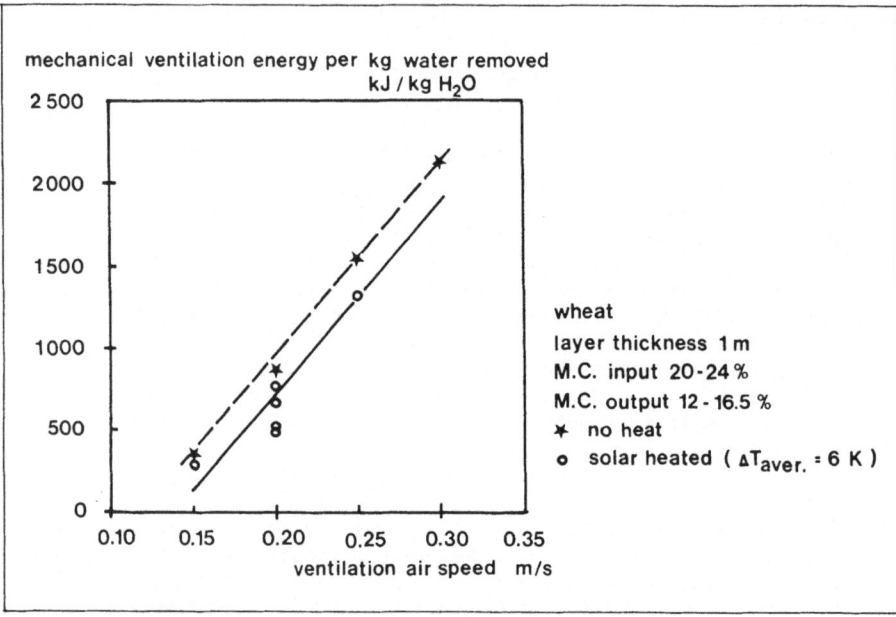

Fig. 4.1 : Electricity input for mechanical ventilation per kg of
          evaporated water. Comparison between solar heated and
          non-heated drying set-up

- Even though the more <u>expensive collectors</u> allow for higher temperature
  performance, it was found that in the working range of the collectors
  in question (less than 10 K temperature increase; high air flow rates)
  the efficiency of the medium-temperature, and cheaper roof type
  collectors are in the same range of 40 - 60 % (see fig. 4.2).
  The cheap roof-collectors therefore emerged as the most economic
  choice for this application. They are reliable, longlasting, and
  low cost : 17-35 ECU/m². In the local application studied in North
  Germany, operation of these collectors over 200-400 hrs/year makes
  them economically competitive with traditional methods.
- Using a diesel-driven fan, the heat production of which can be
  recuperated and used (1), allows for drying operation even at
  night (20 operation hours/day average), and hence can shorten the
  drying period. In the local study-situation in N. Germany, the
  economic feasibility of this solution is found to be comparable
  to that of collector-heating.
  For more sunny climates, collectors would probably be preferable,
  while for less sunny climates the diesel-driven fan is to be chosen.
- Sufficient resource material is now available to adequately design
  economically competitive applications for different situations.

+ Study Center : F. Wieneke/W. Grim
                Institute for Agricultural Engineering
                Gutenbergstr. 33
                3400 Göttingen
                W. Germany

(1) *The lower heating value is intended here since latent heat
    production does not help for drying purposes (N.O.T.E.)*

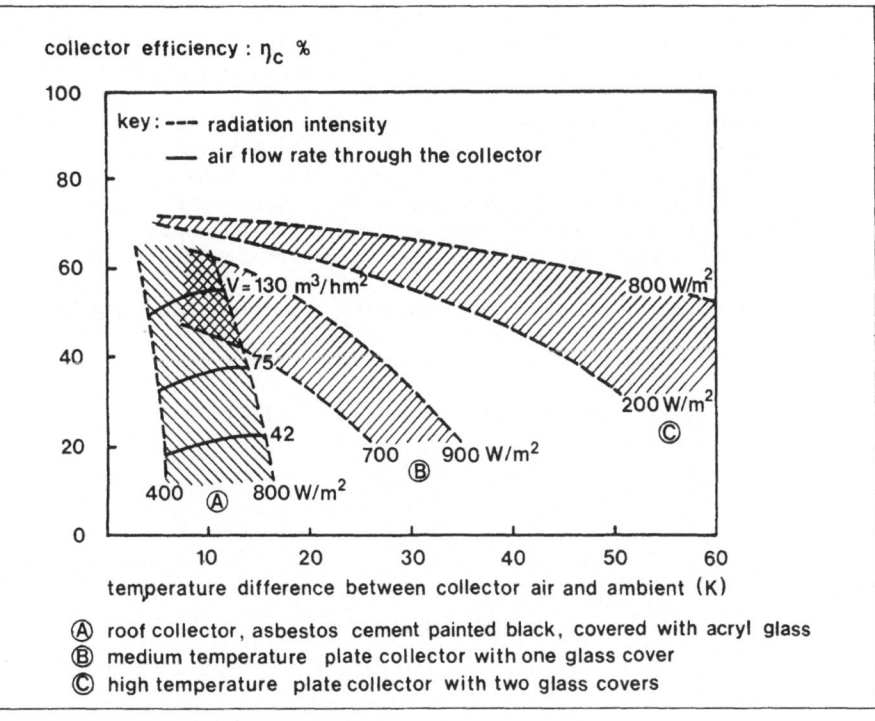

Fig. 4.2 : Collector efficiency as a function of
- air temperature increase
- air flow rate
- radiation intensity

---

5. Solar plant for drying spices

+ Project in Ismaning, near Munich, West-Germany. (since 1982)

+ Description (see fig. 5.1) :
On a spice drying plant in Ismaning 529 m^2 of collector surface are used to help heat up the drying air.
- The collectors are easy to construct, consist of common building materials, and function as the roof cover itself.
- The air, heated by the collectors, is in some cases further heated by a heat exchanger, also connected to a heat recycling system in the exhaust air.
- Finally, if necessary, a 650 kW oil-fired heater brings the air to the required temperature.
- The continuous drier processes 25 tons of herbs over an operation time of 600 hours per year. The spices (especially parsley and peppermint) take about 5-7 hours to dry to a moisture content of 11-15 % using air at 60-100°C and air flow rates of 7-9 m^3/s.
Measurements are being made in order to assess the performance of the solar and heat recuperation system; those measurements will help validate a simulation model, which will allow for optimization of

system components, and calculation of economic feasibility.

+ Conclusions :
  - Over the period of 1982-1983, the solar fraction of the total energy consumption was 32 %. This would correspond to an energy saving of 340 GJ/year.
  - The estimated payback period is 14-15 years, if it is assumed that the cover needs replacement every 8 years. Increasing the operation time per year would significantly reduce the payback period.
  - The polyester glazing transmissivity decreases over time. (This is especially true in cases of high operation temperatures.)
  - Air ducts have to be optimized as to material, length and pressure drop.
  - The contribution of the heat recovery unit has been negligible so far, due to technical problems.

+ Study Center : H. Schulz
    Bayerische Landesanstalt für Landtechnik.
    Vöttinger Strasse 36
    D-8050 Freising
    W. Germany

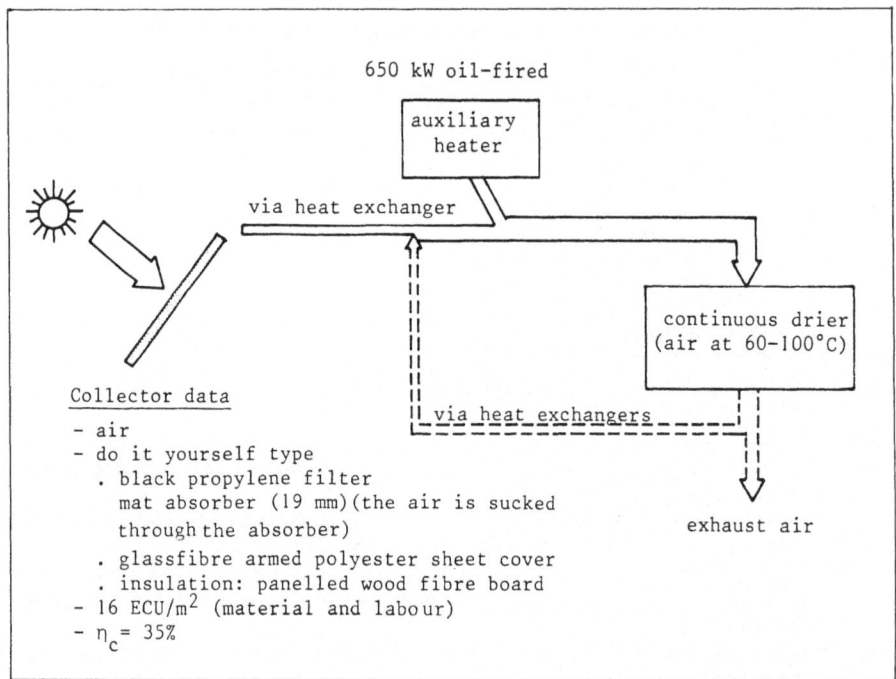

Fig 5.1 : Data on the installation in Ismaning.

6. Inflated solar energy captation tube, developed for drying of agricultural products

+ Project in Montoldre, France.

+ Description (see fig. 6.1) :
  A black polyethylene tube (length 300 m; diameter 1.9 m; polyethylene thickness 250 μm) is placed in front of the air inlet of a traditional in-bin-forage-drier with a capacity of 410 m³ hay (stacked up to 4 m high). A ventilator at the entrance of the plastic duct blows the air into the duct at a rate of 10.6 m³/s and makes it take on its cylindrical shape. This device heats up the incoming air by 7-10 K when exposed to the sun.

+ Conclusions :
  - Advantages :
    - easy to place/low price
    - good performance, even in the absence of direct sunlight.
    - useful for different applications, requiring high energy inputs.
  - Problems :
    - easily caught by the wind (less problematic with diameters smaller than 1.5 m).
    - easily damaged; short lifespan (3 years maximum).
    - takes up a lot of space.
    - needs to be stored away after the season (requires manhours).

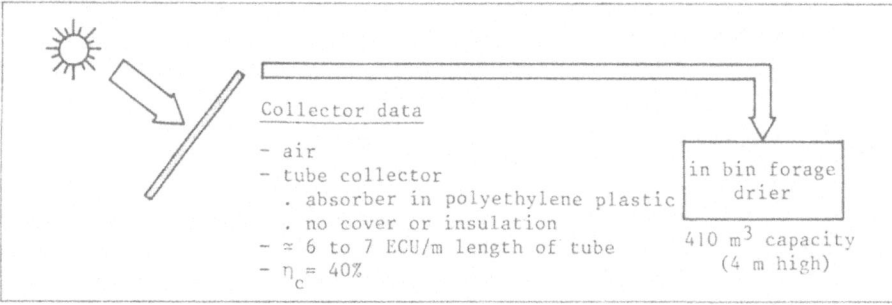

Fig. 6.1 : Data on and sight of the installation in Montoldre.

+ Study Center : P. Feuilloley
             Cemagref
             Domaine des palaquins
             03150 Montoldre-Varennes-s-Allier.
             France

7. Use of a greenhouse for warm air production for hay drying

+ Project in Vélezy - Haute Savoie, France (since 1981)

+ Description (see fig. 7.1) :
  A plastic greenhouse, (80 m long, 4.5 m wide, 2 m high in center),
  with a black plastic sheet covering the ground, is applied as an
  intermediate solution between solar air heating by means of an
  inflated black polyethylene cylinder, and by means of roof collectors.
  The air drawn through the greenhouse is afterwards blown through the
  hay that is to be dried.
  (At least 4 similar set-ups exist in the same region.)

+ Conclusions :
  - In the given set-up 5 T.O.E./year are being saved. (1)
  - Advantages and disadvantages of the system are mainly the same as the
    ones mentioned in BD nr. 6. Another advantage is the absence of
    excessive pressure-drops which often occur in flat plate air
    collectors.
  - Connection with barn drier was problematic.
    (groundwater collection in the ventilation duct).

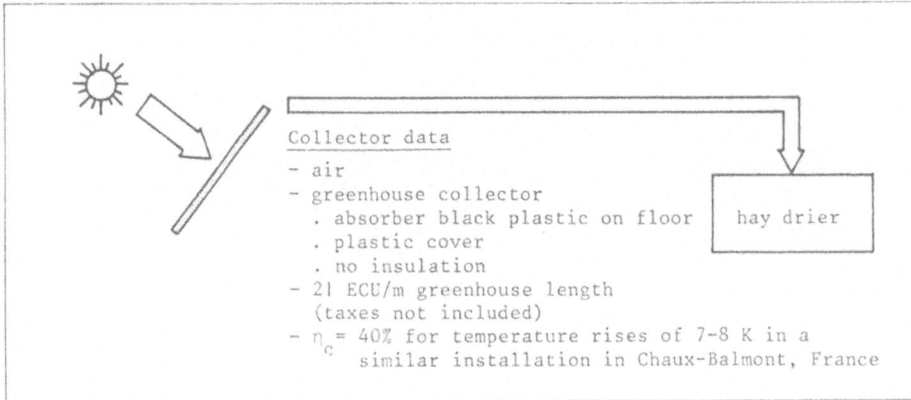

Fig. 7.1 : Data on the installation in Vélezy.

*(1) In a similar but 20 % smaller greenhouse collector, annual energy
savings of 1.3-1.7 T.O.E. are reported. The annual utilization rate
certainly has a role to play in the size of the energy savings.
(N.O.T.E.)*

8. Drying installation for hay drying, by means of a simple roof collector

+ Project in St. Bonnet, France (since 1980)

+ Description (see fig. 8.1) :
An existing asbestos cement roof of a total of 240 m² on a barn hay drier was painted black and covered with a polyester cover. The air is sucked sideways between the 2 layers, and is afterwards blown through the hay at a flowrate of 6.9 - 8.3 m³/s.

+ Conclusions :
- The average collector efficiency is 50-60 %; temperature increase 12-15 K.
- The set-up is solid, long lasting, and has no problems with wind or snow, (unlike the inflatable types); this is an important feature in mountainous areas.
- Compared to the inflatable type collectors, the system is more costly, and requires additional fan power due to significant pressure drops.

+ Study Center : Mr. Veglia
          SERT (Technical Research Center of the firm "Eternit")
          Rue de l'Amandier
          78540 Vernouillet
          France

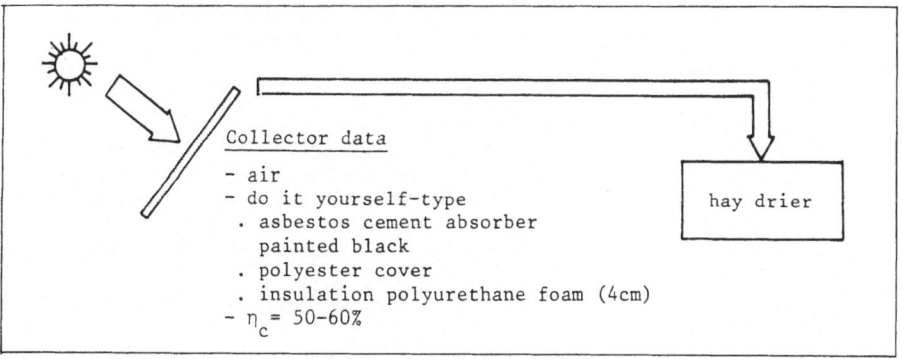

Fig. 8.1 : Data on the installation in St. Bonnet.

9. Comparative study of existing forage driers

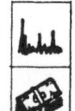

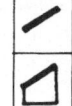

+ Project in Savoie, France.

+ Description :
A comparative study was made of different existing solar forage driers on farms in South-East France.
For this purpose, several installations were visited and thoroughly analyzed, while at the same time interesting publications on the same subject were studied.

+ Conclusions :
1. As regards roof collectors without cover (air flowing between absorber on top and insulation below) :
   - According to the researchers, use of a covering material would not be essential in regions with little wind. In windy places, the collector efficiency is however significantly reduced !
   e.g. : Measurements showed that an increase of wind speed from 2-5 m/s can cause a decrease of the non-covered collector efficiency from 51 % - 38 %.
   - For absorber materials, galvanized steel and coloured asbestos cement are mostly used. It is not clear which one is to be preferred. Some prefer asbestos cement because of its so called "thermal diode" characteristics (not demonstrated), others choose galvanised steel, because of its high heat conductivity (100 x larger than for asbestos cement), as well as its increasing spectral selectivity with increasing age (no need for painting).
   - Even though a 1 cm thick plywood insulation would suffice from solar collecting viewpoint, insulation with for instance 3 cm of polystyrene would be preferable above stables for animals, since this would at the same time prevent condensation of water during wintertime, on poorly insulated roofs. (dual purpose) In addition, polystyrene is cheaper than plywood.
2. As regards roof collectors with cover (air flowing between cover and absorber) :
   - According to the researchers, is suitable for windy regions, or in applications with polyvalent use of the collector.
   (e.g. : Contribution of the collector, via heat exchanger, to water heating during off-drying season or, to house heating).
3. Greenhouse collectors.
   - Some of their problems are quick deterioration, problems with snowfall, need for space, air leaks ...
   - Their efficiency can be improved, by providing wind shields if necessary; by using PVC or EVA instead of polyethylene; by increasing the air speed and the absorber surface; by insulating the north side of the greenhouse ...
   The low cost and possibility for polyvalent use (greenhouse crops during winter) should be borne in mind.
   - The seemingly most efficient greenhouse type would be one oriented south, and leaning against a wall.
4. A possible set-up incorporating the acquired knowledge on roof collectors is given in fig. 9.1.

+ Study Center :
G. Savatier & Y. Frigo
A.S.D.E.R. (Association for the
Development of Renewable
Energies in Savoie, France)
299 Rue de Granier
73230 Saint Alban Leysse
France

A. Anglade
GRET (Grouping of Research and
Technological Exchange)
30 rue de Charonne
75011 - Paris - France

197

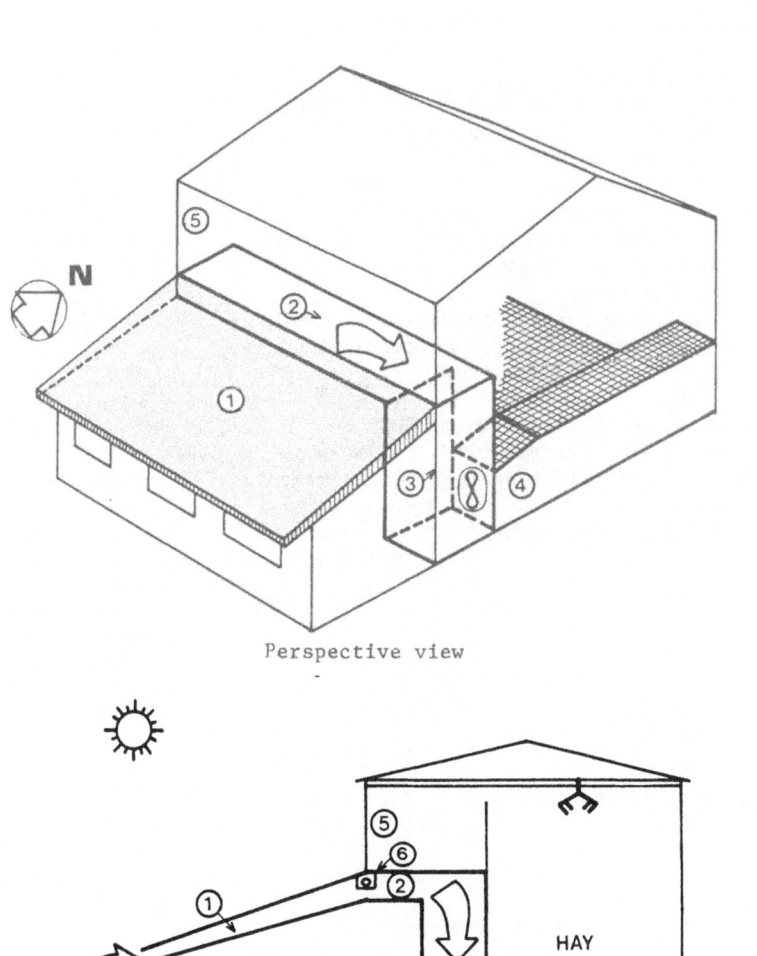

Perspective view

Section

Key
1. Roof collector on top of animal house (cf. comments above)
2. & 3. Air duct (optimum air speed 4 m/s)
4. The distribution duct should diverge right after the ventilator. (Otherwise turbulence of the air could reduce the fan efficiency by 10-30 %).
5. The wall facing the collector should be covered with a reflecting material.
6. Possibly an air-water heat exchanger for hot water production. (multi-purpose use)

Fig 9.1 : Scheme of the proposed optimal installation

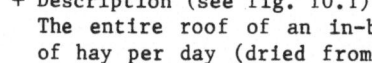

10. Hay drying

+ Project in Vassieux en Vercors, France (since 1980).

+ Description (see fig. 10.1) :
The entire roof of an in-bin-drier with a capacity of about 2.5 tons
of hay per day (dried from 40 % - 15 % m.c.) has been equipped with a
solar collector of 300 m².
The self-made roof-collector consists of a galvanized steel layer on
top that acts as an absorber, and (13 cm beneath it) a plywood plate.
The air is drawn through the roof-collector, towards the top of the
roof, where a well designed duct system leads the air, propelled by a
fan, into the hay.
The air is blown through the hay at a flow rate of 4.4 m³/s.
The installation allows for cereal drying also.

+ Conclusions :
- During an 11 day period in July '82 an efficiency of the collector
system of 47 % was measured. This means an average energy input of 9
kJ/m³ air over given period. A few measured values are given below :

Date	Weather conditions	Q_{ci} (MJ/m²)	Q_{co} (MJ/m²)	η_c (%)	Average energy input per m³ air (kJ/m³)	Average air temper. rise (K)
6/7	Sunny	25.6	11.5	45	9.0	8.4
7/7	Cloudy all day	20.9	9.7	47	7.2	6.8
8/7	Very sunny	33.1	15.5	47	9.7	9.3
9/7	Misty	17.6	9.7	55	6.5	6.2
10/7	Mist in the morning	25.2	13.0	51	9.7	9.3
11/7	Sunny	27.7	13.3	48	10.1	9.5
	Averages	25.2	12.2	49	8.8	8.25

with Q_{ci} = solar irradiation on the collector.
Q_{co} = solar heat extracted from the collector.

- Over the year 1982 about 38 GJ of heat was collected by the solar
system. This should cover for the planned payback period of 15 years,
given the sturdy long lasting character of this installation.

+ Study Center : A.R.E.S.
    39 Rue Croix Baragnon
    31000 Toulouse
    France

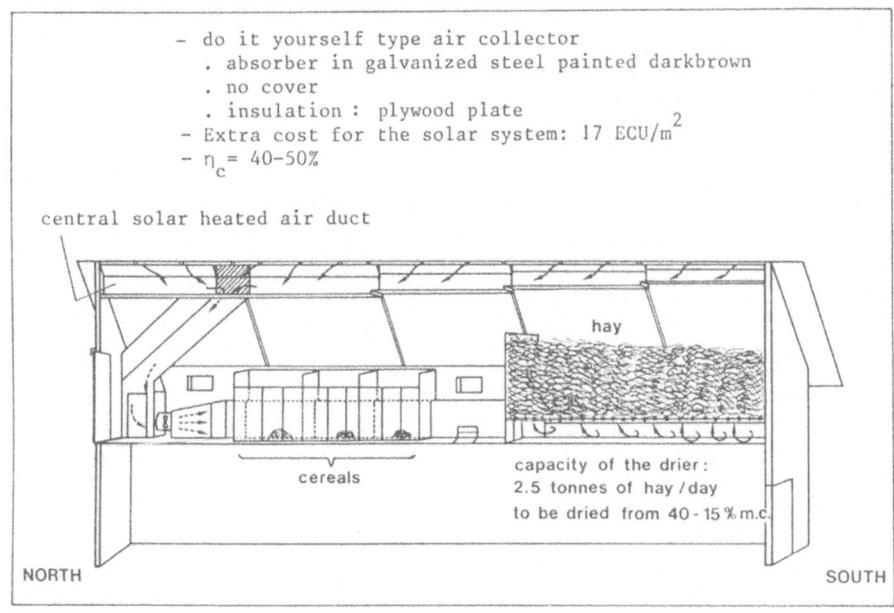

within figure:
- do it yourself type air collector
  . absorber in galvanized steel painted darkbrown
  . no cover
  . insulation : plywood plate
- Extra cost for the solar system: 17 ECU/m^2
- η_c = 40-50%

central solar heated air duct

hay

cereals

capacity of the drier :
2.5 tonnes of hay / day
to be dried from 40 - 15 % m.c.

NORTH   SOUTH

Fig. 10.1 : Scheme of installation in Vassieux en Vercors.

11. Solar drying of grains

+ Project in Bazlège, France. (since 1981)

+ Description (see fig. 11.1) :
  3 different cheap collector types are being tested and evaluated as to their potential in drying grains.
  1. A vertical collector, wrapped around 2/3 of a bin drier.
  2. A cheap roof collector.
  3. A greenhouse-type collector.
  Different parameters of low temperature drying were assessed (batch versus continuous drying, optimization of air flow regulation, quality of resulting product).
  The goal of the experiment is to develop new low cost drying methods, easy to implement on existing drying plants, which can create transition from a dependence on fossil fuels to a dual energy system (solar energy + nuclear/hydro electricity).

+ Conclusions :
  - The average collector efficiency values measured for an air temperature rise of less than 10 K are 50-55 % for collector 1, 40 % for collector 2, and 30 % for collector 3.
  - A simulation model, running over a 95 day reference drying period from September to December predicted a specific electricity consumption of about 2.2 MJ/kg, and payback periods of 4-5.3 years. (lowest payback period for the greenhouse-type collector).
  - The quality test results of the dried grain were positive.
  - This technique is only suitable for climatic regions corresponding to the one in Southern France, and gives good quality results for sun-flower seeds and sorghum.

+ Study Center : J.C. Lasseran
                      Experimental station for grain drying and conservation.
                      G.I.E. SECOGRAINS
                      3, allée des Sylphes
                      31520 Ramonville Saint Agne
                      France

- Ref. : For a more detailed description of this project, cf. p. 301.

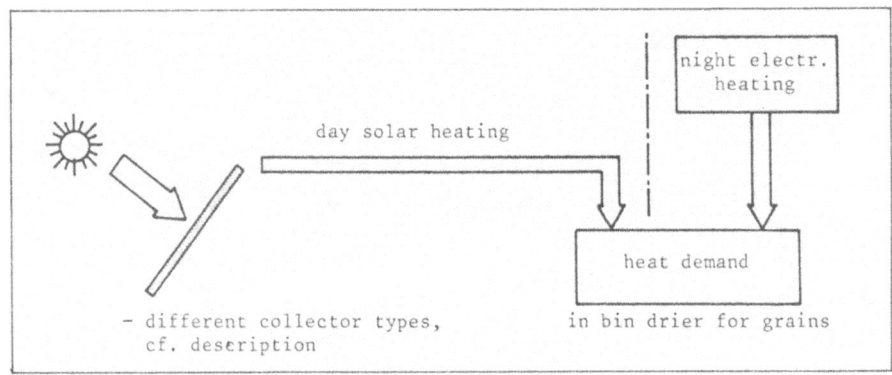

Fig. 11.1 : Scheme of the set-up in Bazière.

12. Development and analysis of rural solar driers

+ Project in Talence, France (since 1980).

+ Description :
In the University of Bordeaux several prototypes of solar fruit driers were developed and tested, with the intention of creating an economically feasible simple installation, needing no other than solar energy (not even energy for operation).

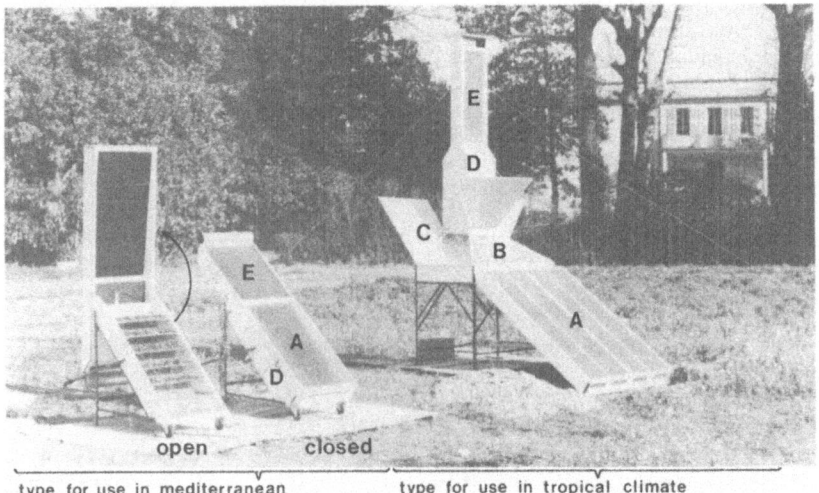

Fig. 12.1 : Picture of rural solar collectors developed in Talence.

Special attention is being paid to make this small scale system accessible to, and useable by local inhabitants of tropical regions.
After testing both modular, and completely integrated set-ups, a more final set-up was created allowing for some of the flexibility of the modular system, but adding to its efficiency.
This system comprises the following elements (cf. set-up on right side of figure 12.1).

A. Air collector : By natural convection, the air is drawn through a porous absorber, mounted diagonally in the collector. (The porous absorber allows for a very good exchange of heat with air.)
B. Heat storage : Water cans inside the storage are heated up directly by the sun through the storage transparent covers, as well as through the transfer of heat from the air coming out of the collector.
C. Reflectors : they reinforce solar radiation on storage during the day, and are closed at night, to provide insulation of the storage.
D. Drying chamber : 6 useful trays, well insulated; contains 10-30 kg of fresh fruit.
E. Solar chimney : helps provide spontaneous draught through the collector, based on natural convection.

Tests, both in the field, and in lab set-up, as well as a simulation model were used to assess the system performance.

+ Conclusions :
  - The system seems economically feasible.
  - A problem with running only on solar energy, is the unpredictability of the performance for each situation (sun is definitely needed; wind-effects can have a large impact).
  - A study, including experiments in the field as well as lab and simulation work, provided a clear picture of the performance of the studied system, as a function of different parameters.
    Each of the elements of the set-up (collector, storage and chimney) proved to have a strongly positive impact on the drying operation.
  - Up to a certain air speed level, the natural convection is amplified by superposing the different thermosiphons (drier, chimney ...).
    Above this level the increase of the driving force of the air is completely compensated by increasing pressure drops.

+ Research Center : J.R. Puiggali/A. Tiguert
                    Laboratory of physical Mechanics
                    University of Bordeaux I
                    351 cours de la Libération
                    33405 Talence Cedex
                    France

  Note : This project is part of a more large scale research including several French organisation (GERDAT, CEEMAT, SIARC, and others). The collector presented on the left of fig. 12.1 is a prototype "Seresol", developed in this same context for mediterranean climates.

13. Drying of fruits and vegetables

+ Project in Prades, Pyrenées-Orientales, France. (1981)

+ Description (see fig. 13.1) :

24 m² of solar air collectors were installed to help heat up the drying air in a semi-industrial drying plant of fruits and vetetables (capacity 5 tons of fresh fruits/month).

The air is preheated by the collector, postheated by an auxiliary heater of 12 kW if needed, and then – at desired temperatures ranging from 20-80°C – blown through a well insulated tunnel of 3 m³, in which the 100-900 kg of produce – lying on 62 m² of trays – are dried. Air flow rates range from 0 - 1 200 m³/h.

The drier can be used as a continuous drier 24 hours per day.

Due to high specific energy input values per kg evaporated water, a recycling air system was added.

+ Conclusions :

- A simulation model was developed, based on measurements done with the installation.

  This model calculated the results for continuous drying of a total of 6 tonnes of fresh apples over the month of June, to a m.c. of 15 %, at which point the residual weight was 748 kg (losses for coring and peeling are about 25 %).

  The drying temperature was 75°C, and air speed 1 m/s (in case of an empty drier).

  This model yielded a solar energy input of 20 % (4400 MJ), auxiliary heater input of 70 %, and ventilation input of 10 %. (Taking the extra required ventilating power for the solar system as opposed to the reference system into account, the net solar energy savings are 19 %.)

  If the collector cost were spread evenly over 7 years, then the cost of the solar supplied energy over this month of June (i.e. the ratio collector cost for 1 month over net energy gain) can be 30 % lower than the electricity cost for ventilation and auxiliary heating.

  How this would look for the situation over a year as a whole has not been assessed vet.

- The program yielded higher productivity and lower production costs for a same set-up, but operating at higher temperatures (and lower air flow rates).

- The model needs further validation and improvement before more definite conclusions can be given.

- The dried fruits are well accepted on the local market, and prove to be of good quality.

+ Study Center : M. Daguenet/J.Y. Quinette
                 Laboratory for Thermodynamics and Energetics
                 University of Perpignan
                 Avenue de Villeneuve
                 66025 Perpignan
                 France

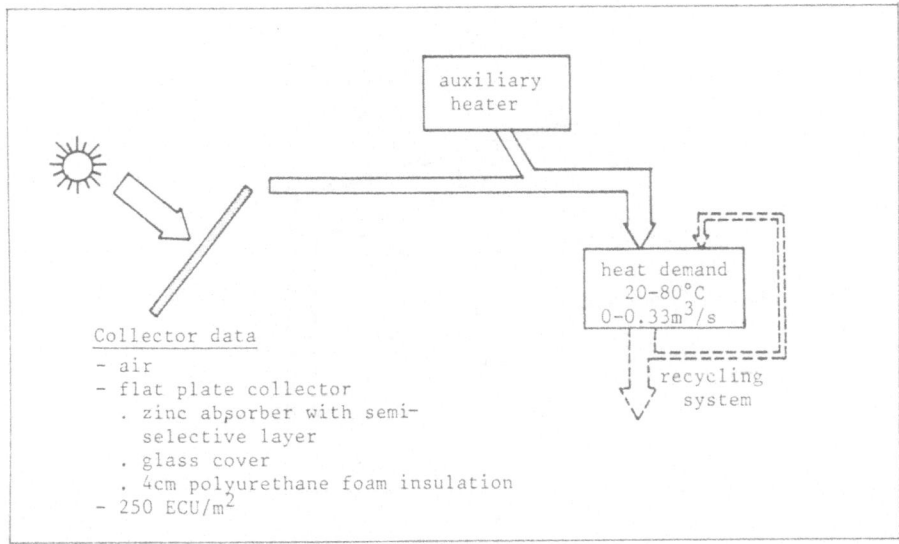

Fig. 13.1 : Data on the installation in Prades.

Note : the incoming air moves up in front, and back down behind the absorber.

14. Wood drying

+ Project in Menestreau - Font Romeu, France.

+ Description (see fig. 14.1) :
  Several drying tests have been carried out on an experimental solar wood drier. (tests with resinous wood, - thickness 34 and 65 mm -; and on hardwood, - thickness 30 and 58 mm).
  The drier is comparable to a somewhat modified greenhouse, and consists of a two layer cover of polymethacrylate of methyl which functions as roof and south facing wall.
  Ventilators draw outside air through adjustable ventilation holes in the top of the insulated opaque North wall.
  This air moves along a black aluminium absorber, through the wood pile and out of the ventilation holes in the bottom of the North wall.

+ Conclusions :
  - For resinous wood, solar drying is only interesting as a finishing process of natural drying.
  - For hardwood, solar drying is always quicker than natural drying :
    . Predrying up to 30 % mc is 1.2-2 times quicker.
    . Via solar drying one can reach moisture content levels of 11-12 % as opposed to 22-25 % for natural drying.

Study Center : Mr. Aleon
               Technical center for wood
               10, avenue Saint-Mandé
               75012 Paris
               FRANCE

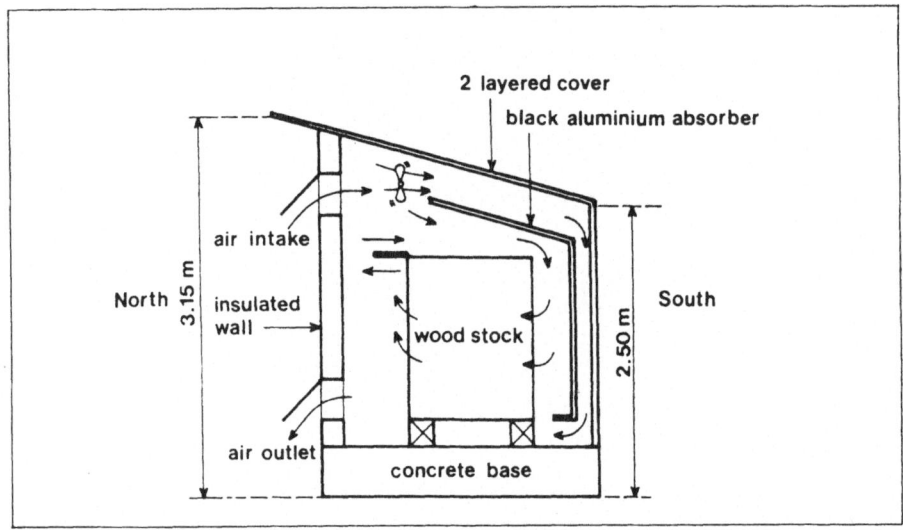

Fig. 14.1 : Cross-section of wood drying plant in Menestreau.

15. Development of a solar drier for wine lees

+ Project location : Canohes, France. (since 1982)

+ Description (see fig. 15.1) :
  Instead of a traditional batch drier for wine less, an experimental greenhouse-type drier was developed.
  In this greenhouse, 8 tonnes of fresh wine lees are stacked on trays exposed to the sun. The air within the closed volume of the greenhouse is continually mixed by ventilation. The inside air is only renewed when the relative humidity becomes higher than 70-80 %.
  As part of the research project, a simulation model is being developed, which will help the optimization of the system.

+ Conclusions :
  - Based on a few experimental data, rather phenomenal energy savings are expected (> 90 %). On the other hand, the drying time is increased from 1 day per batch for the traditional set-up, to 15 days per batch for the new one. This means only six batches of 8 tonnes of fresh wine lees can be dried over the 4 month long drying season. However, compared with the traditional one, this greenhouse-type drier has considerable potential for other drying applications during the 8 remaining months.
  - Experience with the system is gradually gained.
    . To prevent moulding, the drying time needs to be reduced. (Options are the reduction of the batch-size; use of auxiliary heating; conditioning of the incoming air ...)
    . Adaptations of the ventilation and regulation are required in order to reach acceptable system performance.

+ Study Center : M. Daguenet
                Laboratory of Thermodynamics and Energetics
                University of Perpignan
                Avenue de Villeneuve
                66025 Perpignan Cédex
                France

+ For detailed description, cf. p. 263.

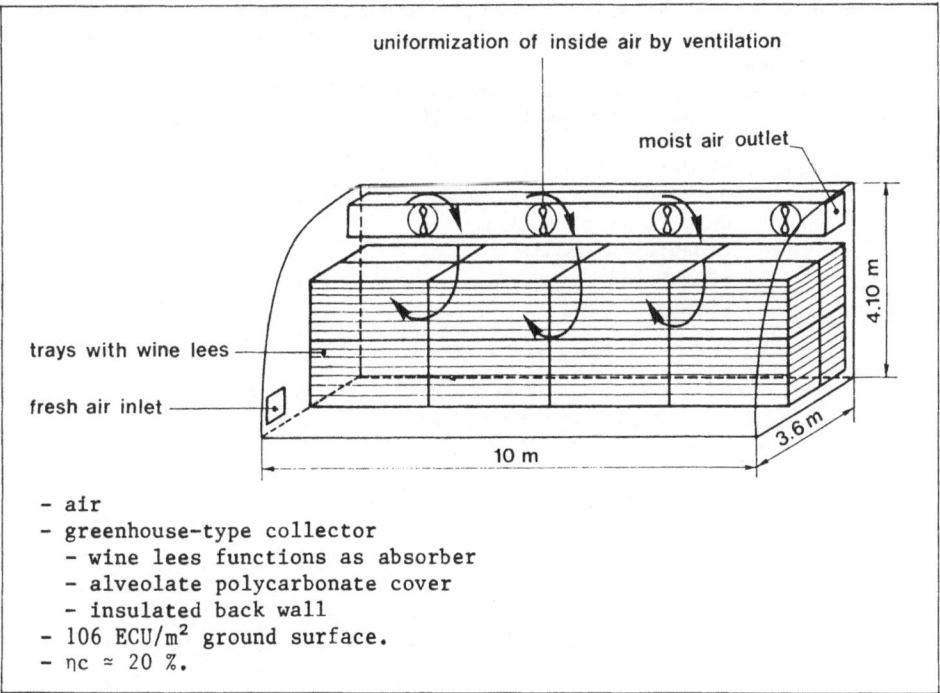

uniformization of inside air by ventilation

moist air outlet

4.10 m

trays with wine lees

fresh air inlet

10 m

3.6 m

- air
- greenhouse-type collector
  - wine lees functions as absorber
  - alveolate polycarbonate cover
  - insulated back wall
- 106 ECU/m^2 ground surface.
- $\eta c \simeq 20$ %.

Fig. 15.1 : Scheme of the installation in Canohes.

16. Construction and evaluation of a pilot scale solar drier

+ Project in Athens, Greece.

+ Description (see fig. 16.1) :
  5 m^2 of flat plate air-collectors are being used to help heat up the
  drying air blown through a drier, filled with 5 shelves of grapes,
  totalling a load of 500 kg (fresh).
  The experience acquired in this experiment as well as a simulation
  model validated by means of the installation, will help design and
  evaluate an efficient system for grapes and other produce on a larger
  scale.

+ Conclusions :
  - The drying test on sultana grapes resulted in good quality grapes,
    but not as good as could be expected. There are reasons to believe
    that this shortcoming, as well as the technical problems experienced
    (too low capacity of collectors and fans, air leaks ...) should be

dealt with adequately.
- The collectors themselves functioned satisfactorily, and one hopes to be able to come up with industrial applications that are economically feasible.
- Practical experience gained :
  - In tray drying, poor uniformity is often reached throughout the layers (top still wet-bottom too dry).
  - Air velocity, more than temperature, seems to be of utmost importance, especially in the beginning stages of the drying process, when important water loss is necessary to prevent development of mould.
- The development of adapted regulating mechanisms is an essential part of making the system work as efficiently as possible.

+ Study Center : M. Tsamparlis
                 Department of Physics/Mechanics division
                 University of Athens
                 Panepistimiopolis
                 Athens 621
                 Greece

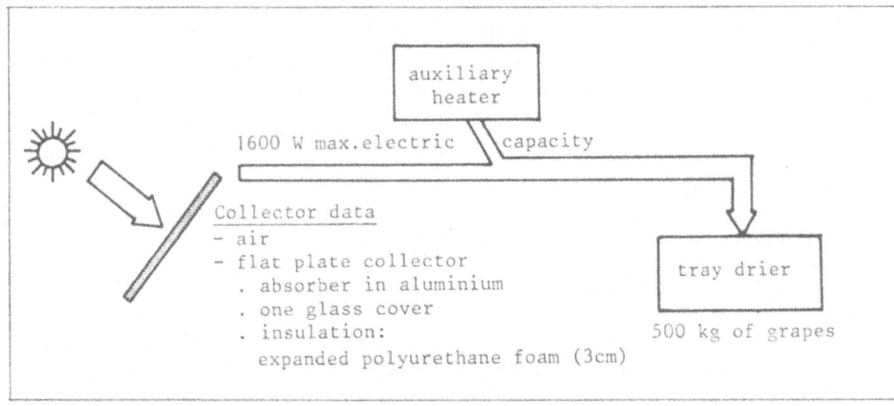

Fig. 16.1 : Data on the installation.

17. Solar drying of Greek fruits

+ Project in Athens, Greece.

+ Description (see fig. 17.1) :
The use of a solar collector system for drying local fruits is being tested and evaluated.
The experimental system consists of a solar collector of 3 m², a storage bed of concrete balls (thermal capacity 250 kJ/K), and a tray drier for 8 kg of fresh fruit.

+ Conclusions :
- A drying experiment was done with alkaline treated Sultana seedless grapes (80 % - 24 % m.c.). This experiment led to a dried product of good quality in 3-4 days only, as compared to more than 8 days in a greenhouse with natural ventilation, after alkaline treatment.

(drying times in this greenhouse were up to 18 days for non-alkaline treated grapes).
- The drying air temperatures ranged between 30 and 62°C. For those temperatures the average collector efficiency was about 40 %.
- The storage bed smoothed out the solar heat input over time. The value of this impact does not seem to outweigh the added cost and heat losses.
- Further research will involve :
  Optimization of the system efficiency (operating conditions, component set-up, multipurpose use) and calculation of the economic feasibility.

+ Study Center : G.D. Saravacos
  Laboratory of Unit Operations
  National Technical University
  Parission 42
  10682 Athens
  Greece

+ For a more detailed description, cf. p. 224.

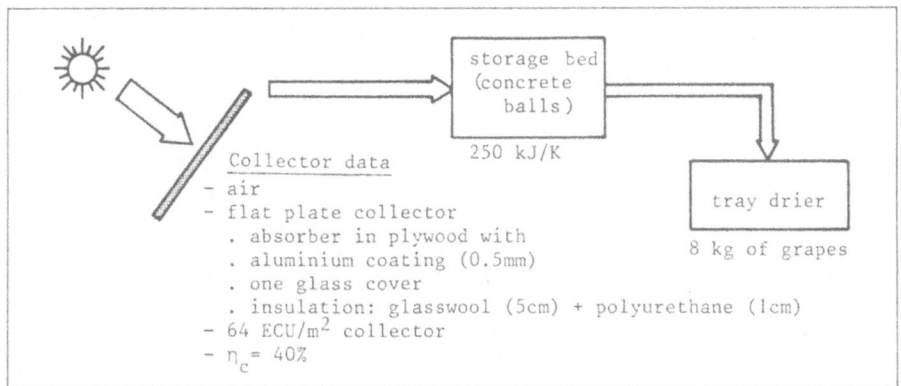

Fig. 17.1 : Data on the installation.

18. Artificial drying of grapes and other fruit

+ Project in Athens, Greece. (since 1980)

+ Description (cf. fig. 18.1) :
  On a pilot scale plant, high efficiency liquid·reflecting collectors,
  - a hybrid form between normal flat plate, and concentrating collectors -, in combination with heat storage and heat transfer from liquid storage to drying air via a heat exchanger, are being tested as to their performance in drying grapes.
  At times, part of the air is recirculated so as to fall within the most optimal range of temperature (70°C max.) and relative humidity (The r.h. value should not be smaller than 10 % at 70°C).

+ Conclusions :
  - The drying period was reported to be significantly shortened (according to the researcher, only 24 hours of drying time instead

of the 20 days required for drying in the open air). *(1)*
- The dried products are of high quality.
- The losses by rain, insects, fouling, etc. ..., occurring in open air drying, are significantly reduced. (by about 10 % of the total production)
- Different size grapes were dried in a non-uniform way. Experience from conventional drying indicates that during storage, uniformization would occur.
- Future work :
  - Economic evaluation of this artificial drying process.
  - Further investigation in post-treatment of insufficiently dried larger grapes.

+ Study Center : A. Deliyannis
                 Plastira st. 3
                 Amarousion - Pefki
                 Greece

Note : Experimentation with collectors and solar installations for drying foodstuffs stored in silos is also planned in cooperation with Prof. Boettcher from W. Germany.

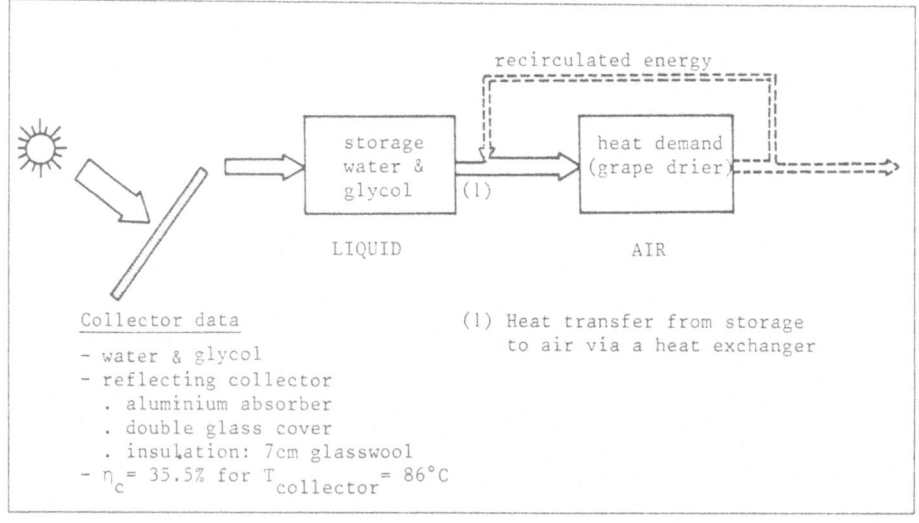

Fig. 18.1 : Data on the installation.

(1) Note from the editor : it is not clear how this statement should be interpreted. Comparing the results with the next project would suggest that the 20 days could refer to open air drying of non alkaline-treated grapes, while the 24 hours most likely refer to 24 daytime active drying hours (2 days instead of 1) on alkaline-treated grapes.

19. Use of solar energy for drying of Cretan sultana grapes

+ Project in Crete, Greece.

+ Description :
   In order to reduce the losses (weight loss; quality loss; high risk) involved in traditional grape drying, a research program was set up.

   This program involved 3 steps :
   1. Fundamental research, studying the grapes themselves, as well as the drying parameters (size of grapes; type of chemical treatment; drying temperature, air velocity).
   2. Based on this research, the development and testing of cheap, easy to make solar drying installations.
   Four main solar set-ups were tested.
   Type 1 : Greenhouse type (3 variants). (See fig. 19.1)
          A. Grapes on the ground, protected by a plastic foil roof (no walls).
          B. Grapes on the ground, protected by a plastic tunnel.
          C. Ventilated tunnel.
   Type 2 : The traditional drying rack, covered with plastic foil.
   Type 3 : Solar drier equipped with a simple collector based on natural convection. (See fig. 19.2)
   Type 4 : Solar drier equipped with more intricate collector using forced convection. (See fig. 19.3 & 4)
   Note : The plastics used are P.V.C. or P.E.
   3. An economic analysis of the feasibility of the studied systems.

+ Conclusions :
   - The fundamental research pointed out that drying temperature and type of chemical treatment are the parameters that have the greatest impact on drying performance.
   - Types 1C and 4 produced the best quality grapes, in the shortest time (about 5 days), with the smallest risk.
   - The other greenhouse variants, as well as type 3 produced somewhat less but still good quality grapes in about 7-8 days. The risk factor in the greenhouse type is rather high.
   - Type 2 produced the grapes of least quality, yet still better than the traditional types; the drying time was 12-16 days; the risk factor rather small.
   - All the above described systems perform significantly better as to risk, quality, and drying time, than the traditional drying-types on the ground (8-10 days) or on the thread-structure drier (15-20 days).
   - The economic analysis pointed out that all the studied alternative systems are economically competitive with the traditional ones.

+ Study Center :
   This study was a joint W. German/Greek undertaking.
           H.D. Kutzbach/W. Mühlbauer
           Institute for Agricultural Engineering
           University of Hohenheim
           Garbenstrasse 9
           D-7000 Stuttgart 70
           W. Germany

210

V. Vardakis
Institute of Vegetables and Ornamental plants
Iraklion
Crete
Greece

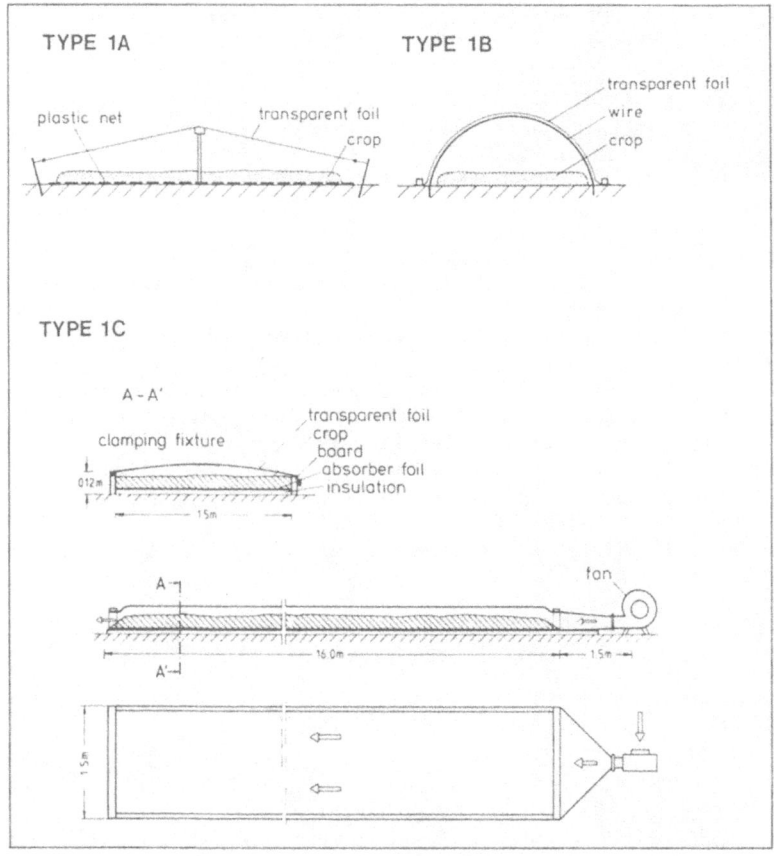

Fig. 19.1 : Greenhouse type collectors.

Fig. 19.2 : Type 3 Solar collector.

plastic cover

black iron absorber

Fig. 19.3 : Type 4 collector.

Fig. 19.4 : Typer 4 drier.

20. Solar energy for produce drying and milking parlour washing

+ Project in Pavia, Italy (since 1978).

+ Descriptions (see fig. 20.1) :
  A simplified water solar collector of 180 m² net surface is being used
  for 2 purposes :
  - to heat up the drying air of a forage drier with batch capacity of
    30 tonnes,
  - to help prepare 1 000 1 of water per day at 40°C for a milking
    parlour on the same farm.
  Next to the collector, the main elements of the solar installation
  are :
  - a 10 m³ water storage tank, mainly used in connection with the
    drier.
  - a 1 m³ water storage tank, in connection with the milking parlour.

+ Conclusions :
  - The solar drier only operates at full capacity during 60-90
    days/year; the heat demand of the milking parlour is very small,
    compared to the capacity of the collecting system.

- If a high utilization rate of the solar system (0.7-0.8) could be achieved by adding extra applications in periods when supply exceeds demand, the predicted payback period would be around 10 years under North Italian weather conditions.
- If the solar system is going to be used for drying only, air collectors are definitely preferable.

+ Study Center : G. Castelli
                   Institute of Agricultural Engineering
                   University of Milano
                   Via Celoria 2
                   Milano
                   Italy

+ For a more detailed description, cf. p. 274.

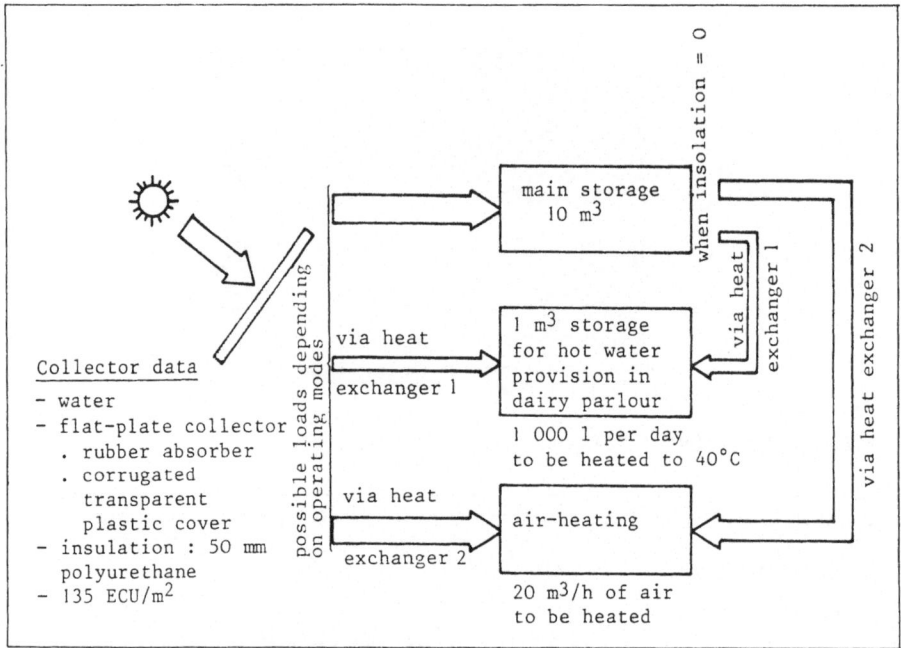

Fig. 20.1 : Data on the installation in Pavia.

21. Solar plant for forage drying in a mountainous area

+ Project in Allessandria, Italy (since 1980).

+ Description (see fig. 21.1) :
The south-facing part of the roof of a 500-head goat shed was painted black and covered with a GRP-clear cover, in order to form a simple 490 m² large collector.
At a rate of 5.6 m³/s, the air is drawn to the top of the roof collector, by means of a 10 kW diesel driven fan, and then blown into a "German-type" drier for predried hay (with capacity of 100 tonnes of dry hay per batch).

The aim of the project is :
- to assess the decrease in energy requirement for hay drying by means of solar energy.
- to verify if such an extremely simple set-up is economically feasible, given the fact that it is only being used during a few months per year (+ 70 days/year of actual drying).
- to gain experience with the system, so as to be able to gauge its suitability for applications on the farm.

+ Conclusions :
- A few short term measurements revealed an average collector efficiency of 36 % for an air flow rate of 40 m³/(h.m²) collector and temperature rise of 15 K.
  This low figure is mainly due to a non-uniform pattern of the air flow through the collector, amplified by the relatively low flow rate.
  The flow rate was actually lower than expected (design flow rate : 8.3 m³/s), since significant pressure drops occurred in the airducts. A better design of those ducts is certainly desirable. Measurements, with part of the collector cut off, and leaving only 160 m² of collector surface, yielded collector efficiency values of + 60 % for practically the same temperature increase at flowrates of 120 m³/(h.m²) collector.
- The heat recovery of the diesel driven fan creates an average temperature rise of 1.8-2.5 K for the current air flow rates.
- The reduction in drying time that can be expected by applying solar collectors is of the order of 30-35 %.

+ Study Center : G. Castelli
                 Institute of Agricultural Engineering
                 University of Milano
                 Via Celoria 2
                 Milan
                 Italy

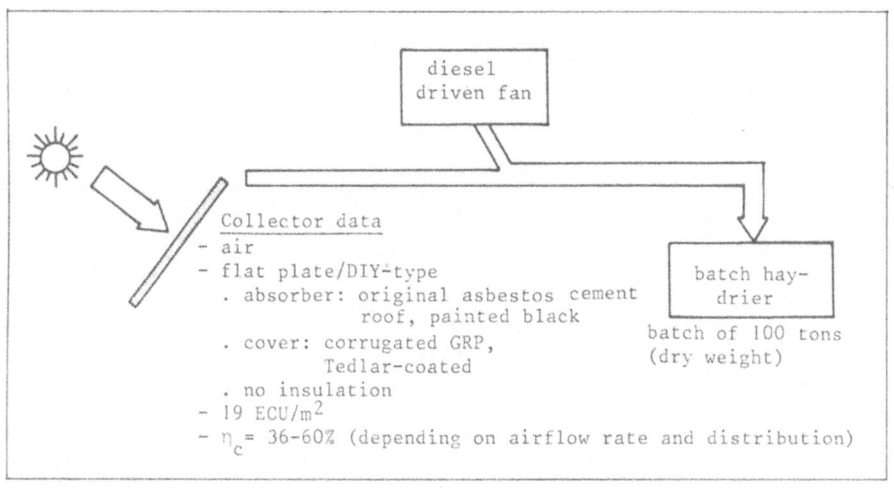

Fig. 21.1 : Data on the installation in Allessandria.

22. Fodder and cereal drying

+ Project in Torino, Italy.

+ Description (see fig. 22.1) :
  80 m² of roof collectors are used to help heat up the drying air for a
  fodder and cereal drying plant with a drying capacity of 1 ton/day.
  The drying air is first preheated by the collectors, then passed over
  a 50 m³ pebble bed heat storage, (with heat capacity of 1.54 MJ/(m³
  K), suitable to meet the energy needs of the drying plant for 3-4
  days, when necessary).
  Depending on the needs, the air leaving the storage bed is sometimes
  mixed with ambient air, or sent back to the collectors, to further
  heat the solar storage.
  After post-heating (if necessary), the air is blown through the drier
  at a rate of about 0.42 m³/s.
  This experiment tries to assess the potential of this application of
  solar energy use in the North of Italy.

+ Conclusions :
  - Of the average 1 400 MJ/day falling on the collectors, 60 % is put to
    use in heating up the air. This amount of energy is about 53 % of the
    total energy need.
  - 80 m² seems to be an optimal collector surface for the given
    drier-capacity.
  - Incorporation of a storage bed proved to be inefficient and
    problematic :
    - The storage requires an accumulation of heat over 7 days in order to
      produce less than 1 day's worth of heat.
    - It proved impossible to keep the pebble bed dry.

+ Study Center : Lisa/ M. Gioco
                 C.N.R. Laboratory for Mechanization in Agriculture
                 Torino
                 Italy

  *Note : The same study center has also conducted experiments with
          different collector types (e.g. : black tube collectors) with the
          intention of reducing the energy costs of forage drying and other
          energy requiring functions.*

23. The use of solar energy for bulb conditioning and soil heating

+ Project in Anna Paulowna, Netherlands (since 1981).

+ Description (see fig. 23.1) :
  700 m² of air collectors are used to help meet the heat demand in a
  hyacinth growing farm. The collected solar heat is transferred via an
  air to water heat exchanger to 2 storage tanks of 6 m³ each.
  This heat is being used :
  1. from April to June for soil heating (18-60°C).
  2. from June to November for heated ventilation of the stored bulbs
     (20-30°C).

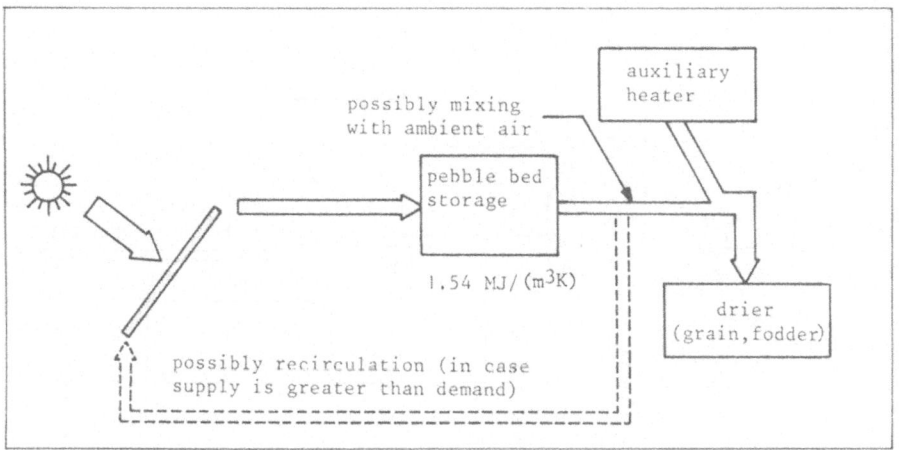

Fig. 22.1 : Data on the installation in Torino.

3. from November to April for floor heating of the work area.
   The project has been monitored over a 1 year period, in order to gain
   experience with the system, and to be able to develop a simulation
   model that should help design and evaluate optimized systems.

+ Conclusions :
  Only soil heating provided positive results, due to its low
  temperature and high load.
  Since the ambient temperature variation is in phase with the solar
  supply, ventilation air heating at such a low temperature could only
  occur in combination with storage; hence low efficiency.
  The supply of solar energy was too small during winter for the solar
  floor heating to operate.
  Even though this application, strictly speaking, does not fully fit
  under the heading of "drying", it does reveal quite a few of the
  problems that have to be dealt with in using collectors for several
  purposes (typical for solar drying).

+ Study Center : D.E. Brethouwer
          Institute of Applied Physics TNO-TH
          Stieltjesweg 1
          2628 CK Delft
          or   P.O. Box 155
          2600 Delft
          The Netherlands

+ For detailed description, cf. p. 289.

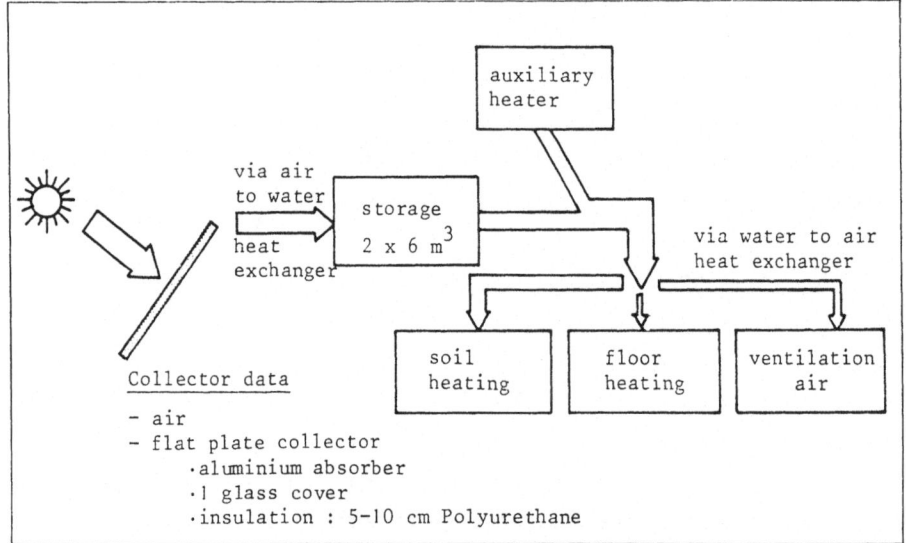

Fig. 23.1 : Data on the installation in Anna Paulowna.

24. Solar heating for barn hay drying

+ Project in Nottingham, United Kingdom (since 1979).

+ Description (see fig. 24.1) :
A comparison is made between 2 identical 30 ton-batch driers for pre-dried hay bales. One drier uses ambient ventilation air, while to the other one, a greenhouse collector of 120 m² ground surface is added for purposes of ventilation air heating.

+ Conclusions :
- The average collector efficiency over the testing period of 1979-1981 was 27 % for air flow rates of 7 m³/s and an air temperature increase of 1-2 K. In a reasonably good summer (1981) the average heat input into the collector during operation was around 10 kW.
- Key factors in future research will be the increase of collector efficiency as well as of its utilisation rate.
  . Collector efficiency can be increased :
    1. by extra insulation of the greenhouse (ground insulation with straw or polystyrene or by using a double cover).
    2. by improving the operating modes : The net radiation into the collector is **negative** from 1 hour before sunset till 1 hour after sunrise. Making use of the collector at those times gives negative results in comparison with the ambient air drier !
    The accurate determination of the end of the need for drying is another factor of importance in making appropriate operation decisions.
  . Utilisation rate :
    Since storage drying is a current practice in the UK the collector is presently only used during 1 drying run (8-10 days) per year.

In order for the collected solar energy to be economically
competitive with energy supplied by electricity, the utilization of
the solar collector in 4 drying runs per year is a strict minimum.
Therefore a study will be made of the potential of distribution
ducts or portable collectors in helping to use the solar heat in
different storage driers, possibly for different products. (The
greenhouse could also be used as a storage area for machinery.)

+ Study Center : J.A. Clark
                University of Nottingham
                School of Agriculture
                Sutton Bonington
                Loughborough
                Leics LE 12 5RD
                Great Britain

+ For more detailed description, cf. p. 252.

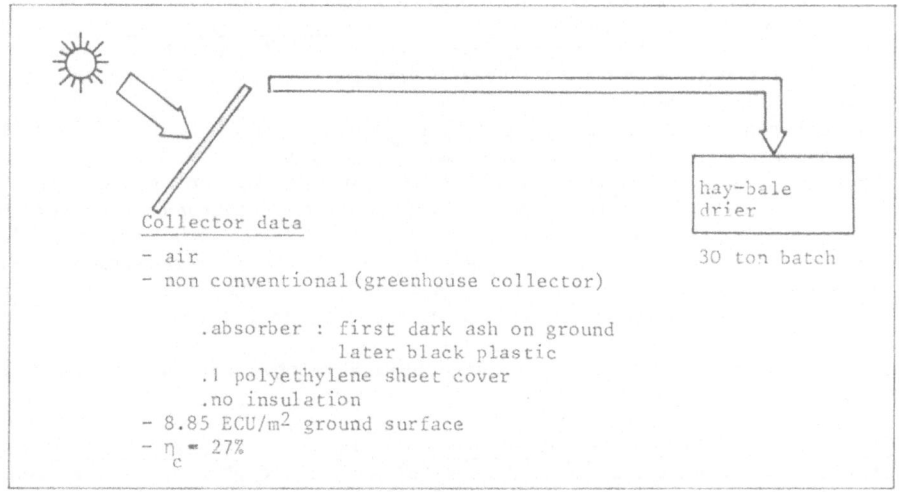

Fig. 24.1 : Data on the installation in Nottingham.

3. GENERAL CONCLUSIONS FOR SOLAR DRYING

3.1. Considerations

The character of current research on solar drying is certainly worth a
few comments. The quality and content of the basic descriptions would
suggest a different researcher than for instance in the application field
of solar water heating. Indeed, it is not in the first place the
rigorously scientific, fundamentally oriented university type researcher,
but the more practical, application oriented developer from an industrial
and rural background, who dominates the scene.
General involvement of this type of person strongly suggests a very high
potential for economic feasibility and for quick development of
marketable solar drying installations.

Typical of low-temperature drying is the simplicity of the applied solar
set-ups :
- In the desired ventilation air temperature increase range of the low
  temperature drying technique (5 - 10 K), cheap do it yourself collectors
  proved to yield efficiency values (40 - 60 %) comparable with the
  significantly more expensive "medium temperature plate collectors". (cf.
  figure in Basic Description (B.D.) nr.4).
- In different experiments where a heat storage was built in, this extra
  expense was found to be uneconomical (cf. B.D. nr.17 and 22).
- The simple air collectors are definitely preferable for drying over the
  more complex and costly water collectors (water-tightness requirement,
  frost danger, extra weight -load on roof...) (cf. B.D. nr.20 ; 21).
  Nevertheless an air collector has some specific design requirements (in
  relation with pressure drops, air patterns and leaks) which have often
  been overlooked in the past (cf. B.D. nr 3; 5; 8; 16; 21).

Different simple collector types have been developed for "low temperature
drying technique" applications.

- A variety of cheap roof collectors with plastic transparent covers were
  tested and often found to be successful both as to physical and economic
  feasibility (cf. B.D. nr. 1; 4; 5; 8; 11; 21).
- According to some (cf. B.D. nr. 9) roof collectors without cover, but
  instead with bare plate absorbers in galvanized steel or painted
  asbestos cement (cf. B.D. nr. 10) under which the heating air is
  conducted, would be less costly and preferable in regions with little
  wind. In this case, the plate, placed a little distance underneath
  the absorber in order to help form the collector air duct, could be
  made out of cheap insulating material and so prevent condensation on
  roofs of animal houses during winter.
- Greenhouse and tube collectors are another area of experiments (cf. B.D.
  nr. 6; 7; 9; 11; (15); 19).
  These low price, easy to place collectors with reasonable performance
  and at least theoretical potential for multi-purpose use (e.g.
  possibility for horticulture' or storage of equipment outside the drying
  season), reveal some significant drawbacks however:
  + large space requirement
  + fragility: the plastics are easily torn (resulting in air-leaks), and
    have a short life-span (± 3 years).
  + problems with wind and snow: these collectors are sensitive to wind,
    and in areas with snowfall they need to be taken down; hence high

labour inputs).
Ways to improve the greenhouse collector efficiency are suggested in
B.D. nr.9; 24.
- Other interesting new and simple collector types can be found in B.D.
nr. (3); 11; 12; 14; (15); 19; 20.

*Note: The heat production of a diesel driven fan placed behind the
collector was found to be a significant asset, allowing for drying,
even at night. (cf. B.D. nr. 4).*

The simplicity, characteristic of the above solar drying systems seems to
blend well with the socio-economic environment in agro-climatic zones I
and II, a privileged area as far as solar energy supply is concerned.
The task of finding several adapted applications for a solar system
(covering different periods of the year; with heat demarus in the same
range; each one compatible with the solar system as well as located
near the collectors ...) is very complex and in many cases even
impossible (cf. B.D. nr. 20; 23).

This limitation will probably have a significant impact on how far north
the application of solar drying will remain economically competitive.
Going north means less sun and a shorter drying season. However, it is
precisely in the more Northern parts of Europe, with their variable and
often poor weather conditions, that an artificial (solar) drying system
could be a great asset. Indeed, with solar drying - of hay for example -
the need for good weather for proper quality drying is reduced from a
period of 1 week to a few days only).

Note: In the given working ranges of temperatures and air flow rate,
      experts estimate there is little health danger in the use of
      asbestos cement plates as collector materials.
      For absolute safety, use of coated plates, or application of
      coating is however advised. *(1)*

3.2. Product specific evaluation

3.2.1. Hay

In - bin solar post - drying by means of slightly heated air is one of the
most promising application fields for solar energy. It incorporates all
the above-mentioned advantages of the low temperature drying technique
and its application can significantly reduce the weather dependence of
the hay production.
The potential is situated especially in agro climatic zone II (mainly in
the mountainous areas of eastern France, southern Germany, Switzerland
and northern Italy) where controlled in-bin drying is currently
practised, and where the number of operational solar installations is
constantly rising : witness the great variety of described "on farm
operational systems" in France (cf. B.D. nr. 6; 7; 8; 9; 10).
The construction of 150 solar installations in Switzerland over the period
1976-1983 is another clear indication.

Different elements contribute to the success of solar hay-drying in this
zone :

(1) Lammerant, Eternit - Personal communication, 1986.

- As opposed to agro climatic zone I, the weather is sufficiently variable to produce a risk of poor harvests when open air drying is practised.
- Nevertheless the solar supply is relatively high in comparison with that in agro-climatic zones III and IV.
- Furthermore in zone II (e.g. in certain areas, based on prescriptions connected with quality label) general practice is directed towards hay rather than silage production (a trend, opposed to what happens in zones III and IV).
- The developing practice of hay making at an increased frequency (up to 10 times per year) not only allows for a better utilization of the drier capacity, but is also accountable for an increased hay quality. Indeed, grass mowed in an early stage of the maturation process is very rich in nutrients. In practice, the high feeding value of this short cut grass, combined with optimal processing conditions, result in a hay quality that can lead to production values of 6 000 litres per year for cows fed exclusively on hay ! *(1)*
- The easy integration of cheap solar collectors in barn roofs is another factor adding to the potential of this application.

Even though in agro-climatic zones III and IV a trend towards silage making exists, hay-drying is still practised on several farms, where the integration of cheap solar systems could be profitable. In Sweden, with a zone IV type climate for instance, about 100 solar barns are actually operative.

3.2.2. Raisins

Simple solar driers for raisin production (from $\pm$ 80 % mc to $\pm$ 24 % and in some cases even below 14 % mc), in combination with alkaline treatment of the grapes, are reported to be economically competitive with traditional drying methods, due to increased quality and lower risk for reasonable costs. (cf. B.D. nr 19).

3.2.3. Grain

Also solar low temperature grain drying (from in the worst cases more than 30% m.c. to $\pm$ 15% mc) is reported to have some good chances of success in certain areas (cf. B.D. nr 11).
Its application might however remain limited to certain climatic regions (mainly agro climatic zone II) as well as grain species.
Indeed, unless sufficiently heated air is provided in a consistent way (importance of climate), and unless the initial moisture content of the grains is sufficiently low (impact of grain species and climate), it will be difficult to prevent mould development when using the low temperature drying technique. Given also the limited time span of the drying season, "multipurpose use" needs to be made of the system in order to maximize the utilization of the drier over time.
The extra attention required for such a system at a time when manpower is short is a significant drawback of this option, especially considering the fact that the drying cost only amounts to 1-5% of the total grain production cost.

(1) How this solution compares with the yield, nutritional value and cost per hectare of well fertilized green crops, like maize for instance, is a different question, certainly worth being addressed.

In agro-climatic zone I cereals are generally at good moisture content
levels when harvested, except for rice and maize which are worth-while
objects of study in the solar application context.
As for agro-climatic zone III, the impredictability of the solar
conditions is expected to be a prohibitive obstacle in many cases.

3.2.4. Others (apricots, apples, spices, wood, wine lees, tobacco,...)

Certain other applications seem to be promising. However, given the fact
that less research has been done, and at the same time, the above
mentioned conditions are often less fully met, the potential of these
applications is not always so evident.

4. PROSPECTS FOR FUTURE RESEARCH

First of all, there is clear value in the acquisition of more fundamental
knowledge on the drying process itself.

As to those products for which significant questions still remain, (E.g.
wood, tobacco, wine lees, etc. ...), the non-covered areas of research,
corresponding to the criteria to be considered, need further
investigation.
Especially for wine lees drying (cf. B.D. nr 15) and bulb conditioning
(cf. B.D. nr 23), one would think that before further research in this
area is done, a clear assessment should be made of the importance of the
energy demand on a macro-economic basis.

Furthermore, as to the more clearly promising applications (hay, raisins
and grains) several steps still need to be taken.

1. Research in the area of spreading the use of the solar system over a
   longer period of the year seems worthy of further investment.
   - Investigation of the applicability throughout the E.C. of the highly
     promising technique of hay-drying 7 - 10 times per year, a technique
     in which the drier is put to use intensively throughout the haymaking
     season.

     Note:
     *Even given the labour saving possibilities of pneumatic hay
     transportation - made possible by the short length of the grass -
     the required labour for this application is still relatively
     high!*
     *However, given the current rates of milk-overproduction in the E.C.,
     authorities are taking measures of production control. Such measures
     inevitably imply limitations on the labour requirements of the milk
     producing farms. The above described more labour intensive technique
     could put the freed-up labour to valuable use, since it enables the
     farmer to produce more of his feedstocks himself, and thus also to
     reduce his dependence on other food producers.*

   - Analysis of the impact that cooperatives (multi-users) can have in
     extending the potential of optimized multipurpose drying
     installations (so called "drying cabins").
   - Evaluation for European conditions of the "desiccant system"
     (cf. p. 173).
     The advantages of the desiccant system need to be weighed against

factors like :
- higher investment costs (E.g. the investment in grain stored over 1 year;
- quality loss of the stored overdried and rewetted grain.
2. Even though support, enabling finalization of different currently running research projects as well as enabling more specifically oriented investigations like the ones suggested above, is certainly worth while, at this point it seems that especially for hay- quite a wealth of both practical experience and fundamental knowledge is already available.
The integration of all this experience and knowledge into adapted and optimal systems seems to be one of the next steps worth taking. (The project in Savoie, France (B.D. nr. 9) is an example of a step in this direction).
3. It seems that sufficient data are available, or will be soon, to allow for a detailed and quantitative analysis of the climatic and socio-economic regions in which the different drying applications - with their specific feasibility characteristics - would be economically feasible.
Such a study would certainly provide a more reliable and even quantitative basis for market development.

In this context, it would not be out of place to consider the potential benefit of including the developing countries. Indeed, given certain adaptations to allow for differences in products to be dried, as well as in socio-economic environment, this might prove to be a significant means of market expansion, as well as a source of further experience in this area. At the same time, simple solar drying techniques could help the local inhabitants produce higher quality agricultural products with less risks ...
(cf. project in Talence, France, B.D. nr. 12).
4. The final step of solar drying development seems to be that of communicating the acquired information to the actual user.
This step could include several activities.
E.g. - development of computer aided design programs, incorporating the acquired knowledge, in such a way as to be usable by study-centre personnel, without requiring detailed insight into the contents of the program itself.
- counselling
- effective information packaging
(at the University of Perpignan, France, several of these activities are already being developed, cf. table 2 on p. 181)

APPENDIX TO PART II
Some specific research projects on solar drying

In this appendix, 7 solar drying projects are described in detail. They were chosen to represent as far as possible the wide range of the research on this subject.

1. SOLAR DRYING OF GREEK FRUITS - p. 224
   (A Greek project, representing the 'physical feasibility' stage of the research)

2. THE USE OF SOLAR AIR COLLECTORS FOR FORAGE AND CEREAL DRYING - p. 239
   (A Belgian project, as an example of the search for cheaper collector types for drying of agricultural products)

3. SOLAR HEATING FOR BARN HAY DRYING - p. 252
4. DEVELOPMENT OF A SOLAR DRIER FOR WINE LEES - p. 263
   (Examples in the U.K. and France of quite original collector types)

5. SOLAR ENERGY FOR PRODUCE DRYING AND MILKING PARLOUR WASHING - p. 274
6. THE USE OF SOLAR ENERGY FOR BULB CONDITIONING AND SOIL HEATING - p. 289
   (Italian and Dutch projects, representing the area of "multi-purpose" use of solar collectors)

7. SOLAR DRYING OF GRAINS - p. 301
   (A French project more explicitly covering several of the above mentioned areas : collector development, experiments with multi-purpose use, optimization of the solar drying system as a whole)

The reporting formats used to compile the detailed descriptions are explained in the ADDENDUM, cf. p. 467. The numbering of the subheadings in the detailed descriptions corresponds to the reporting format.

1. Solar drying of Greek fruits

Project location

ATHENS, GREECE.

Study Center :
  - Project leader : Prof. G.D. Saravacos
  - Institution :   Laboratory of Unit Operations
                     National Technical University
                     Parission 42
                     106 82 Athens
                     Greece

Running since 1983

Project description

An experimental solar drier has been designed and constructed at the Technical University of Athens. It consists of a 3 m² flat-plate solar collector for air heating, a storage bed of concrete balls of 250 kJ/K thermal capacity, and a cabinet tray-drier with a capacity of up to 8 kg of fresh fruit.
Two modes of operation were tested :
- Direct connection of the solar collector to the air drier.
- Connection of the solar collector, the storage bed and the air drier in series.

Experiments were performed during the drying season (August to October) of 1983. These experiments gave useful data on the efficiency of the collector and the storage bed, and on the feasibility of solar drying of Sultana grapes.

Goals of the experiment :

1. To evaluate solar drying of typical Greek agricultural products. In this first stage, Sultana seedless grapes (raisins) were considered.
2. To evaluate the air-collector efficiency under various operating conditions.
3. To evaluate the dynamic behavior of the heat-storage bed and its effect on the overall efficiency of the installation.

Scheme

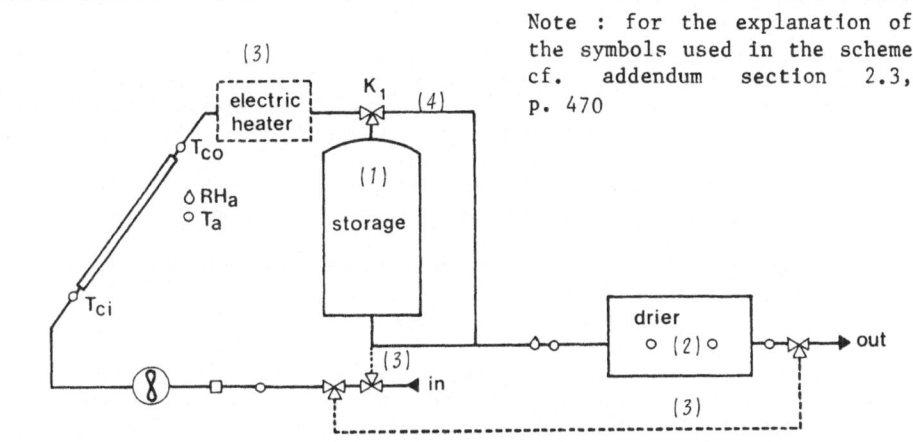

Note : for the explanation of the symbols used in the scheme cf. addendum section 2.3, p. 470

(1) A total of 10 temperature sensors were placed in the storage bed. Half of them measure the air stream temperature, while the other half measure the temperature of the concrete balls. For temperature sensor locations in the storage bed, cf. section 2.4.of this description.

(2) The dry solids content of the product was determined ; at regular time intervals, the weight of the product was also measured.

(3) The components of the system given in dotted lines were not put to use in the described experiments.
  - The electric heater was installed, to enable the simulation of certain patterns of the heat input into the storage, in order to assess storage performance and behaviour.
  - The other dotted components would allow for storage loading (when no load is present in the drier), or air recirculation.

(4) The 3 way valves shown in the diagram are symbolic. In fact, the system is set up in such a way that the different options for the flow circuit can easily be materialised. This action would however involve more than just the switching of a few valves.

Operating modes :
In some experiments, the air is blown through the storage bed (cf. sections 5.2. and 5.7.) ; in others, the storage bed is by-passed (cf. section 5.5).

Reference solution :
There are different other options for drying grapes.
- Open air drying.
- Driers, applying forced-fuel heated-air ventilation (temperature + 50°C)
- Drying in greenhouse with natural air circulation (drying time 8-18 days even when dipped in alkaline solution).

In the given project, no quantitative comparison was made between the solar drier and one of the above mentioned solutions.

Conclusions

- The solar drying experiments with alkaline treated sultana seedless grapes, over the grape drying season of August to September 1983, yielded good quality raisins within 3-4 days (30-40 daytime drying hours for a change in moisture content from ± 80% to ± 25% m.c.).

- The drying air temperatures ranged between 30 and 62°C. For those temperatures, the average collector efficiency was about 40%.

- The storage bed smoothes out the temperature variations, and because of its storage characteristics, the daily duration of time over which heated air is being blown into the drier is being extended by a few hours.
  2% of the solar heat is being lost in the storage. The advantages of the storage bed, listed above, do not seem to outweigh this loss-factor as well as the added cost for the storage bed, so that leaving out this extra element seems desirable.
  (A possible alternative would be to use the storage in parallel with a by-pass, in such a way that it is only loaded when solar supply exceeds the demand, and discharged when solar supply is insufficient).

- This same system can be applied to other fruits that require drying over summertime. E.g. : apricots (drying season June-July).

- Future work :
  - More experimental work will be done, in order to establish the optimum regulation of the main parameters of the drying process.
  - A larger scale project will be set up, and will help assess in a more accurate way the economic feasibility of such an installation.

PROJECT ANALYSIS

---

1. Site and climate

1.1. Site

Latitude :	37°58' N
Longitude :	23°45' E
Altitude :	107 m
Nearest main city :	Athens
Distance from main city :	0 km
Obstructions :	None

1.3. Annual long-term averages *(1)*

1. Prevailing wind direction	W *(2)*	NNE/NE
	S	NE/SSW/SW
2. Average wind speed	W m/s	2.3
	S m/s	1.9
3. Total precipitation	mm/yr	408.2
5. Global irradiation on horizontal plane	MJ/(m^2.yr)	*(3)*
8. Hours sunshine	hrs/yr	2 871
9. Ambient temperature	C	17.5
10. Average max. temp. (July)	C	27.0
11. Average min. temp. (Jan)	C	8.9

12. Meteorological station :	Meterological Institute National Observatory of Athens
13. Latitude :	37°58.3' N
14. Longitude :	23°45' E
15. Altitude :	107 m
16. Distance from site :	1.5 km

(1) The long-term averages were calculated from data collected by the Greek Meteorological Institute, and published in the "Climatological Bulletin". The averages represent the years 1974-1980.
(2) W = Winter (October-March)
   S = Summer (April-September)
(3) The long-term annual average according to the map on p. 5 is $\pm$ 6200 MJ/(m^2.yr) (N.O.T.E.).

---

2. Plant

2.2. Drying-place

1. Drying system : Experimental cabinet tray-drier, connected to the solar air-heater.
2. Material : Sheet-iron, insulated with glass-wool, 3 cm thick.
3. Dimensiosn : cf. diagram

4. Capacity : 8 kg of fresh product (maximum)
5. Ventilation - Power : 1.2 kW
              - Flow rate : 0.66 m³/s
6. Conventional heating source *(1)* : Electric resistances
   calorific power : 3 000 W (max)
7. Diagram : Cabinet tray-drier (all dimensions in mm)
   cf. figure 2.2.7.

(1) The conventional heater was not used in the given experiments.

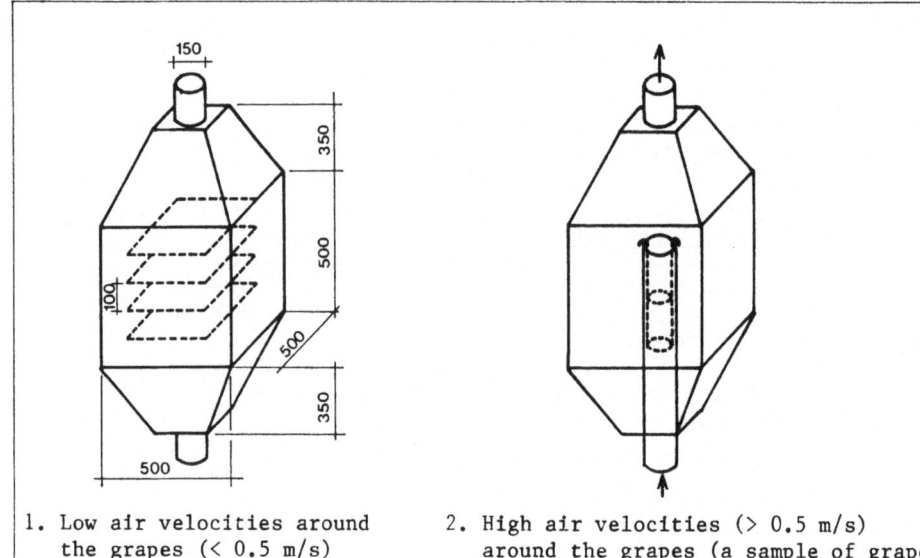

1. Low air velocities around   2. High air velocities (> 0.5 m/s)
   the grapes (< 0.5 m/s)   around the grapes (a sample of grapes
                                     is kept in the central tube only)

Figure 2.2.7. : cabinet tray drier

2.3. Collector (Flat-plate collector)

1. Absorber
   - Coolant :   Air
   - Material :   0.5   mm thick aluminium foil
   - Aperture area :   3.0   m²
   - Absorber area :   3.0   m²
   - Working pressure : 10^5 Pa
   - Flow rate :   3.3-20 1/(sm²)
   - Emittance of plate :   0.88-0.90
   - Coef. of absorption : 0.94-0.98
     of visible light

Note : The air that is being heated up, flows between absorber and
        insulation. The air in between absorber and cover functions as
        insulation.

2. Glazing
   - number of glass covers : one
   - material : glass
   - special characteristics : window glass with greenish edges

(3 mm thick)
3. Insulation : Glass-wool, 5 cm (bottom)
   Polyurethane, 1 cm (sides)
   Frame : plywood with aluminium coating.
4. Mounting
   - Orientation : North-South
   - Tilt : adjustable (26° during the drying test of 1-9/9/'83)
5. Collector performance (experimental data) cf. figure 2.3.5.

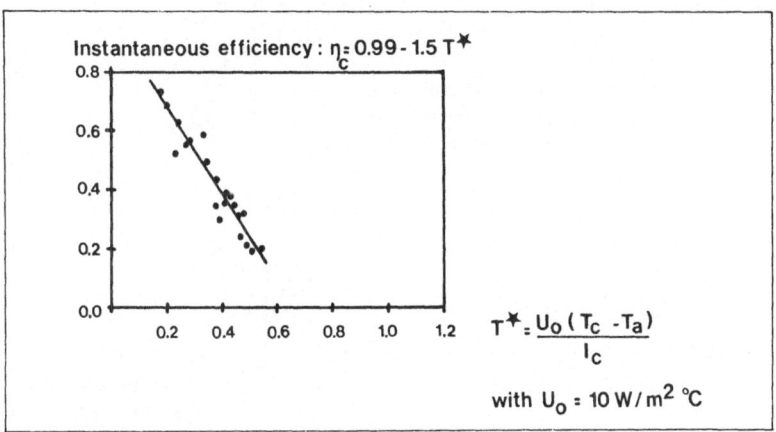

Figure 2.3.5. : Collector performance

The instantaneous collector efficiency η_c is defined as

$\eta_c = \dfrac{Eco}{S_c I}$ as a function of T^*

with Eco : useful power extracted from the collector (W)
   I_c : incident solar radiation on collector (W/m^2)
   S_c : aperture area (m^2)

$$T^* = Uo \; \frac{Tc - Ta}{I_c} = 10 \; \frac{Tc - Ta}{I_c}$$

   Tc : average temperature of fluid in collector (°C)
   Ta : ambient temperature (°C)

η_c : 0.99 - 1.5 T*

2.4. Heat storage

1. Medium of heat storage : concrete (cement) balls. Concrete mix :
   Portland cement 100 g, sea sand (particle size 1-2 mm) 100 g, and
   water 45 g.
2. Description, location in the system : storage bed, next to
   collector. 2 900 concrete balls, 4 cm in diameter. Total weight
   210 kg. Bed of balls supported by perforated metal plate.
3. Insulation :
   Armaflex sheet, 10 mm thick. Thermal conductivity 0.035 W/m K.
4. Dimensions :
   Cylindrical bed of volume 200 l. Base diameter 0.5 m, height
   1.0 m. Cylindrical plenum of diameter 0.5 m and height 0.15 m.
5. Thermal capacity : 250 kJ/K
6. Overall heat loss coefficient : 5.0 W/K (calculated)

7. Diagram : cf. figure 2.4.7.

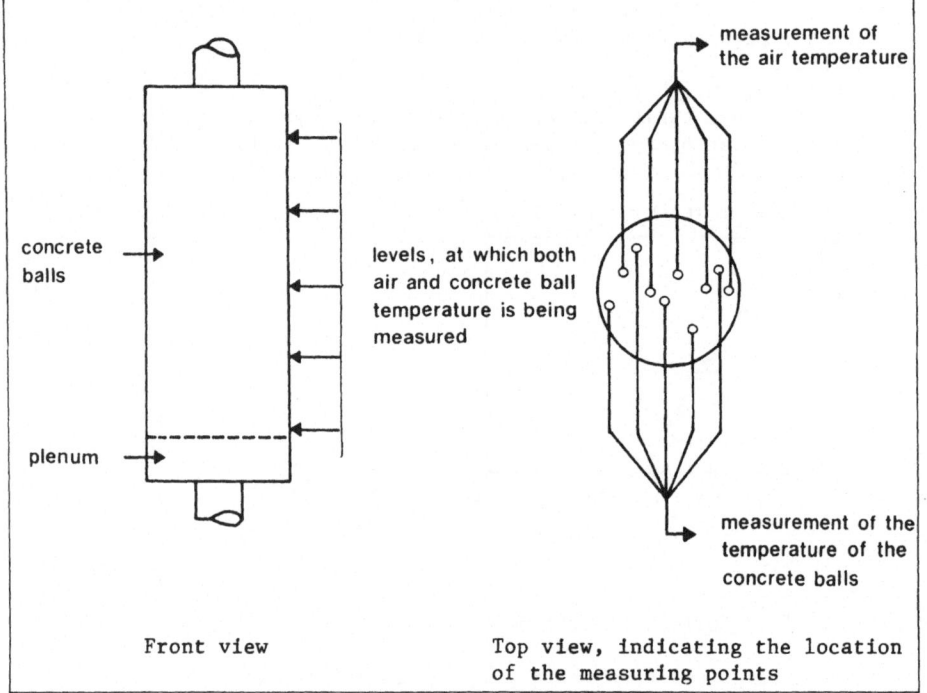

Front view   Top view, indicating the location
of the measuring points

Figure 2.4.7. : Diagram of heat storage

2.5. Installation parts and measuring devices

1. Technical description of the installation parts :
   - Electrical fan :   Capacity 0.65 m³/s
                    Power 1.2 kW
                    Static pressure : 640 Pa
                    1287 RPM
   - Electric heater : 3 electrical resistances
                    (1) 220 V - 1 000 W
                    (2) 220 V - 1 000 W
                    (3) 0-220 V - 0-1 000 W

                    total maximum : 3 000W

2. Description of the sensors
   - Temperatures were measured with two types of thermocouples :
   - Copper-constantan

or   - Iron-constantan
        Precision of temperature measurements : $\pm$ 0.1 K
        Recording of temperature in a Data logger (Autodata Ten/5, Acurex Corp).
        Measuring frequency varying from 10 to 60 min.
   - Flow rates of air were measured with a Pitot tube, with direct reading of the pressure on an inclined manometer.
        Measuring frequency, every 60 min. Precision of measured air

velocity + 5% in the range of 0 - 12 m/s.
- Solar radiation was measured with an Eppley Precision Pyranometer, model PSP. Measuring frequency varying from 1 to 10 min. Precision of 0.5% at radiation fluxes up to 1 400 W/m^2.

---

3. Products

3.2. Drying-products

1. Designed for special products : During the first operating season, Sultana grapes (raisins) were dried.
2. Required conditions of inlet air : Air velocity 0.5 - 2.0 m/s, temperature 45-55°C, low relative humidity.
3. Properties of the products : Sultana grapes (seedless raisins) from the Corinth area. Samples of grape berries, separated from the stems, 600 g per tray (2 trays per experiment).
   Sugar content of fresh grapes 18° Brix (by refractometer).
4. Other products which could be dried :
   Fruits : prunes, figs, apricots. Vegetables : potatoes, peppers, onions. Cereals : maize.
5. Any special treatment of the crop ?
   The Sultana grapes were dipped in a hot solution (90°C) of sodium hydroxide (0.5%) for 10 seconds. Afterwards they were rapidly cooled and washed with cold water.

---

4. Economic aspects

1983 prices

4.1. Economic environment *(1)*

1. Energy price of oil	ECU/l	0.32
2. Energy price of electricity	ECU/kWh	0.064
3. Energy prices evolution prospects		
– of oil or gas	%/yr	15
– of electricity	%/yr	15
5. Interest rate of loans for farm equipment (rate without subsidy)	%/yr	14
6. Subsidies available for solar and competitor farm equipment (as capital grant or interest rate reduction)	% or ECU	none

(1) The given values were adopted from a report of another researcher in the same city (N.O.T.E.).

4.2. Investment expenses *(1)*

2. Cost of equipment :		
- cost of collector per m^2	ECU	64
- expenses for buildings	ECU	none
- cost of regulation equipment	ECU	none
- other material cost	ECU	229
3. - Time of personnel for the conception, installation and start-up of equipment (time unit)		1 man year
- Cost of personnel for the conception, installation and start-up of equipment	ECU	6 400
4. Lifetime of the equipment (evaluation)	years	10
5. Residual value of the investment	ECU	None

(1) Data available from the actual experimental set-up

4.5. Rough estimation of the potential economic performance of a hypothetical operational scale solar drying installation

(Based on the experience with the experimental solar drying system at the National Technical University of Athens during 1983). The assumed set-up is a solar installation, operated by the owner of a small farm, drying his own grapes.

Solar Drying Installation
- 10 flat plate solar collectors, positioned in parallel (3 m^2 surface each). The solar heated air is directly blown into the tunnel drier by a centrifugal fan (no thermal storage).
- The tunnel drier has a cross-section of 2 m x 2 m and a length of 5 m. It has a capacity of 1 000 kg of fresh grapes.
  (40 trays of 1.5 m^2, carrying 25 kg of fresh fruit)
- The drying time is 4 days (40 hours of active solar drying).
- The initial moisture content of the grapes is 78% m.c. (wet base), the final moisture content is 16% m.c. (wet base).
- Production of dried raisins : 262 kg/4 days.
- The installation is operated 2 months/year (August/September, which is the raisin processing season in Greece).
- Assuming operation of 28 days/month, the toal amount of processed grapes is 14 000 kg, resulting in 3 660 kg of raisins.

Estimation of cost and benefits
- Gross annual income :	3 660 kg of raisins sold at + 1.1 ECU/kg	3 960 ECU/yr
- Investment :	Solar collectors 30x63.7	1 919 ECU
	Tunnel drier (brick construction)	1 910 ECU
	Trucks and trays	640 ECU
	25 kW fan and electricals	1 270 ECU
	Ductwork/insulation	640 ECU
	Total investment	6 370 ECU
Total investment costs per year (depreciation in 5 years)		1 270 ECU/yr

- Operating cost :

	2 months operation/year	
	Electrical energy	635 ECU/yr
	Treatment of grapes	255 ECU/yr
	Maintenance	380 ECU/yr
	Total operating cost/year	1 270 ECU/yr

- Total solar drying system costs/year   2 540 ECU/yr
  (64% of gross
  annual income)

---

5. Results

5.1. Method of calculation

1. Calculation of the energy flows
   Solar radiation on the collector

$$Q_{ci} = {}_{t1}\int^{t2} S_c I_c(t)\,dt$$

Input solar energy

$$Q_{co} = {}_{t1}\int^{t2} \psi_g \dot{V}_g C_g (T_{co} - T_{ci}) + dt$$

Collector losses
$$Q_{cl} = Q_{ci} - Q_{co}$$

Input energy storage

$$Q_{si} = {}_{t1}\int^{t2} \psi_g \dot{V}_g C_g (T_{si} - T_{so}) + dt$$

Output energy storage

$$Q_{so} = {}_{t1}\int^{t3} \psi_g \dot{V}_g C_g (T_{so} - T_{si}) + dt = \text{Useful energy}$$

Explanation of the symbols :
- All the energy flows (Q) are given in Joule.
- S_c is the collectorsurface area (m^2)
- I_c is the solar radiation intensity on 1 m^2 of collectorsurface (W/m^2).
- t stands for time (s).
  t1 = 8 AM
  t2 = 8 PM
  t3 = 12 AM
- T stands for temperature of the air at the locations that the different indexes refer to (°C).
  the first index c refers to the collector.
                     s refers to the storage bed.
  the second index i refers to the air inlet of the element, referred
                     to in the first index.
                     o refers to the air outlet of the element, referred
                     to in the first index.
- $\dot{V}_g$ is the air flow rate (m^3/s).
- ψ_g is the specific mass of the air (kg/m^3).
- C_g is the specific heat of the air (J/kgK).

- The symbol $(........)^+$ is used when the integral is only being computed in case the temperature difference between brackets is greater than zero.

2. The moisture content, given in the second part of section 5.5., was determined by the vacuum oven method (0.5 Torr, 70°C for 24 hrs). Accuracy : $\pm$ 1%.
   For good storage stability, the final moisture content of the raisins should be close to equilibrium moisture content in the storage atmosphere (approximately m.c.= 0.20 (d.b.)).

5.2. Weather conditions and energy

1.	1/9	2/9	3/9	4/9	5/9	6/9	7/9	8/9	9/9			from: 1/9	PERIOD
2.	1983	1983	1983	1983	1983	1983	1983	1983	1983			until: 9/9/1983	
3.	27.1	28.4	29.0	28.9	28.7	27.4	29.6	29.5	28.4	Day time	°C	average ambient TEMPERATURE	
	23.8	23.5	23.1	25.5	24.6	24.5	24.6	24.7	24.6	Night time			
5.	66.4	69.6	65.7	64.9	58.7	69.7	66.0	64.4	62.5		MJ	SOLAR RADIATION on collector	
6.	18.6	20.4	20.0	28.3	23.0	25.2	35.9	33.6	32.3		MJ	input SOLAR energy	
7.	14.5	15.9	15.0	24.1	16.2	19.7	28.0	25.9	24.6		MJ	input energy STORAGE	
8.	13.1	10.9	11.4	18.5	14.4	17.3	25.4	22.9	21.3		MJ	output energy STORAGE	
9.	47.4	49.0	49.6	51.0	46.2	47.7	47.4	46.6	44.5	Day time	°C	average TEMPERATURE STORAGE	
	26.2	25.7	28.3	28.4	26.7	25.6	25.9	25.8	25.5	Night time			
10.	13.1	10.9	11.4	18.5	14.4	17.3	25.4	22.9	21.3		MJ	useful energy	

Note : The air flow rates in this drying round were varied from 72 m³/h for the first 3 days, to 106 m³/h for the second set of 3 days, to 178 m³/h for the last set of 3 days.

5.5. Drying results *(1)*

1.	14/9	23/9	29/9		from: 14/9 1983	PERIOD
2.	17/9	26/9	2/10		until: 2/10/1983	
3.	0.32	2.5	2.5	m/s	flowrate	INLET AIR
4.	30-60	25-50	30-55	°C	temperature	
5.	32	33	33	%	relative humidity	
6.	77.9	76.8	75.6	% w.b.	moisture content (initial)	PRODUCTS
7.	1.2	1.2	1.2	kg	useful quantity	
8.	v.g.	v.g.	v.g.		quality	

- Quality of dried raisins is very good (v.g.) both as to structure and colour.
- Drying data for the first 2 rounds - with very different air flow rates - (14-17/9 and 23-26/9) are given in figures 5.5.1. and 5.5.2.

(1) The periods, referred to in this section, do not correspond to the one given in section 5.2.

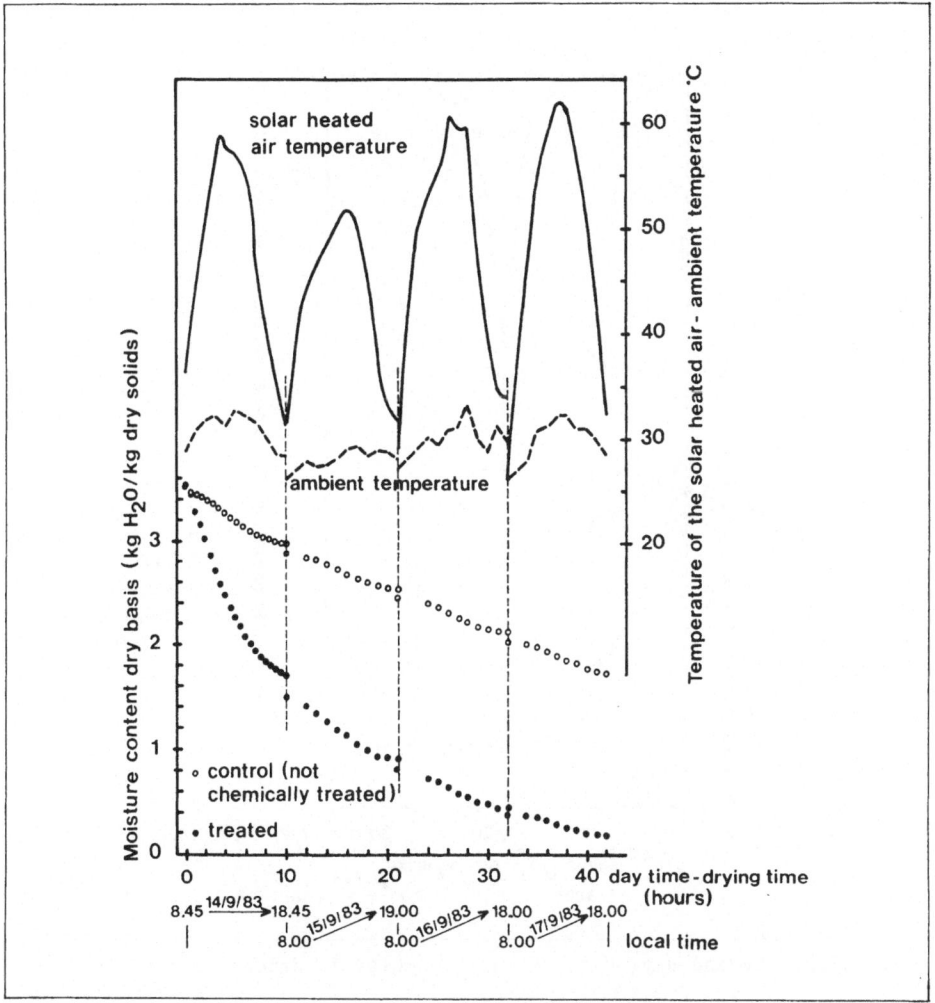

Fig. 5.5.1 : First drying round (air velocity in drier 0.3 m/s)

Note :
1. As can be seen in figures 5.5.1 and 5.5.2, the drying rate can be
   increased significantly when the fresh grapes are first dipped in an
   alkaline solution (e.g. Potassium Carbonate).
2. Even though it is not mentioned in the report, it seems obvious from
   the pattern of the solar heated air temperature, that in this case the
   storage bed is being bypassed. Indeed, if this were not so, the drying
   air would reach its maximum temperature level much later in the day,
   and would remain significantly above the ambient air temperature level
   during the first part of the night. (Compare with the course of T_{so} in
   section 5.7) (N.O.T.E.)

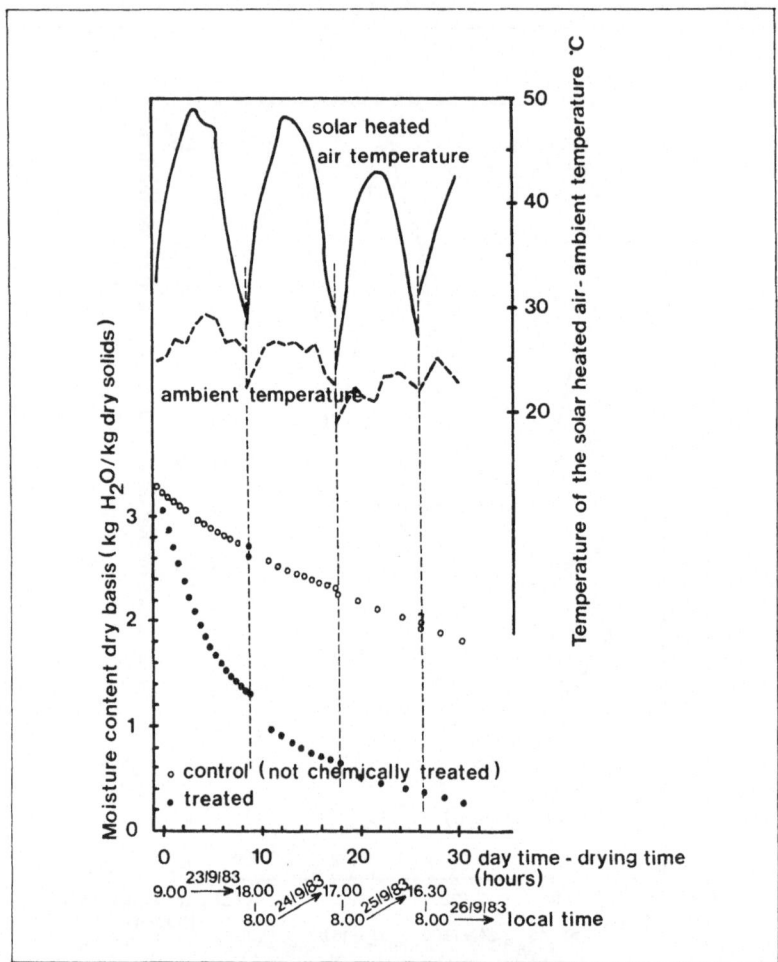

Fig. 5.5.2 : Second drying round (air velocity in drier 2.5 m/s)

5.7. An example of some measured data

The evolution over time of some values, measured on 9/9/83, are given.
(for explanation of symbols other than T_a (ambient air temperature in °C,
cf. section 5.1).

1. Variation of collector and storage temperature as a function of solar
   radiation over time (cf. figures 5.7.1. and 5.7.2).

2. Temperature distribution over time in the storage bed (cf. figures
   5.7.3 and 5.7.4)

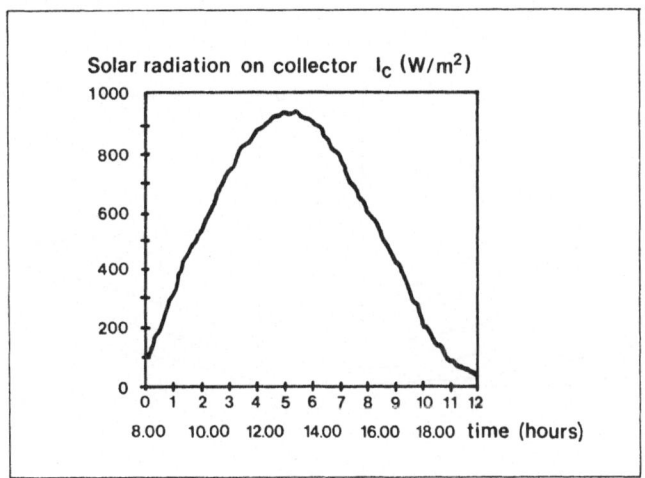

Fig. 5.7.1 : Solar radiation on collector

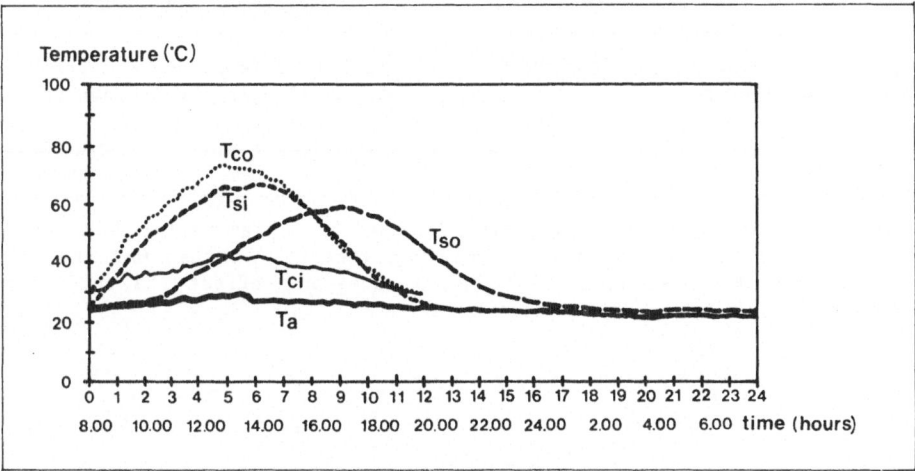

Fig. 5.7.2. Solar system temperatures over time

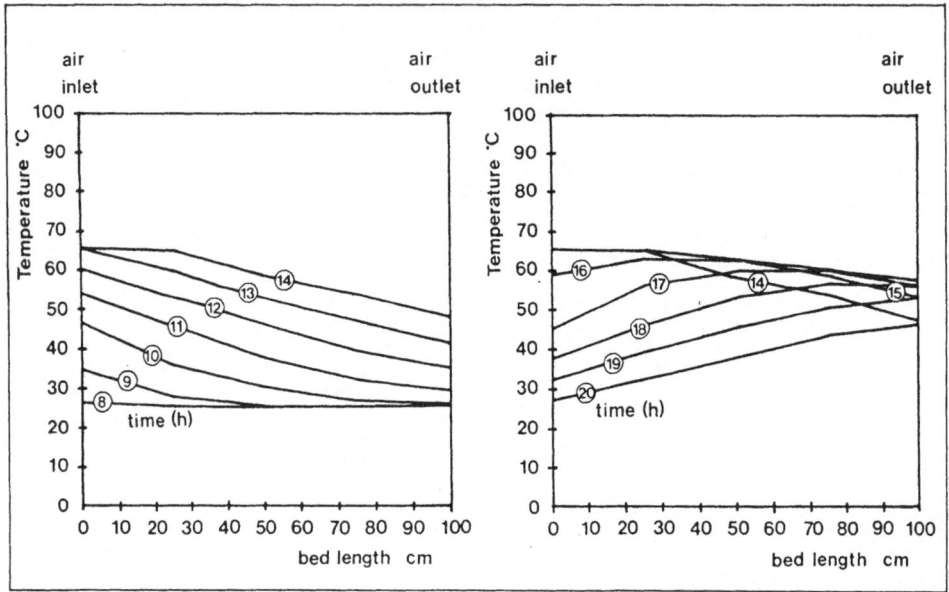

Fig. 5.7.3. Temperature distribution in charging storage bed

Fig. 5.7.4. Temperature distribution in discharging storage bed

6. Remarks

- An adaptation which would slightly increase the system performance would be to place the fan <u>behind</u> the solar collector. In that way the air temperature increase caused by the fan does not affect the collector efficiency.

2. The use of solar air collectors for forage and cereal drying

Project location

GEMBLOUX, BELGIUM

Study Center :
- Project leader : A. PLETINCKX
- Institution :   Station de Génie Rural
                    Chaussée de Namur 146
                    B-5800 Gembloux
                    BELGIUM

Research running since 1982.

Project description

This study aims at assessing the feasibility of solar grain and hay drying (in particular for the Belgian situation).

The first step of the research involved the development of air collectors, in which efficiency, cost and durability are optimally combined. With this goal in mind, 5 different collectors of 2 m² each, using cheap conventional building materials, were manufactured. (For an explanation of the composition of the different collector types, cf. section 2.3)
Their performance (efficiency and durability) was then compared under the same operating conditions.

Scheme

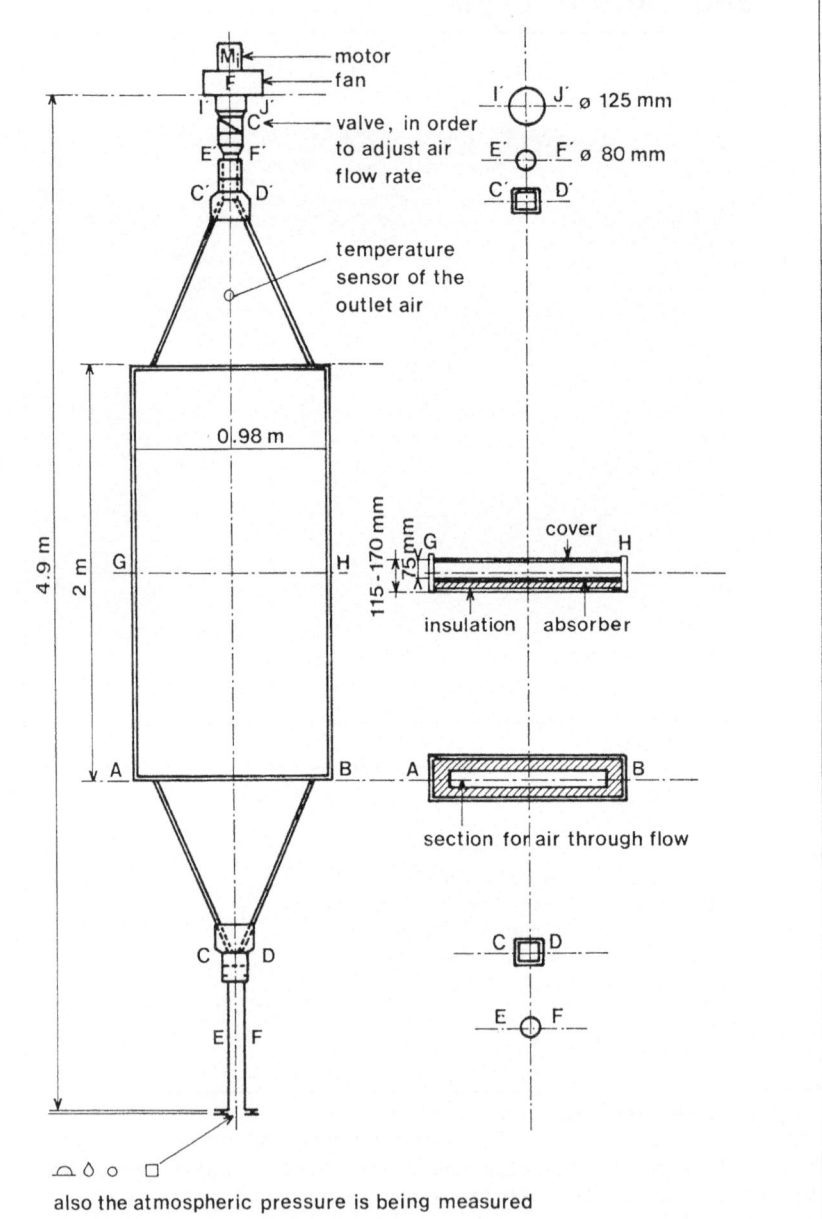

motor
fan
valve, in order to adjust air flow rate

ø 125 mm
ø 80 mm

temperature sensor of the outlet air

0.98 m

4.9 m
2 m

115 - 170 mm
75 mm

cover
insulation   absorber

section for air through flow

also the atmospheric pressure is being measured

Explanation of the symbols:
- o temperature sensor
- ◊ relative humidity sensor
- △ solar radiation intensity sensor
- □ orifice plate for air flow rate measurement

Among others, the influence of collector orientation and slant, radiation intensity, air speed and insulation ... were assessed.

A second step will involve the assessment of the performance of the collectors integrated in the drying set-up (energy savings, quality of dried products ...).

This report is concerned with the first step only.

The different collectors are all mounted on the same swivelling support.
Except for the building materials all the other parameters (e.g. air flow rate, inclination, collector dimensions) were set identically for the 5 types.
In order to be able to calculate the collector efficiency as a function of the different climatological data, the solar radiation on the collectors, the atmospheric pressure, the ambient air temperature and humidity, as well as the air flow rate and air temperature at the outlet of the collector are being measured.
Data of those variables - each one of them based on an average value of 3 to 4 readouts - are available for every 5-15 minutes, depending on the weather. Over the period between end of May and end of August, 11 to 60 of such measurements were made per day.

Conclusions

- The average collector efficiency ranges from 56-73% for the different collectors. Collectors 1 and 2 show highest efficiency values while 3 and 4 show the lowest. The temperature increase ranges from 3.5-13 K, depending on the weather. The saturation deficit of the air is increased by 50-200%.
  (All these values were measured, given a specific air flow rate of 150 $m^3/h.m^2$ collector).
  The efficiency can even be raised, when larger air flow rates of up to about 190 $m^3/h.m^2$ are applied.
  Both as to efficiency and cost, these collectors, which can be built separately or integrated as part of a roof structure, seem perfectly adapted for forage and cereal drying.

- Durability of the material is a significantly limiting factor for collectors 1 and 2, since deterioration of the absorber/insulation can already be detected after 4 months.
  This leaves collector 5 as the best choice. The price of this collector can be reduced - without significantly affecting collector efficiency - by replacing the 10 mm polycarbonate cover by a 3.5 mm one.

- A (rather low key) economic analysis predicts a positive outcome as to economic feasibility of the application of the given solar collectors for forage drying.

- (The accurate measurement of the temperature of the air coming out of the collector posed some problems :
  Good positioning of the sensor as well as insulation of the convergent channel are necessary to exclude interference from the collector wall temperature).

PROJECT ANALYSIS

1. Site and climate

1.1. Site

Latitude :   50°34' North
Longitude :   4°4' East
Altitude :   180 m
Obstructions :   No shading caused by obstructions between
                        8 AM and 5 PM.

1.3. Annual long-term averages

1. Prevailing wind direction		S.W.-W.S.W.
2. Average wind speed Winter	m/s	2-8
Summer	m/s	2-5.5
3. Total precipitation	mm/yr	816
5. Global irradiation on horizontal plane	MJ/(m².yr)	3 600
6. Diffuse proportion	%	60
8. Hours sunshine	hrs/yr	1 604
9. Ambient temperature	°C	10.9
10. Average max. temp(July/August)	°C	22.7
11. Average min. temp(January)	°C	-2.3

12. Meteorological station : Gembloux
16. Distance from site : 2 km

1.4. A few data on temperature and relative humidity variation over time

Season	Temperature during the day (°C)			Average r.h. from 8 AM to 8 PM(%)
	Average	Maximum	Minimum	
Spring	8.5	13.4	4.0	88
Summer	15.9	21.7	11.0	73
Fall	9.4	13.9	5.6	71
Winter	2.9	5.6	-0.1	83

2. Plant

2.3. Collector (Flat plate)

1. The 5 collector types consist of following materials :

Collector type	Materials (cover, absorber, insulation)
1	Glass (4 mm) ; polyurethane (40 mm) painted black ; chip board (12 mm)
2	Polycarbonate (3.5 mm) ; Eurothane B 12 (25 mm) painted black ; chipboard (12 mm)
3	Flat PVC (Greca Cristal)(1.5 mm) ; corrugated asbestos cement, painted black ; expanded polystyrene (20 mm) ; chipboard (12 mm)
4	PVC "Peveclair" with trapezoidal corrugations ; corrugated bituminized cardboard ; chipboard (12 mm)
5	Polycarbonate (10 mm) ; flat asbestos cement, painted black (5 mm) ; expanded polystyrene (20 mm) ; chipboard (12 mm)

2. Dimensions : cf. scheme
3. Flow rate : 140-180 m^3/h and per m^2 collector
5. Collector efficiency : cf. section 5.2

2.5. Installation parts and measuring devices

1. Technical description of the installation parts :
   - fans : Each centrifugal fan has a capacity of 460 W (can yield 250 m^3/h, given a pressure drop of 775 Pa)
2. Description of the sensors
   Note : Due to lack of funding the acquisition and processing of the measurements was not automated.
   - Temperature measurements ($T_{ambient}$; $T_{collector\ out}$)

   Precision mercury meters (± 0.1°C)
   - Air flow rate through collector
   Orific meter (according to ISO5167 norms)
   Pressure drop measured by a "Schiltknecht" micromanometer
      - Range : 0-2 450 Pa
      - Precision : ±2 Pa
   - Radiation on the collectors
   Kipp en Zonen solarimeter (parallel to collectorsurface)
      - gauged at the Belgian Royal Meteorological institute
      - the voltmeter reads the electrical signal from the solarimeter with an accuracy of ± 10 mV, which corresponds to ± 1 W/m^2.
   - Relative humidity
   psychrometer "HAENNI" based on 2 temperature measurements, each with precision of ± 0.1°C.
   - Atmospheric pressure
   "CASELLA" precision barometer.

4. Economic aspects

A barn drier with a capacity to dry all the forage growing on 5 ha, by means of ambient air, could yield a higher capacity, and better quality of the dried product if the incoming drying air is preheated by e.g. 4 K.

The expected economic performance of such a system, using solar energy as a heat source, is compared with different other traditional systems. (cf. table).

These results are based on following <u>hypotheses</u>.

- Installation for forage drying equipped with a 7.36 kW fan.
  Air flow under load : 12 m^3/s
  Quantity of heat, required to increase the temperature by 4 K :
  1.25 kJ/(m^3K) x 4 K x 12 m^3/s = 60 kW
- Required area of collector
  with an average daytime radiation intensity during drying season of 500 W/m^2 and an efficiency of the collector of 65%
  $$\frac{60 \text{ kW}}{0.5 \text{ kW/m}^2 \times 0.65} = 184 \text{ m}^2$$
- Cost of the collector : 22.5 ECU/m^2 (13.5 ECU for materials and 9 ECU for manpower - 1 h/m^2) *(1)*

- Energy price : oil   310 x 10^{-3}ECU/1
           electricity   100 x 10^{-3}ECU/kWh
- Oil savings : 35%
- Subsidies available : 35% of the investment for solar equipment
- Drying capacity :
  Installation with ambient air
       40 t of dried forage/year at 15% m.c.
       24 kg of evaporated water/h
  Installation with oil or oil and solar
       68 t of dried forage/year at 15% m.c.
       40.6 kg of evaporated water/h

Note : 1) The cost of the different collectors, for materials only, is
  given in table 4.1..
  Table 4.1. : Material costs for collectors.

Collector	Cost ECU/m^2
1	23
2	15
3	26
4	17
5	31

  2) A comparison of the performance of hay processing systems is
  made in table 4.2.

(1) Compared with the collector cost table(4.1.), the chosen collector cost is a very optimistic one ... (N.O.T.E.)

Table 4.2. : Comparison of the performance of different hay processing
systems.

Drying system	Yield of hay at 15 % m.c.(w.b.) (kg/ha)	Feeding value of the hay U.F.[1]/kg	Real output U.F.[1]/ha	Total[2] cost (1982 values) 10⁻³ ECU/U.F.
Field drying				
- bad conditions	4 000	0.3	1 200	491
- mean conditions	6 000	0.4	2 400	246
Barn drying				
- ambient air	7 500	0.5	3 750	178
(capacity 5 ha/yr)				(energy costs 19)
- heated air				
(capacity increased to 8 ha/yr)				
- oil	8 500	0.6	5 100	172 (energy costs 44)
- oil + solar	8 500	0.6	5 100	174 (energy costs 34) [3]
Pre-dried ensiling	8 000	0.48	3 840	167
Ideal situation (no losses)	10 000	0.7	7 000	not applicable

(1) 1 U.F. = Feeding value of 1 kg Barley
(2) The total cost includes the cost for production of grass, harvesting,
drying and foddering
(3) For an 8 ha/yr capacity drier the annual oil savings of the solar
assisted model compared to the conventionally heated one are expected
to be about 1 400 l/yr.

5. Results

5.1. Method of calculation

1. The instantaneous collector efficiency (η_c) is defined as

$$\eta_c = \frac{I_c}{e_{co}}$$

with - I_c the solar radiation intensity
on the collector, measured by the
solarimeter (W/m²)

- e_{co} the solar energy input into the
air flowing through the collector
per unit of collector surface and
per unit of time (W/m²)

- In order to solve this equation e_{cu} can be calculated as :

$$e_{co} = \vartheta_c \rho\, c_g \Delta T$$

with ϑ_c the specific air flow rate through the collector

$$\left(\frac{m^3}{s.m^2 \text{ collecting surface}}\right)$$

ρ the specific mass of the air (kg/m³)
c_g the specific heat of the air (J/kgK)

ΔT temperature rise of the air over the collector (K)

- The specific air flow rate is indirectly assessed, by measuring the static pressure change over an orifice.
The calculation of air flow rate from the measured pressure drop is given below :

$$\vartheta_c = \frac{\varepsilon \; \alpha \; S_c' \sqrt{\dfrac{2g.C.\Delta p}{\gamma}}}{S_c}$$

with ε the coefficient of compressibility of the air (-)
($\varepsilon \simeq 1$ for $\Delta p < 2940$ Pa)

α coefficient of air throughflow (-)
(is a function of the type of orifice and location of the Δp measurement)

S_c' surface of the orifice opening (m^2)

S_c net collecting surface of the collector (m^2)

g 9.81 m/s^2

C coefficient which eliminates falsification of the calculated air flow rate by variations in air temperature, humidity and atmospheric pressure (-)

Δp static pressure difference measured over the orifice (Pa)

γ specific weight of the air at 20°C and pressure ± 1 bar (11.8 N/m^3)

- The correction coefficient C can be calculated from an experimental equation :

$$C = \frac{56.48 \; (273 + T_g)}{0.164 p_a - 0.062 \; (p_v' - \dfrac{(T_g - T'_g) \; p_a}{1.548 - 1.425 T'_g})}$$

with T_g dry bulb temperature of the air (°C)

T'_g wet bulb temperature of the air (°C)

p_a atmospheric pressure (Pa)

p_v' saturated vapor pressure at temperature T' (Pa)

- p_v' can in turn be calculated, using following experimental equation :

$$p_v' = (4.543 + 0.36979 T' + 0.00707 T'^2 + 0.0003706 \; T'^3)^{133}$$

2. Finally the average daily efficiency of each collector ($<\eta_c>$) can be calculated as

$$<\eta_c> = \frac{\int_c I \; dt}{\int e_{co} dt}$$

5.2. Weather conditions and energy

I. Collector efficiency and temperature rise for air flow rates around

$$150 \ \frac{m^3}{h \ . \ m^2 \ collector}$$

For 15 typical measuring days the average daily solar radiation on the collector (I) and the corresponding collector efficiency values (η_c) for the 5 different collectors 1-5 are given in figure 5.2.1..
Those data were obtained, based on 11 to in some cases 59 measurements per day. (Depending on the climatological conditions, in order to allow for a reliable calculation of the average efficiency - cf. calculation method, section 5.1.).

The long term average of the efficiency value, as well as the average temperature rise of the air is given below.

Collector	Long term η_c	Average ΔT air
1	67%	7.6 K
2	73%	8.1 K
3	61%	7.0 K
4	56%	6.4 K
5	63%	7.0 K

Note : Depending on radiation intensity, the temperature rise can vary from 3.5 - 13 K.

II. Influence of air flow rate on collector efficiency :

Measurements, assessing the effect of air flow rate on collector efficiency, are given in figure 5.2.2.

In this experiment the specific air flow rate (ϑ_c) was changed every 5 minutes in steps of about 15-20 $m^3/(h.m^2)$. (This means the stabilisation time before each measurement was at most 5 minutes.)

The efficiency clearly increases up to air flow rates of 187 $m^3/(h.m^2)$. The last step to 200 $m^3/(h.m^2)$ seems to have a rather negative effect on the efficiency. More measurements under different conditions should be made in order to confirm these results.

III. Pressure loss in collectors

The pressure loss (Δp) in the collector parts, given normal operation, was calculated and is given in table 5.2.3.

$$\Delta p = \xi \left(\frac{v^2}{2g}\right) \gamma$$

with Δp in Pascal
ξ loss coefficient (decimal)
v air speed in given part (m/s)
g 9.81 m/s^2
γ specific weight of air ($\simeq$ 11.8 N/m^3)

IV Durability

Problems with durability in collectors 1 and 2 : absorber -insulation materials show significant signs of degradation after 4 months.

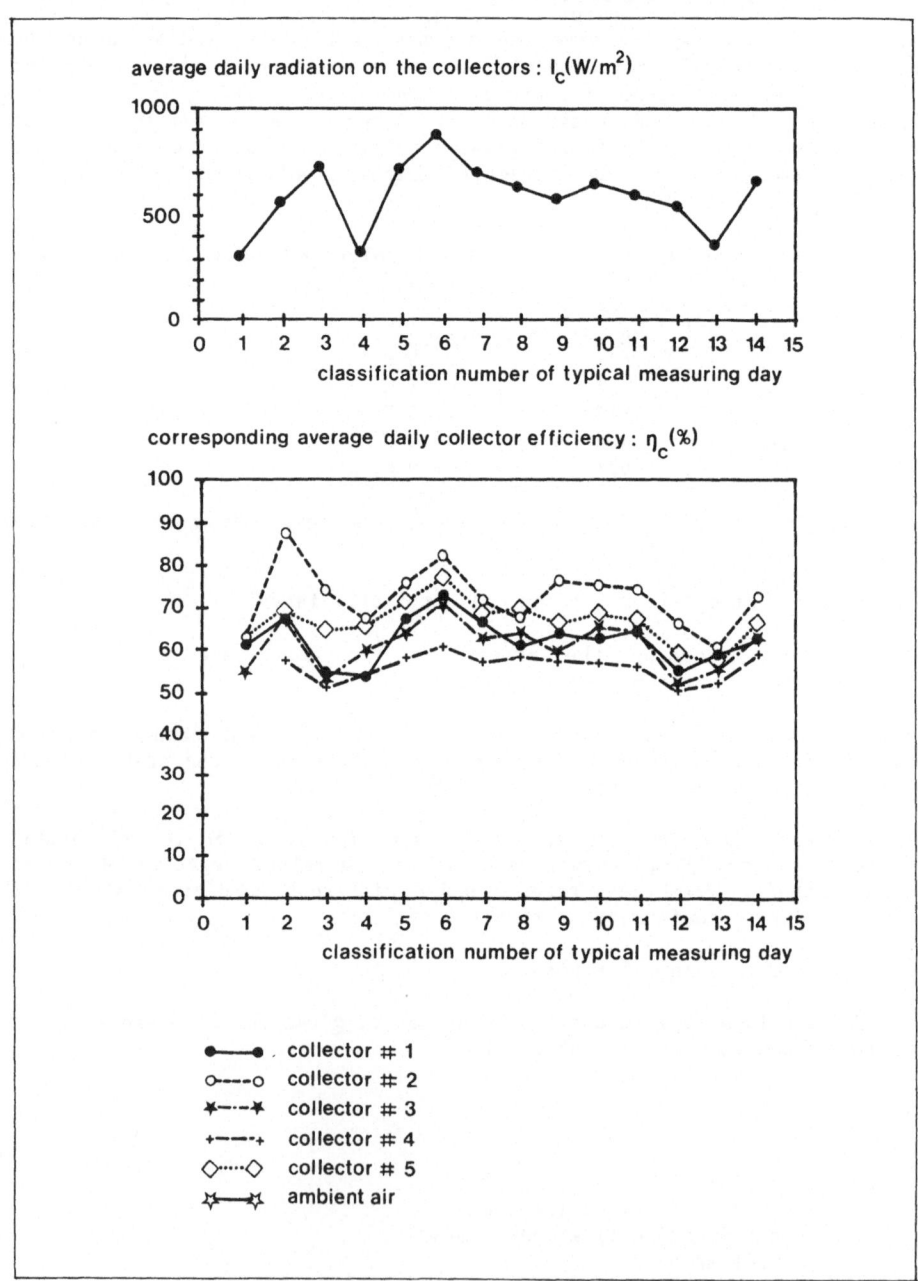

Figure 5.2.1 : Measured collector efficiency values as a function of average daily solar radiation.

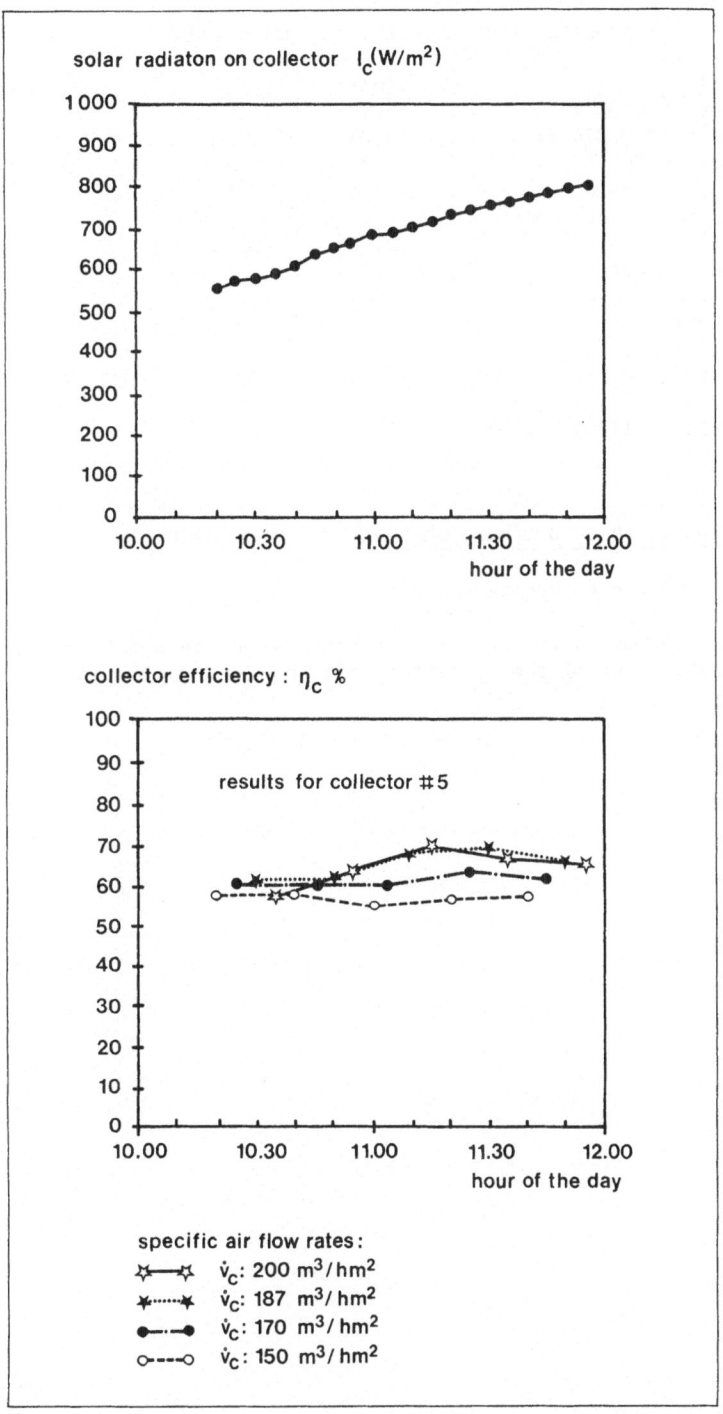

Figure 5.2.2. : Influence of air flow rate on collector efficiency.

Table 5.2.3. : pressure losses in the collector parts.

ELEMENT	v(m/s)	ξ	Δp (Pa)
orifice (∅ 0.07 m)	28.97	0.5	250
PVC duct (∅ 0.074 m)	25.83	0.14	56
Connection CD	25.83	0.41	164
gradual expansion	9.18	1.0	51
transition to collector			
(at AB)	2.035	0.13	≈ 0
in the collector	1.33		≈ 0
transition collector-			
contraction	1.33		≈ 0
gradual contraction			≈ 0
connection (C'D')	9.18	0.39	20
expansion 0.074 to 0.110	25.83	0.3	120
expansion 0.110 to 0.125	11.69	0.08	7
total losses			668

5.5. Influence on saturation deficit of the drying air

(strongly related to drying capacity)

The effect of the solar radiation intensity on the saturation deficit of the air coming out of the collectors can be seen from the figure 5.5.

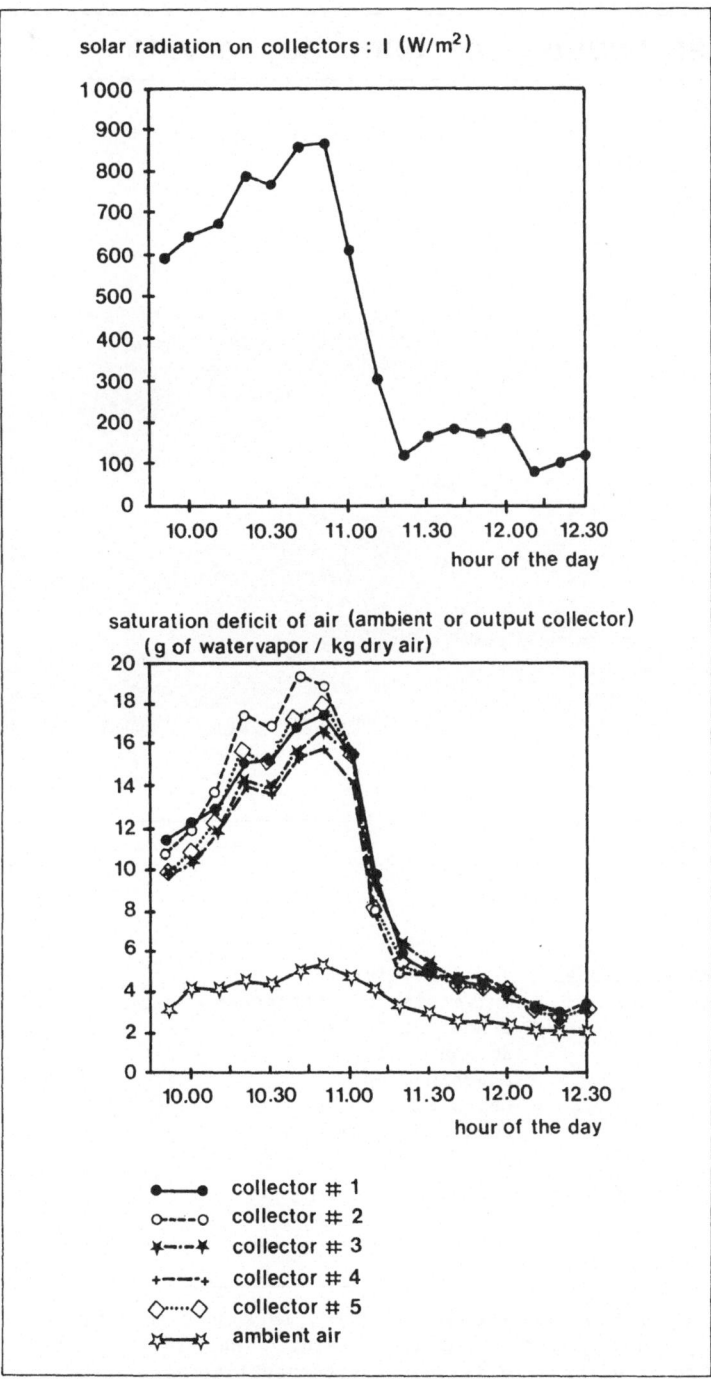

Fig. 5.5. : Effect of solar radiation intensity on the saturation deficit of the heated air. For cloudless situations the saturation deficit can increase by 200 % of its original value. When overcast weather occurs, the increase in saturation deficit is only about 50 % of its original value.

3. Solar heating for barn hay drying

Project location

NOTTINGHAM, UNITED KINGDOM

Study Center :
- Project Leader : Dr. J.A. Clark
- Institution : University of Nottingham
  School of Agriculture
  Sutton Bonington
  Loughborough
  Leics LE 12 5 RD
  United Kingdom

Running since 1979

Project description

For reasons of risk and product quality, in some cases hay is baled after a short predrying period on the field. The still moist bales (m.c. 25-35%) are dried artificially by forced ventilating air.

- In this project, 2 identical forced ventilation air driers are compared, one of which is equipped with a greenhouse-type collector of 120 m^2 ground-surface for air heating.
  In both cases, an 8.5 kW fan forces 25 200 m^3/h of (ambient or solar

heated) ventilation air through a 30 tonne batch of hay bales.

- The aim of the research is to assess the heat balance and energy yield, as well as the costs and benefits of this low cost solar heating device.

Scheme

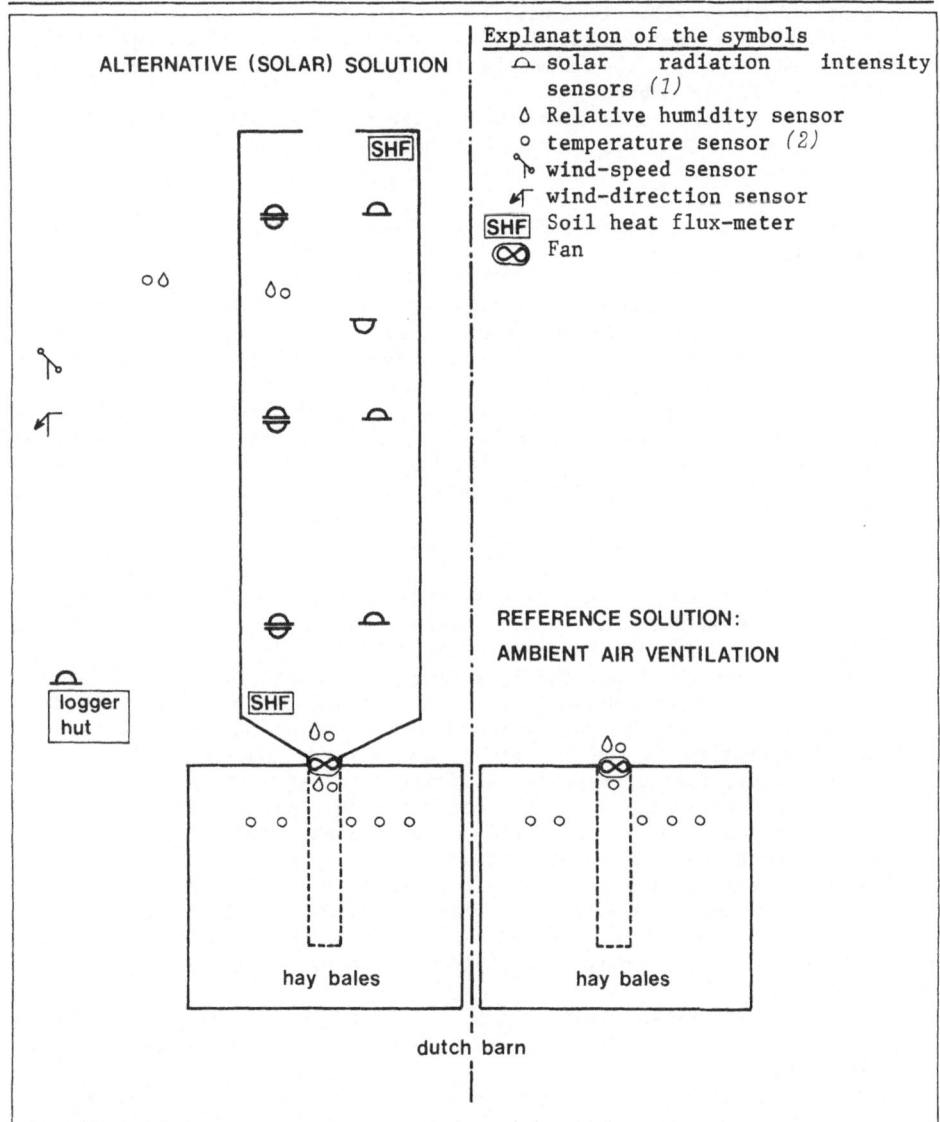

Note :
1. *The solarimeters with right side up measure the incoming solar radiation. The ones upside down measure the amount of energy that is dissipated back to the atmosphere through radiation losses; e.g. reflection ... (= albedo)*
   As such, the double side ones measure the net solar irradiance.
2. *The combination ○ ◊ is measured by means of a wet and dry bulb temperature sensor.*

Operating modes :
- Continuous operation for the first 48 hours (in order to prevent heating of the moist hay-stack).
- Afterwards operation during daytime only (until desired moisture content for permanent storage is reached).

Note :
1. The solarimeters with right side up measure the incoming solar radiation. The ones upside down measure the amount of energy that is dissipated back to the atmosphere through radiation losses ; e.g. reflection ... (= albedo)
   As such, the double side ones measure the net solar irradiance.
2. The combination o ◊ is measured by means of a wet and dry bulb temperature sensor.

Conclusions

- Even in a less favorable climate, this cheap installation is able to collect solar energy in a somewhat effective way. The collector efficiency was about 27% over the testing period of 1979-1981. This yielded an average temperature increase of 1-2 K for an air flow rate of 7 m^3/s. During operation the energy yield varied between 4 kW for a bad summer (1980) to 10 kW for a reasonably good summer (1981). Over the same drying period, 2 identical batches were dried at least 1% point more in the solar than in the other system (13.7% m.c. as opposed to 14.9% m.c. in 1981).
- The collector efficiency could be increased in several ways :
  - ground insulation with straw or polystyrene : calculations would predict that this could increase the collector efficiency by 25% of its current value. This would however cause problems as to potential multi-purpose use of the greenhouse collector.
  - The introduction of adapted operating modes could increase the long term collector efficiency, since the net radiation is negative from about 1 hour before sunset, until 1 hour after sunrise. Thus the air temperature in the collector can be even lower than the ambient temperature, causing negative efficiency values at those times.
- Even though this is a cheap installation, an increased rate of utilisation, up to more than 4 times the original value, is essential in order to reach economic feasibility.
  - Since storage drying of hay bales is most common in the U.K., the collector needs to either be made portable, or supplemented with distribution ducts in order to be able to help dry more than 1 batch of hay per year, or to be applicable for other products in another location.
    If e.g. additional grain drying should become possible, the drying season of the collector would be extended from June-July to June-September.
  - The greenhouse collector could possibly also be used as a storage area for machinery and other materials, and even in some cases for animal housing.
- Future research will involve development of ways to determine more accurately the end of the need for drying. This will mean solving the problems encountered with non-uniform air distribution in the bales, and developing more accurate ways of measuring the moisture content of the batch.

Future research will also involve experimentation with the potential of greenhouses with double cover, which should help reach the air temperature increase, necessary to make grain drying possible by means of this installation. Finally research will also be directed towards development of a portable collector.

PROJECT ANALYSIS

1. Site and climate

1.1. Site

Latitude :	52°50' N
Longitude :	1°15' W
Altitude :	40 m
Nearest main city :	Nottingham
Distance from main city :	15 km
Direction from main city :	NE
Obstructions :	Experimental Barn to E.

1.2. Site micro-climate

Generally sheltered, except from NW

1.3. Annual long-term averages

			Project Southwest
1. Prevailing wind direction			
2. Average wind speed	W	m/s	2.9
	S	m/s	2.3 (both at 2 m above the ground)
3. Total precipitation		mm/yr	603
5. Global irradiation on horizontal plane		MJ/(m².yr)	3 100
8. Hours sunshine		hrs/yr	1 243
9. Ambient temperature		C	9.4
10. Average max. temp. (July/Aug)		C	20
11. Average min. temp. (Jan)		C	2

12. Meteorological station :	Sutton Bonington (Agrometeorological station)
13. Latitude :	52°49' N
14. Longitude :	1°15' W
15. Altitude :	40 m
16. Distance from site :	1 km South

Note : W = Winter
       S = Summer

2. Plant

2.2. Drying-place

1. Drying system :
   - Batch drying
   - Pressure driven air, blown into tunnel by 8-blade axial flow fans
   - Haybales are stacked around the tunnel.
2. Material :
   Steel mesh tunnel.
3. Dimensions : 2 identical batch driers of 4.7 m x 7 m each.
4. Capacity :
   Approximately 30 tonnes per batch.
5. Ventilation :
   - Power : 8.5 kW
   - Flow rate : 7 m^3 s^{-1}
6. Conventional heating source : none

2.3. Collector (semicylindrical horticulture tunnel)

1. Material :
   - cover : Polyethylene Film 600 gauge ; UV stabilized ; lifetime 3 years.
   - supports : 60 mm galvanized steel pipe
   - absorber : dark colored ash on ground ; later replaced by black polyethylene sheet.

2. Dimensions :
   - Ground surface : 120 m^2 (6 x 20 m, approx.)
   - Air inlet : 1.8 m x 4.3 m

3. Flow rate : 7.0 m^3 s^{-1}, during drying runs.

5. Collector performance
   A diagram of the average hourly energy yield of the collector as a function of the incident solar radiation is given in figure 2.3.5.
   - The slope of line A, obtained through a least squares fit linear regression corresponds to a collector efficiency of about 27%.
   - Line B is added to show the additional effect of the fan. (6 kW of extra heat input)
   - The scatter of the data points is due to the fact that certain variables, influencing collector efficiency, have not been considered.
   Indeed, wind speed for instance has a significant impact on the collector efficiency via the heat transfer coefficient (k).

   $k = 4.26 + 0.634 \, v$ with k given in $\dfrac{W}{m^2 \, K}$

   v windspeed in m/s

Note : Over the 3 years of operation, a slight reduction of the cover transmissivity from the original 80% to 78% was observed.

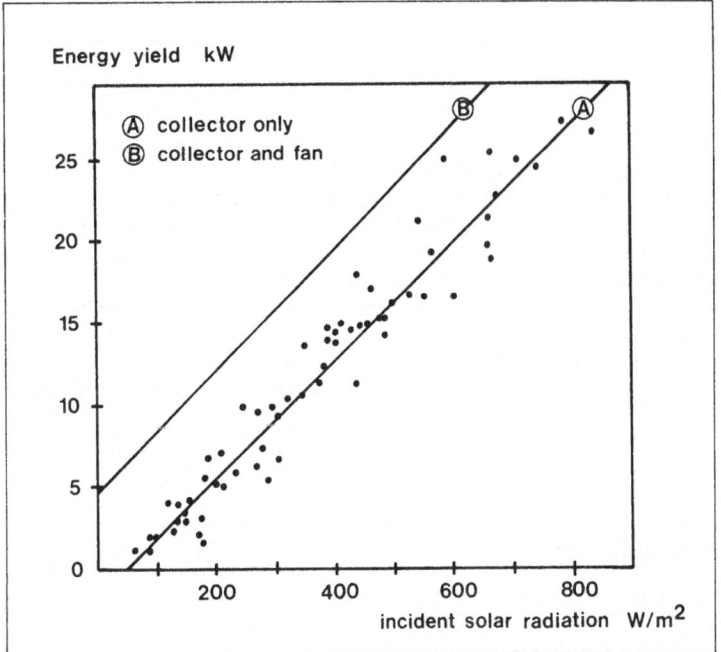

Figure 2.3.5. : Energy yield of collector as a function of solar radiation.

2.5. Installation parts and measuring devices

1. Technical description of the installation parts :
   - Fan, 8 blade axial flow, 8.5 kW (at 2 950 r.p.m.)
     Manually operated switches.
   - Fan is mounted in plywood ducts.
2. Description of the sensors
   Sensors and Measuring Equipment.
   Data Logger : Solartron 3 430 Compact Logger, Cassette Output, ECMA34
                 format
   Data Processing : Using university mainframe computer via PDP11 based
                     tape reader.
   Solarimeters : Reference Kipp and Zonen
                  Secondary instruments, tube solarimeters to Monteith
                  and Sceicz design, similar to the commercial available
                  ΔT design.
   Net Radiometers : Polyethylene covered tube net radiometers.
                  Similar to the commercially available ΔT Instrument
                  design.
   Wet and dry bulb thermometers : Thermistor thermometer unit in
                                    ventilated housing.
   Stack thermometers : Cu-Constantan thermocouples.
   Flux plate : Electroplated thermopile on glass substrate.

Note : *Logger scan interval, 5 or 10 minutes.*
          Data processed to give hourly averages stored on file.

3. Products

3.2. Drying products

1. Designed for special products :
   Designed for baled hay. Installation designed to standard UK criteria for barn hay drying. Batch size matched to fan capacity.
2. Required conditions of inlet air : 7 m^3 s^{-1} flow rate. Temperature not critical. Desired relative humidity < 70%.
3. Properties of the products : product dried to less than 18% moisture content.
4. Other (related) products which could be dried :
   Collector principle could be applied to other products and to grain drying.
5. Any special treatment of the crop :
   Field dried for 2 to 3 days before solar drying.

4. Economic aspects

prices in January 1983

4.1. Economic environment

2. Energy price of electricity	0.066 ECU/kWh
3. Energy price evolution prospects of electricity	10%/yr
5. Interest rate of loans for farm equipment (rate without subsidy)	15%/yr
8. Subsidies available for the solar and competitor farm equipment (as capital grant or interest rate reduction.	0

4.2. Investment expenses

Approximate figures

	Solar only	Solar + support energy	Reference solution
1. Total investment budget (materials) (ECU)	1 065	2 048	983
2. Cost of equipment : (ECU)			
– cost of collectors per m^2 ground surface	8.85		
– cost of regulation equipment		820	820
– cost of other materials		163	163
3. Time of personnel for the conception, installation and start-up of equipment.	Total approximately 20 man days.		
4. Lifetime of the equipment. (evaluation) (years)	Collector cover : 3 frame : 10	Fans : 10	
5. Residual value of the investment (ECU)	328	655	655

4.3. Operating expenses

	Solar only	Reference solution
1. Energy cost approx.	0.54 ECU/hour	0.54 ECU/hour
2. Maintenance cost	Labour not costed	
- spare parts	164 ECU/yr	
	(for cover replacement)	

4.4. Hidden cost and benefits

Installation used to dry produce consumed on the university farm.

4.5. Analysis of the economic feasibility of this application

Following situations are being considered :

1. - The greenhouse collector is only used over 1 drying run per year.
     (8 days of 12 hours/day operation time).
   - A writing-off period of 3 years (the cover gave way during winter
     of 81-82, three years after installation in 1979).
   - An average collector yield of 10 kW over operation time is being
     reached (cf. situation for the best year 1981 in the local
     situation of Nottingham).
   For this situation the calculated solar energy price was more than
   4 x the energy price for electricity (1982 prices).
   Over the years, following this first 3-year-period, the solar energy
   price would become comparable to the price of electricity, in case no
   extra expenses for deterioration of other than covering materials are
   incurred. (The price of a new cover is about 25% of the total
   collector price).

2. - Figure 4.5 based on different assumptions (cf. (1) in figure),
     shows how utilization rate is a crucial factor in the economic
     feasibility of the solar system.

5. Results

5.1. Method of calculation

- Collector yield and efficiency were calculated as follows :
  Collector yield = Airflow rate x temperature rise x volumetric
  specific heat.

$$\text{Collector efficiency} = \frac{\text{Collector yield}}{\text{Incident Solar Radiation} \times \text{Collector Area}}$$

- 'Prices' estimated using published inflation accounting procedures.

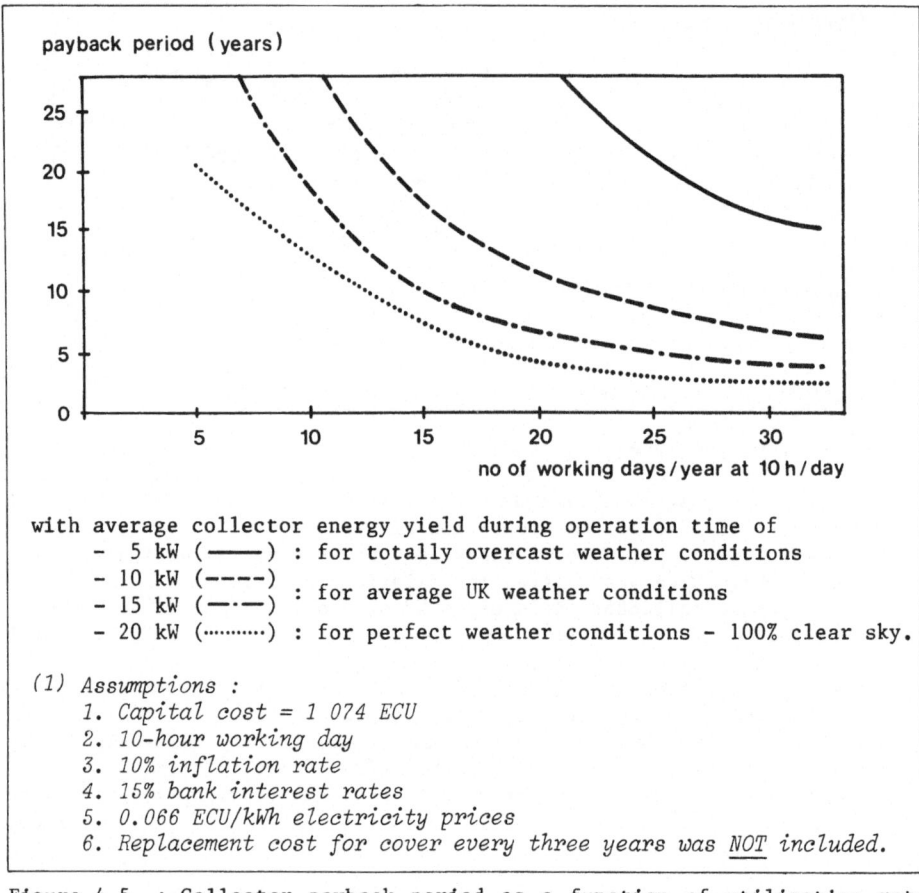

with average collector energy yield during operation time of
- 5 kW (———) : for totally overcast weather conditions
- 10 kW (————)
- 15 kW (—·—) : for average UK weather conditions
- 20 kW (··········) : for perfect weather conditions - 100% clear sky.

(1) Assumptions :
  1. Capital cost = 1 074 ECU
  2. 10-hour working day
  3. 10% inflation rate
  4. 15% bank interest rates
  5. 0.066 ECU/kWh electricity prices
  6. Replacement cost for cover every three years was NOT included.

Figure 4.5. : Collector payback period as a function of utilization rate and climate.

5.2. Weather conditions and energy

	1979	1980	1981		PERIOD during haymaking season (June, July)	
3.	13.5	13.9	13.4	°C	average ambient TEMPERATURE	
4.	2.1	2.6	3.1	m/s	average ambient WIND	
5.	176	138	310	Wm^{-2}	average SOLAR RAD on collector (during operation)	solar
10.	5.2	4.1	10.0	kW	USEFUL energy	
12.	8.5	8.5	8.5	kW	energy for operation	
14.	8.5	8.5	8.5	kW	ELECTRICITY consumption	
16.	8.5	8.5	8.5	kW	energy for operation	ref.
18.	8.5	8.5	8.5	kW	ELECTRICITY consumption	

5.5. Drying results

	1979	1980	1981		PERIOD during haymaking season (June, July)		
3.	25 200	25 200	25 200	m^3/h	flowrate	INLET AIR	
4.	1.0	1.0	1.6	°C	average temperature rise		
6a.	32	24	28	%	moisture content a. original		solar
6b.	16.0	17.0	13.7		b. final	PRODUCTS	
7.	Approx 30 Tonne per batch				useful quantity		
8.	good	good	good		quality		
10.	25 200	25 200	25 200	m^3/h	flowrate	INLET AIR	
13a.	32	24	28	%	moisture content a. original		reference
13b.	18.0	18.0	14.9		b. final	PRODUCTS	
14.	Approx 30 Tonne per batch				useful quantity		
15.	good	good	good		quality		

Note : drying time kept the same for both solar and reference solution.

5.7. An example of some measured data (cf. figure 5.7)

Note : The average soil heat flux over the 2 measured days amounted to about 20% of the net radiation.

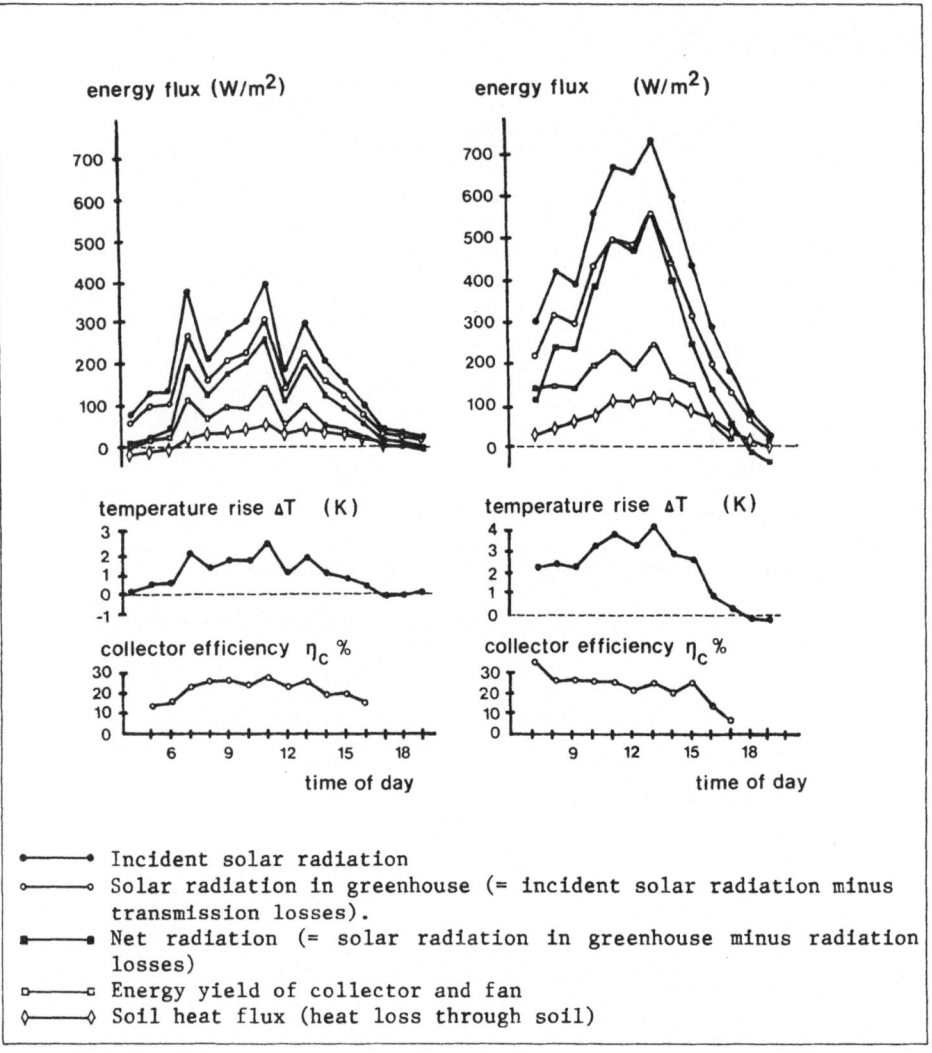

Fig. 5.7. : An example of some measured data

4. Development of a solar drier for wine lees

Project location

CANOHES, FRANCE.

Study Center :
- Project leader : Prof. M. DAGUENET
- Institution : Laboratoire de Thermodynamique et Energétique.
  Université de Perpignan
  Avenue de Villeneuve
  66025 PERGIGNAN Cédex
  France

Running since 1982

Project description

A southfacing greenhouse (10 m long ; 3.6 m wide ; 4.6 m high), leaning against an insulated wall, is being used to dry wine lees from 50% to 5-10% m.c.. The drying of wine lees, which occurs during a 4 month season, is part of the production of tartar and tartaric acid.

In this experimental greenhouse, 8 tons of fresh wine lees are stacked on 360 m² of trays, exposed to the sun. Both the absorption of the solar radiation by the wine lees, as well as the forced ventilation by means of several fans, should normally contribute to an effective evaporation of the water that is to be removed.

In order to make maximum use of the absorbed energy, the greenhouse is closed off all the time, except when the rh of the air inside the greenhouse grows larger than a set value of around 70-80%, at which point the inside air is renewed.

The goals of this research are :
- To assess the feasibility of such a set-up.
- To develop a mathematical model, that will help determine and optimize the parameters of influence in the installation.
- To look for ways to put this drier to use during the off-season for forage, wood, or medicinal plant drying ...

Scheme

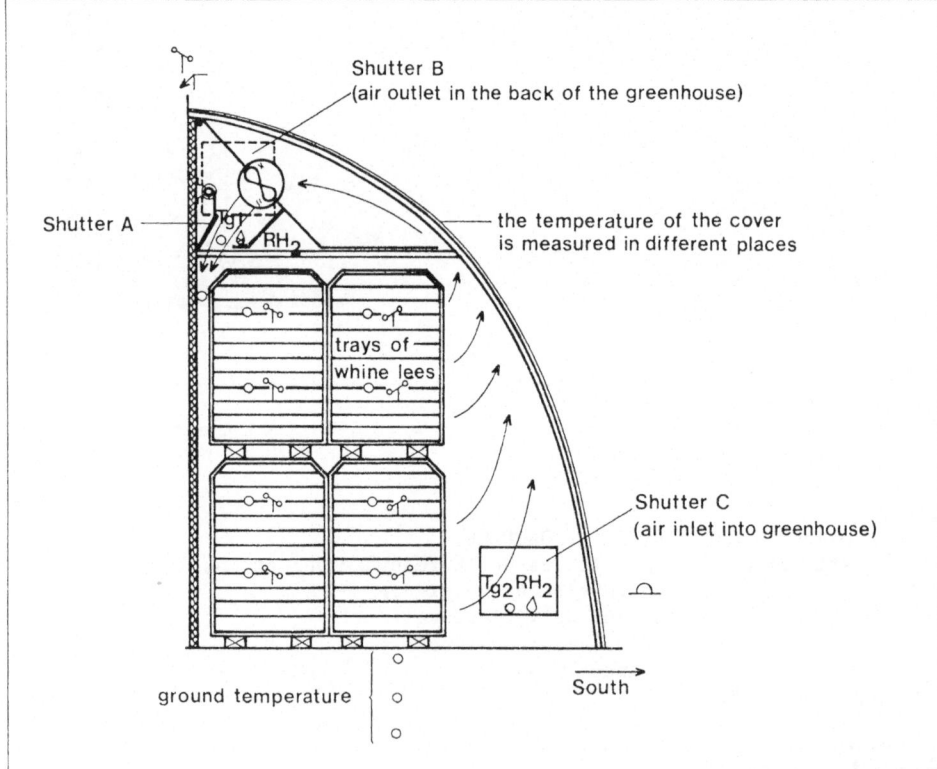

Note :
1. T_{g1} *and* RH_1 *are measured inside the greenhouse,* T_{g2} *and* RH_2 *are measured outside.*
   For our explanation of the other symbols, cf. pp. 470-471.
2. *Also measured are :*
   - *The temperature inside of certain wine lees particles (by means of thermo-couples).*
   - *The weight of representative samples.*
   - *The importance, and the frequency of the air-renewal.*

Operating modes
- As long as the inside relative air humidity is smaller than a set value
(between 70-80% r.h.), the air inside the drier is ventilated within the
closed greenhouse.
For this, the shutter A is opened wide, while shutters B and C are
closed. The goal of the ventilation is to homogenize both temperature
and r.h. of the air throughout the greenhouse, in order to reach a
uniform and effective drying.
- As soon as the set-point value (70-80% r.h.) is exceeded in measuring
point RH_1, a control device opens the power circuit of a servo-motor,
which closes the shutter A.
Consequently, shutter B comes under extra pressure from within, and
opens up. Since the inside air is being blown through shutter B, this
air will be replaced by fresh air sucked through shutter C.
- Even after the humidity in RH_1 has fallen below the set-point value,
shutter A still remains closed for a certain period of time, imposed by
a retardation relay.

Reference solution
- The traditional wine lees drier of the firm in question was a batch
drier with capacity of 5-6 tonnes of fresh wine lees, requiring about
100 litres of fuel per initial tonne of fresh product, for heating
purposes. (7 500 kJ/kg evaporated water).
- As part of its operation, this drier also required long and tiresome
manual agitation.

Conclusions

- A rough estimation, based on very limited experimental data (cf. section
6), would predict an energy saving of 50% compared to the reference
solution. (3 750 instead of the usual 7 500 kJ/kg of evaporated water).
Of the remaining energy need, 92.5% is being supplied in the form of the
collected solar radiation energy, while only the other 7.5% - which is
required for fan operation and regulation purposes - has to be provided
in the form of electrical energy.
- The drying time is however quite a bit longer (15 days, instead of 1 day
for the reference solution). This means only 6 batches of 8 tonnes of
fresh wine lees per drier can be dried over the 4 month long drying
season. On the other hand, this greenhouse-type drier has considerable
potential for other drying applications over the period of the year in
which wine lees is not being dried. This is not the case for the
reference solution.
- The measurements readily revealed some problems with the current set-up:
  - Given the dimensions of the greenhouse, both the limited surface
  which takes part in solar energy collection, and the extensive heat
  losses, especially through the cover, make it impossible to reach the
  desired temperature of about 30°C.
  The resulting low temperature does not allow for sufficiently quick
  drying, and this leads to mould development on the wine lees.
  This problem can be solved either by -1- decreasing the heat demand
  for drying - by reducing the quantity of the batch to be dried - or
  -2- by increasing the drying potential of the air, using an auxiliary
  heater, or -3- an instrument that reduces the water content of the
  incoming air. The unfortunate occurrence of renewed absorption of
  moisture by the wine lees at lower moisture content levels should

also be dealt with by means of the 2nd and 3rd option.
Experiments with the first 2 options are planned in the future.
- The <u>ventilation</u> and its regulation need improvements in many ways. The main problems and the planned solutions are given below :
  - devices will be installed to direct the air flow, in order to adjust for the current non-uniform pattern of the flow. They will be painted black, so they can also contribute to the absorption of solar heat.
  - The renewal of the air, controlled by the hygrometer, falls short, since the obtained air flow rate is insufficient. The installation of an extra fan, controlled by the hygrometer, is planned in order to overcome this problem.
- The described experiment has been considerably delayed due to a combination of unfortunate factors, ranging from availability of measuring equipment and product to be dried, to the unforeseen prohibition of cutting down trees in the neighborhood, which throw shade on the greenhouse after 2 pm, and jeopardize the value of the measurements ...
While these and other problems are being solved, considerable progress is being made in the development of a simulation model.

PROJECT ANALYSIS

1. Site and climate

1.1. Site

Latitude :	42°42' North
Longitude :	2°52' East
Altitude :	40 km
Nearest main city :	Perpignan
Distance from main city :	6 km
Direction from main city :	West
Obstruction :	None

1.3. Annual long-term averages

		Project
1. Prevailing wind direction		NE
2. Average wind speed	m/s	5.1
3. Total precipitation	mm/yr	585.7
4. Relative humidity	%	67
5. Global irradiation on a horizontal plane	$MJ/(m^2.yr)$	5 500
6. Diffuse proportion	%	40
8. Hours sunshine	hrs/yr	2 567
9. Ambient temperature	C	15.7

12. Meteorological station :   Rivesaltes
13. Latitude :   42°42' N
14. Longitude :   2°52' E
15. Altitude :   42 m
16. Distance from site :   10 km

2. Plant

2.2. Drying-place

1. Drying system : The driers consist of 2 south-facing greenhouses, leaning against a wall, only one of which is being monitored.
2. Material : – Alveolate Polycarbonate cover (6 mm).
   - Arched supports, made of galvanized steel ($\emptyset$ 60 mm – 2 metres apart)
   - Back wall insulated with 4 cm of polystyrene.
3. Dimensions : length : 10 m
   (1 drier)   width : 3.60 m
              height : 4.60 m
              surface of cover : 70 m^2
4. Capacity : 8 tonnes of fresh wine lees, exposed on 360 m^2 of trays. (Each tray : 1.50 x 1.07 x 1.07 m^3)
5. Ventilation – Power : 1 080 W
                – Flow rate : 3.5 m^3/s
6. Conventional heating source : none
7. Diagram : cf. "Scheme".

2.3. Collector

The drying place can be considered as a modified form of collector, where that part of the wine lees which is exposed to the sun functions as an absorber, while the heat lost by these absorbing wine lees is transported by air ventilation to the other part of the wine lees which is not accessible to direct solar radiation.
For a description of the drying place, cf. section 2.2.

2.5. Installation parts and measuring devices

1. Technical description of the installation parts : (data of 1 drier)
   - Ventilation : 6 helicoidal fans
               ref. : H x B 4/400/24 trademark S & P
               airflow : 3 600 m^3/h
               power consumption : 170 W   } free running
               diameter of the propeller : 400 mm
   - Hygrometer : ref. HBC A001 trademark Sauter
               range : 0 – 100%
   - Motor : servo-motor, 2 work directions
            torque : 140 Nm
            power consumption : 35 W

2. Description of the sensors
   - Global irradiation : THIES Solarimeter
                     Precision + 1%
                     Range : 0 – 1 500 W/m^2

- Wind :   Speed : precision mechanical anemometer
                         ref. APM 22
                         precision + 0.1%
                         range : 0 - 150 km/h
             Direction : ref. GPM 23, precision-mechanical
- Temperature :   Sensor Pt 100
                         precision + 0.2°C
                         range : 0 - 200°C
- Humidity :   (wet bulb temperature measurement)
                         Sensor Pt 100
                         precision + 2%
                         range : 0 - 100%

4. Economic aspects

4.1. Economic environment

		Solar only
1. Energy price of oil	ECU/l	0.40
2. Energy price of electricity	ECU/kWh	0.06
3. Energy prices evolution prospects	%/yr	15
4. Manpower cost (relevant for the needs of functioning of the installation)	ECU	(1)
5. Interest rate of loans for farm equipment (rate without subisdy)	%/yr	18 (2)

(1) *Salary for labourer, charges included : 5.6 ECU/hr*
    Salary for engineer, charges included : 11.1 ECU/hr
(2) *The owner is not a farmer.*

4.2. Investment expenses

		Solar only (1)	Reference solution
1. Total investment budget	ECU	23 322	28 000
2. Cost of equipment :			
- expenses for buildings (drying greenhouse)	ECU	3 810	
- cost of regulation equipment	ECU	777	
- other material costs	ECU	15 625	
3. - Time of personnel for the conception, installation and start-up of equipment	(months)	3	
- Cost of personnel for the conception, installation and start-up of equipment	ECU	3 110	
4. Lifetime of the equipment (evaluation)	(years)	7	20 (2)
5. Residual value of the investment	ECU	none	none

(1) It is important to note that in this section, the cost is given for both of the existing greenhouses together, while in section 4.3 and 4.4 the data are transformed to correspond to a capacity of 6 tonnes of fresh wine lees, in order to provide a basis for comparison with the reference solution. In the results section (5), the data refer to the performance of only 1 of the 2 driers, the one in which measurements are being made.

(2) As opposed to the solar drier, this reference drier can only be put to use during 4 months per year.

4.3. Operating expenses

for drying 6 tons of fresh **wine** lees

		Solar only (ventilation)	Reference solution *(1)*
1. Energy cost	(ECU/batch)	12.3	323
4. Manpower costs		16h x 5.6 ECU/h	12h x 5.6 ECU/h
6. Raw material cost	(ECU/batch)	91	91

(1) The comparison between the solar and reference drier is based on equal quantity of dried product. It is important to note that the drying time is significantly different : 1 day only for the traditional drier, while 10-15 days for the solar drier.

4.4. Hidden cost and benefits

based on 6 tons of fresh **wine** lees

		Solar only	Reference solution
1. Production obtained (dried substance)	kg/batch	3 000	3 000
2. Average sales price per unit produced	ECU/kg	0.056	0.056
3. Total income obtained	ECU/batch	168	168
4. Storage costs due to the installation	ECU/batch	134	11 *(1)*

(1) Maximum operation time : 4 months/year.

---

5. Results

Note : The given values apply to the 1 greenhouse in which measurements were made.

5.2. Weather conditions and energy

- Values of the first 3 days when the full scale measurement program was started, are given below.

| date | local time | OUTSIDE CLIMATE | | | | RH_l | T_b | T_p | T_{gb} | T_{gp} | ΔT | ρ_g | C_g | E_i |
		I_h	T_{ao}	RH_a	v									
29/11/1983	8	77.5	11	70	8	72	14.3	13	10.2	9.8	0.4	1 244	999.4	597
	10	77.7	12	69	10	74	24.6	14.2	17	14.8	2.2	1 212	996.17	3 187.4
	12	121.88	13.5	67	10	73	35	18	24	19.6	4.4	1 180	992.02	6 181
	14	132.96	14.5	66	2	73	29.7	19.1	20	17.9	2.1	1 198	994.17	3 000
	16	113.57	14	66	1	73	19.1	16.4	15	14.2	0.89	1 221	997.26	1 300.5
	18													
30/11/1983	8	19.39	13	70	3	75	16.5	16	11	10.9	0.1	1 240	998.77	148.6
	10	58.17	13	68	3	74	31.7	17.5	24.5	21.7	2.8	1 178	991.2	3 923
	12	74.79	15.5	64	0	72	34.9	18.2	26	21.1	4.8	1 171	990.4	6 680
	14	77.56	17	57	0	68.5	32.7	18.4	23	20.5	2.5	1 186	993.35	3 534.4
	16	49.86	16	60	0	70	17.8	18.3	16.5	15.8	0.7	1 217	996.9	1 019
	18													
01/12/1983	8	11.1	9.5	65	0	72.5	7.5	7.9	8	7.8	0.2	1 254	999.95	301
	10	254.84	11	65	0	72.5	21.4	9.4	17	15.1	1.9	1 212	996.32	2 753.2
	12	407.19	16	54	0	67	31.7	12.6	21.5	20.6	0.9	1 192	994.37	1 280
	14	354.56	17	46	0	63	31.9	16.2	23.5	21.8	2.7	1 184	993.94	3 813
	16	121.88	14	45	0	62	10.4	9.5	16	14	2	1 218	997.96	2 917.3
	18													

with I_h = global solar irradiation on a horizontal surface (W/m^2)

T_{ao} = ambient outside temperature (°C)

RH_a = relative humidity outside (%)

v = windspeed (m/s)

RH_l = relative humidity inside (%)

T_b = temperature of a black suface (°C)

T_p = temperature of the wine lees (°C)

T_{gb} = temperature of the air around the black surface (°C)

T_{gp} = temperature of the air around the wine lees (°C)

$\Delta T = T_{gb} - T_{gp}$

ρ_g = specific mass of the drying air (kg/m^3)

C_g = specific heat of the drying air $(J/(kg.K))$

E_i = energy absorbed by the drying air (Watt)

- It takes this new type of drier about 15 days, to dry 8 tonnes of fresh wine lees from 49% to 5% m.c..
- A rough extrapolation, based on these few measurements and on the short experience acquired in the meantime with the system, yielded following results : (for correct interpretation cf. section 6).
  Over the 4 month-long season, 6 batches of 8 tonnes of fresh wine lees can be dried in one of the solar greenhouses.
  The energy use over this whole season is about 83 GJ/year, 92.5% of

which is supplied by the sun and used for air heating, while only 7.5% of this value involves actual electricity consumption for fan operation and regulation purposes.

This total energy use corresponds to a specific energy input of about 3 750 kJ/kg evaporated water, which is only 50% of the energy required in the traditional drier.

Figure 5.2 gives an idea of the amount of solar energy that can be incorporated for air heating purposes during the different months of the year. (It is clear that applications need to be sought in order to put to use this installation for other drying applications during the 8 remaining months).

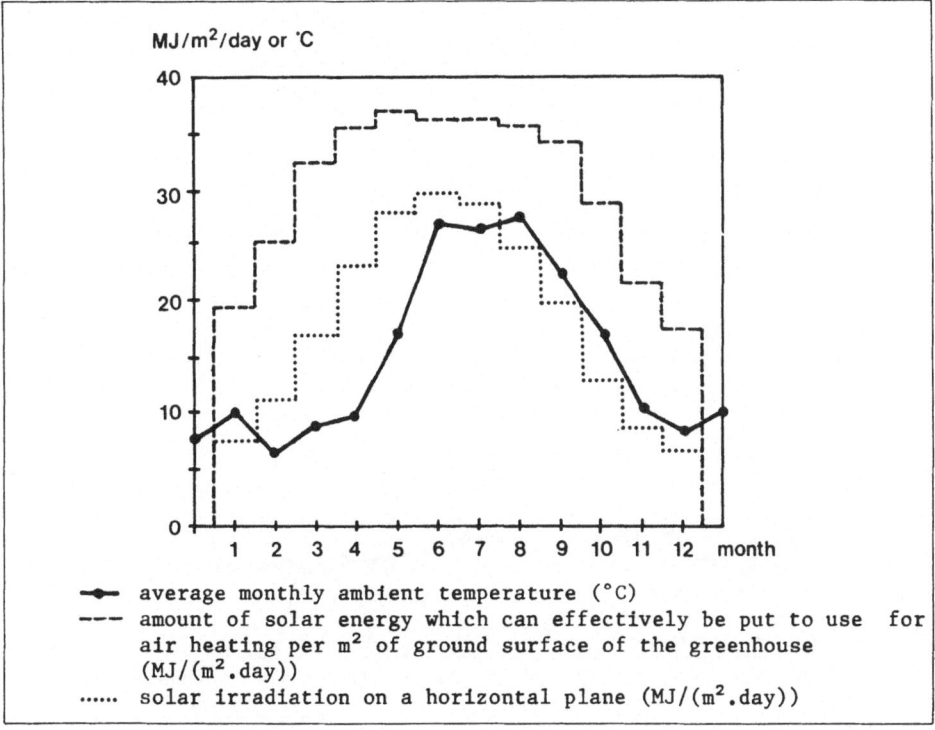

— average monthly ambient temperature (°C)
-- - amount of solar energy which can effectively be put to use for air heating per m² of ground surface of the greenhouse (MJ/(m².day))
...... solar irradiation on a horizontal plane (MJ/(m².day))

Fig. 5.2. : Average solar energy and temperature data for each month of the year.

Note : At the same time the development of the simulation model is well on its way ;
The following steps have already been set
- establishment of the equations, describing the operation of the drying installation.
- literature study, in order to determine certain physical values.
- experimental determination of the influence of the main system parameters on the drying speed (in a lab set-up)
The next step, currently being worked on, is :
- simulation of the operation of the drier as a function of the main parameters.
Future steps will involve :
- validation of the simulation model.
- optimization of the drier.

5.7. An example of some measured data

In figure 5.7 an example is given of a few preliminary measurements, before the planned full scale measuring equipment was available.
The oscillations of temperature and relative humidity correspond with the periodic renewal of the inside air.

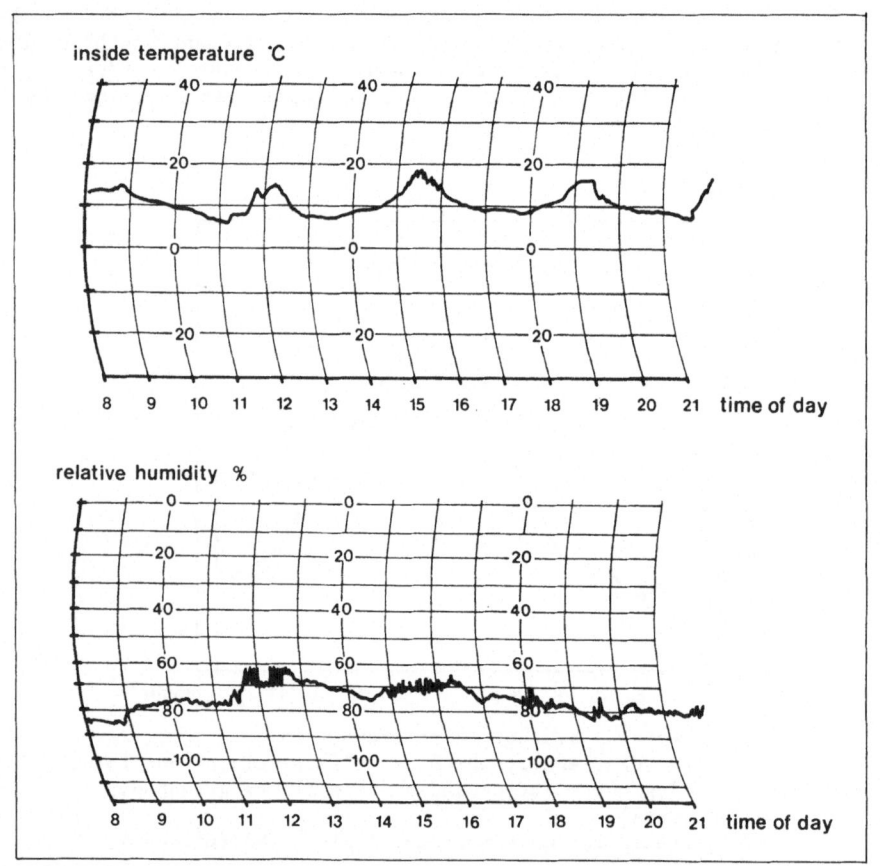

Figure 5.7. : Some preliminary measurement data.

---

6. Remarks

It should be clear that the results and conclusions, especially the given numerical values, are of a very provisional nature, since they are based on a short experience with the system.
In fact it seems that the only basis for these results and conclusions has been the 3 measuring days, given in section 5.2., together with possibly some earlier experience, using limited measuring devices and in the absense of the wine lees itself (cf. 5.7.).

If the measured values of E_i (in section 5.2.) are considered to be representative for the 2 hour-long period in which they were measured, one could get an idea of the value of the absorbed solar energy (Q_{total}) over each of those days :

$$Q_{total} = \frac{\sum\limits_{i=1}^{5} E_i * 2 \text{ hours} * 3\ 600 \frac{\text{seconds}}{\text{hours}} * 10^{-6}}{36m^2}$$

with Q_{total} in MJ/m² ground surface/day
E_i in Watts

The values calculated in this way are Q_{total} = 2.8 MJ/(day.m²) for the first day ; corresponding with a measured solar radiation on a horizontal surface of 3.6 MJ/(m².day) (1).
The corresponding values for the 2 other days are given in table 6 :

Table 6 : Data on solar radiation and Q_{total} for each of the 3 measuring days.

	Solar radiation (MJ/(m².day))	Q_{total} (MJ/(m².day))
day 1	3.6	2.8
day 2	2	3.1
day 3	8.3	2.2

How the results and conclusions have been calculated, based on the given measurements, is not at all clear. Probably some earlier measurements were included, as well as a quantitative interpretation - based on several, unmentioned, assumptions - of the experienced problems and their influence on the potential system performance.

(1) The given data of solar radiation are calculated, following the same train of thought as for the calculated value of Q_{total}.

5. Solar energy for produce drying and milking parlour washing

Project location

PAVIA, ITALY

Study Center :
  - Project leader : Prof. G. CASTELLI
  - Institution :   Institute of Agricultural Engineering
                    Via Celoria 2
                    Milano
                    Italy
Running since 1978

Project description

A simplified solar water collector, with net surface of 180 m², was built on a medium size farm in northern Italy, in order to provide heat for a batch forage drier, containing 30 tons of predried hay. Since the drier only works at full capacity for 60-90 days/year (over the period May-September), the solar heated water is also used for washing and sanitary purposes in the milking parlour. This parlour, which takes care of about 100 cattle, requires 1 000 l/day at 40°C for this purpose.

The research aims at comparing the performance of air and water type solar collectors. It also aims at assessing the economic feasibility of simple water collectors (< 135 ECU/m²), applied on medium size farms.

Note : This project is part of a larger scale research, the first part of which consisted in developing cheap collectors, adapted to the farm situation (a brief discussion of the development of such collectors for the production of warm water is given in section 5.8.).
The second part consists of the study of different set-ups and the feasibility in Northern Italy of several solar heating applications in the agricultural realm (biogas production, forage drying, ...).
A few examples are given in the overview chapters of the current research within the EC.

Scheme

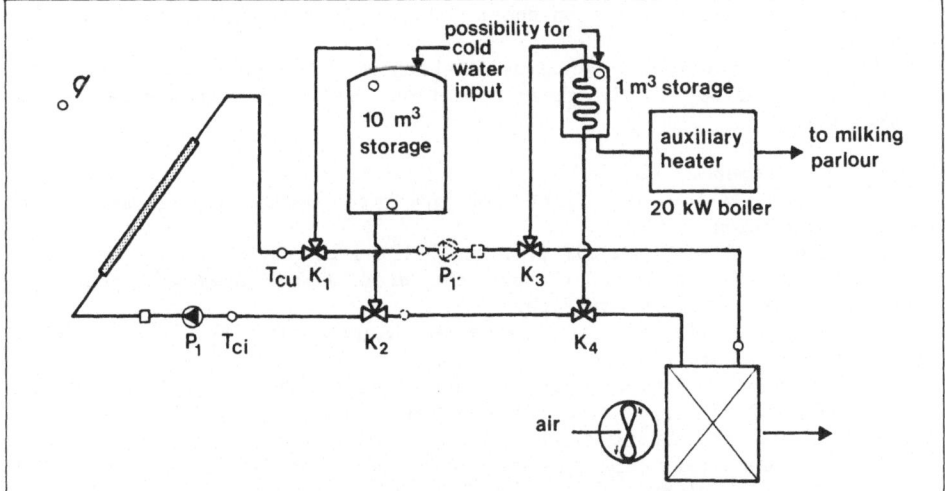

Note : an explanation of the symbols can be found on pp. 470-471.

Pump P'1 and the measuring points around it are presented in dotted lines, since in practice, they are identical to P1 and its surrounding measuring points.
To simplify the drawing, the intricacies of the actual set-up have been deleted. (N.O.T.E.)

Operating modes
1. Manual controls
   + the collector as the heating element :
     - heating the 10 m³ storage → K1&K2 in position ⊣⊳⊲⊢ (nr. 1)
     - heating the 1 m³ storage → K1&K2 in position ⊣⊳⊲⊢ (nr. 2)
                                  → K3&K4 in position ⊣⊳⊲⊢
     - heating the ventilation air → K1,K2,K3 & K4
                                      in position ⊣⊳⊲⊢
                                      (nr. 3)
   + The 10 m³ storage as a heating element
     - heating the 1 m³ storage (at nil insolation)
                                  → K1&K2 in position ⊣⊳⊲⊢ (nr. 4)
                                    K3&K4 in position ⊣⊳⊲⊢
     - heating the ventilation air (at night)
                                  → K1&K2 in position ⊣⊳⊲⊢ (nr. 5)
                                    K3&K4 in position ⊣⊳⊲⊢

The pattern of the manual control during the drying period is as follows :
- Before drying, the collector is connected so as to heat the 10 m^3 storage tank (nr. 1)
- On the first drying day, the drying air is directly heated by the collectors (nr. 3)
- On the first night, the heat in the 10 m^3 storage is used for the ventilation air (nr. 5)
- In the following days the collectors are connected either to heat up the air directly (nr. 3) or to load the 10m^3 storage (nr. 1), depending on relative humidity of the ambient air and moisture content of the hay (low rh $\rightarrow$ (nr. 1)) while drying is being done by means of ambient air.
- At some time during the day, the 1 m^3 storage is also loaded
  - (nr. 2), possibly nr. 4) -. Loading this storage tank requires about 1 hour of collector operation under good insolation conditions.
2. Automatic controls
A differential thermostat determines whether the pump operates or not, depending on whether in the given situation, heat can actually be gained or transferred.

Reference solution
A forced ventilation air drier for predried forage was chosen as a reference solution.
- The forage is dried from 50% m.c. to 10-15% m.c.
- Ventilation air is blown through the product at rates of about 0.1 m^3/(s.m^2) of ventilated surface.
- In most cases the air is heated at night only, by 5-10 K (High r.h.-values at night).
Per ton of resulting hay, this process requires about 2 300 MJ for air heating and 250 MJ for ventilation purposes.
N.O.T.E. : *Only the drying aspect of the plant has been monitored. For a rough estimation of what can be expected from this part of the installation, cf. section 6.1.*

Conclusions

- Solar drying definitely produces better quality hay than natural drying in open air, while the required energy input is lower than that of conventionally heated artificial driers.
- Given the low utilization rate (only 70-80 days of full capacity use of the solar system per year), the given application is not economically feasible. The importance of the energy demand of the milking parlour is too small to adjust this problem. In order for this system to become economically feasible, ways have to be found to increase the utilization rate (feed preparation for other animals, domestic water, etc...).
For high utilization rates of about
0.7 - 0.8 ($\dfrac{\text{energy actually put to use}}{\text{total energy that can be made available by the system}}$)
the given plant with total solar system cost of 190 ECU/m^2 collector would require a payback period of about 10 years under Northern Italian weather conditions.
- If the solar system is used for drying only, air collectors are definitely preferable.
- Even though frost is not a problem for the rubber absorber itself, the system collector-circuit-pipes is at risk. Hence, the solar system

could not be used in December and January.
- (Even though the collectors were designed to be fitted by unskilled labour, the use of unskilled labour posed some problems).
- For the reliability of the given conclusions, cf. section 6.2.

Further references
L. Bodria, G. Castelli, G. Riva, U. Facchini : "Primi risultati di prova di um impianto per l'essiccazione in due tempi del foraggio, utilizzando collettori solari piani". Proceedings of the 3rd AIGR National Convention, Catania, 16-19 May 1979.

PROJECT ANALYSIS

1. Site and climate

1.1. Site

Latitude :	45° N *(1)*
Longitude :	9° E *(1)*
Altitude :	115 m
Nearest main city :	Milan
Distance from main city :	10 km (as the crow flies)
Direction from main city :	Southeast
Obstructions :	None

(1) Approximate

1.2. Site Micro-Climate

The climate is typical of the Milan province.
Main characteristics :
- Poor insolation because of fog in the winter.
- Good insolation in summer. The diffuse component is constantly high ; generally, the sun can be seen only for a few hours after sunrise even on a clear summer day.

1.3. Annual long-term averages

			Project	Country in general *(1)*
1. Prevailing wind direction	W		E	
	S		NW	
3. Total precipitation		mm/yr	1 195	870
4. Absolute humidity	W	g/kg	3.8	
	S	g/kg	6.2	
5. Global irradiation on horizontal plane		MJ/(m².yr)	3 960	
				(2)
7. Degree days (Base temp.=18°C)		C days	2 340	
8. Hours sunshine		hrs/yr	1 848	
9. Ambient temperature		C	12.2	14.5
10. Average max. temp. (Jul/Aug)		C	27.4	27.8
11. Average min. temp. (Jan)		C	- 4.6	3.0

12. Meteorological station :   Milan Linate
13. Latitude :   45°28' N
14. Longitude :   9°12' E
15. Altitude :   120 m
16. Distance from site :   7.5 km (as the crow flies)

Note : W = Winter
S = Summer

(1) Refers to the values of the Po plain in general.
(2) According to the long term data on the map on p. 5 the annual solar radiation intensity on a horizontal plane is around 5 000 MJ/(m².yr). (N.O.T.E.).

2. Plant

2.2. Drying place

1. Drying system : Two cells with central hot air duct. Aulendorf-type drying system.
2. Material : Wood (grating and ducts)
3. Dimensions : 50 m² ground surface/cell
4. Capacity : 20 t of produce at 15% moisture per cell.
5. Ventilation - Power : 3 - 6 - 11 kW (three speeds)
   - Flow rate : 5.5 - 8.3 - 11.0 m³/s
6. Conventional heat source : none
7. Diagram : cf. fig. 2.2.7.

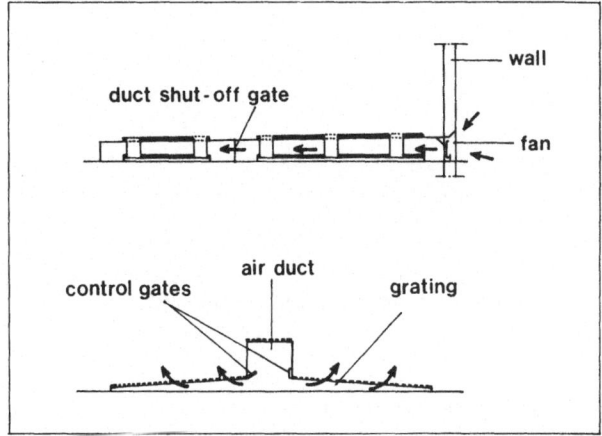

fig. 2.2.7. : diagram of drying system.

2.3. Collector (flat plate)

1. Material :
   - Absorber :   Rubber
   - Coolant :   Water
   - Working pressure :   80 kPa
   - Collector fluid content :   3 l/m²
   - Emittance of plate :   0.9
   - Coefficient of absorption
     of visible light :   0.95
2. Dimensions :
   - Aperture area :   228 m²
   - Absorber area :   180 m²

   The plant is formed by five custom built collectors of 36 m² net each. 4 collectors are covered by a corrugated GRP sheet (corrugation height 18 mm). 1 collector is covered with alveolar polycarbonate (thickness 10 mm)

3. Flow rate : 15.4×10^{-6} m³/(s m²)

4. Diagram of collector : cf. figure 2.3.4.
   Orientation : South   Tilt : 40°

5. Collector performance
   The equations of instantaneous collector efficiency are based on outdoor test-runs on 6 m² collectors built according to the same technology.
   Test conditions : diffuse radiation 25% ; wind, 1 m/s ; fluid : water ; flowrate : 33×10^{-6} m³/(s m²).
   The obtained formulas are :

   $$\eta_c = \begin{cases} (0.7 - 7.1 \frac{\Delta T}{I_c}) & \text{for a collector with Polycarbonate cover.} \\ (0.7 - 7.5 \frac{\Delta T}{I_c}) & \text{for a collector with GRP cover} \end{cases}$$

   with ΔT : temperature difference between collector and ambient air (K)

   I_c : radiation on collector surface (W/m²)

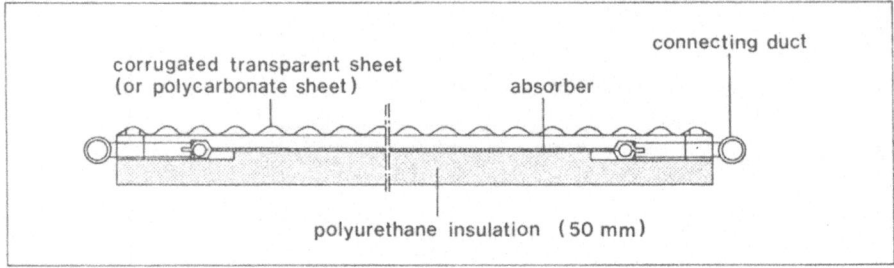

Figure 2.3.4.: Horizontal section of collector.

2.4. Heat storage

1. Medium of heat storage : water
2. Description, location in the system :
   - metal tanks with waterproof lining
   - the tanks are located outside
3. Insulation : glasswool
4. Dimensions : The storage tank, connected to the experimental drier, contains 10 m^3 of water.
   The one used for hot water preparation for the dairy farm has a volume of 1 m^3.
5. Thermal capacity : 45.9 MJ/K for the total of 11 m^3 of water.
6. Overall heat loss coefficient : assessed at 920 W/K for the 2 tanks together.

2.5. Installation parts and measuring devices

1. Technical description of the installation parts :
   Pumps : Two (one standby), 1 kW each.
   Air/water exchanger : Effective power (with 55°C water-35 kW ; copper exchanger with aluminium fins
   Fan : Refer to 2.2.5. The fan has a helicoidal impeller.
   Differential thermostat : Standard ON/OFF type.

2. Description of the sensors
   - Ambient temperature : Weather station by SIAD, Bologna, with bimetallic thermometer.
   - Fluid temperature : Pt 100 thermoresistors by LSI, Milan, to DIN 43760 (0.2°C tolerance).
   - Solarimeter : Kipp & Zonen, Model CM5
   - Flow rate : Floater-type flowmeter by ASA, Milan, precision 2%.
   - The measured data were recorded every 2 minutes by a galvanometric recorder on graph paper.

3. Products

3.2. Drying-products

1. Designed for special products : Mixed meadow forage
   Lucerne
2. Required conditions of inlet air : Common practice recommends approximately 400 m^3/h air flow per m^2 of perforated wood floor, air temperatures 5 to 10 K higher than the ambient and air humidity

below 50%.

3. Properties of the products : Produce is fed to the drier at average 50% moisture ; ventilation continues until the moisture content drops to 12 - 15%. Each batch is of 3 - 3.5 t dry.

4. Other (related) products which could be dried :
The plant is designed for forage drying only.

5. Any special treatment of the crop :
After mowing, the forage can be conditioned (by pressing with rubber rollers) to make drying faster and more uniform.

4. Economic aspects

1983 prices

4.1. Economic environment

1. Energy price of oil	0.31 ECU/1
2. Energy price of electricity	0.075 ECU/kWh
4. Manpower cost (relevant for the needs of functioning of the installation)	9.3 ECU/hr (labour cost genetic filter)
5. Interest rate of loans for farm equipment (rate without subsidy)	22%/yr
6. Subsidies available for the solar and competitor farm equipment (as capital grant of interest rate reduction)	50-60% (1) 30-50% (2)

(1) Interest - also for reference solution
(2) Capital - only for solar plant

4.2. Investment expenses

		Solar + support energy	Reference solution (Forage drier cf. p. 276)
1. Total investment budget	ECU	34 830 *(1)*	3 225 *(1)*
2. Cost of equipment :			
– cost of collectors per m² net surface (placement cost included)	ECU	135	none
– expenses for buildings	ECU	750	1 500
– cost of regulation equipment	ECU	300	150
– cost of 20 kW boiler (aux. heater)	ECU	1 120	
– other material costs	ECU	8 360	1 575
3.– Time of personnel for the conception, installation and start-up of equipment	hours	390 *(2)*	30
– Cost of personnel for the conception, installation and start-up of equipment	ECU	4 550	450
4. Lifetime of the equipment (evaluation)			
– solar collectors	years	15 *(3)*	15
– Boiler	years	10	
5. Residual value of the investment	ECU	?	37

(1) Material costs only
(2) This was the first installation (university personnel)
(3) Replacing the transparent cover is budgeted after 7.5 years.

4.3. Operating expenses

		Solar		Reference solution (Forage drier cf. p. 276)
		forage drying	hot water prep.	
1. Energy cost	ECU/yr	750	150	2 100 *(1)*
2. Maintenance cost				
– spare part	ECU/yr	150	37	60
– salaries	ECU/yr	300	52	225
4. Other manpower costs	ECU/yr	1 400 *(2)*	225 *(3)*	837 *(4)*
5. Insurance cost *(5)*	ECU/yr	750	none	750

(1) The reference solution considered is one supplying the same amount of heat as the solar plant.
(2) 0.5 h/day during 300 days/yr for controls, adjustments, and daily insolation readings.
(3) Labour cost involved in operating the auxiliary hot water supply to the milking parlour during the months when the solar plant cannot provide the required energy.
(4) Labour cost for operation of the energy supplying part of the reference solution : 1 h/day over the 90 day/yr drying season.
(5) Generally, insurance either covers the dried produce and the building where the produce is stored in the drier, or the building in which the thermal plant is located.

4.4. Hidden cost and benefits

	Solar		Reference
	Forage drying	hot water prep.	solution
1. Production obtained (tons/drying season)	40(1)	250(2)	40(1)
2. Average sales price per unit produced ECU/ton	150		150
3. Total income obtained (3) ECU/drying season (4.4.1. x 4.4.2.)	6 000		6 000
4. Storage costs due to the installation ECU/m³	0.82		0.82

(1) Hay at 15% moisture
(2) Hot water at 40-45°C
(3) Hay is fully consumed on the farm

5. Results

5.1. Method of calculation

- The solar radiation on the collector, during its operation time (line 5 in section 5.2.) is calculated as follows :

$$Q_{ci} = S_c \int_o^t I_c \, dt$$

with : Q_{ci} = incident energy (J)
S_c = net collector surface (m²)
I_c = instantaneous insolation, read every 2 min (W/m²)
t = time, over which the collector is actually working (s)

- The solar energy input (= the energy that the collector can actually collect) is calculated as follows :

$$Q_{co} = c_w \, \mathring{V}_w \, \rho_w \, (T_{co} - T_{ci}) \, t$$

with : Q_{co} = thermal energy supplied by the collectors (J)
c_w = specific heat of water (4 180 J/kg°C)
$\mathring{V}_w$ = mass flow rate of water in the collector circuit
(flow meter readings) (m³/s)
ρ_w = specific mass of water (kg/m³)
T_{co} = mean temperature at collector outlet (°C)
T_{ci} = mean temperature at collector inlet (°C)
t = operating time under given conditions (s)

- The other energy flows are calculated in the same way, based on measured values of liquid flow rates and operating temperatures.
- The calculation error of the tabulated data is expected to be smaller than 15%.

Note : Only the drying part, in combination with the 10 m³ storage, has been monitored. No data are available of the performance of the solar system as to hot water preparation for the milking parlour.

5.2. Weather conditions and energy

The results which follow refer to the four drying cycles which were performed in the hay season of 1980.

1st Drying cycle (3-6 June)
- 5.94 tons of forage, with 55% initial moisture content were dried to a final 20% m.c. (water removal 2.6 tons)
- The storage had been heated to 83°C prior to starting the drying cycle.
- The air was blown through the hay during a total of 73.5 hours.

3/6	4/6	5/6	6/6		Date
15-19	8.45-19.30	9-19	8.30-19	hour of the day	Time of operation of the collectors
1296	2714	4277	3758	MJ	Incident solar radiation on the collectors over given collector operation time (Ω_{ci})
61	42	45	47	°C	Mean operating temperature of collectors
421	1404	2282	2016	MJ	Thermal energy (Q_{co}) supplied by the collectors [1]
32	52	53	53	%	Mean collector efficiency (η_c)
421	400	778	886	MJ	Part of Q_{co} transferred directly to the drying air (cf. operating mode stand # 3)
-	1004	1505	1130	MJ	Part of Q_{co} to storage (cf. operating mode stand # 1)
2426	1044	1382	-	MJ	Energy output from storage (cf. operating mode stand # 5) [2]

(1) *Since T_{co} and T_{ci}, used for the calculation of Q_{co} (cf. section 5.1.), are being measured quite far away from the collectors itself, Q_{co} and η_c are strictly speaking an understatement of the real collector performance, given the pipe losses that are incorporated in the calculated values.*

(2) *Since the storage was loaded before the drying operation, its energy output is not correlated with Q_{co}, the value of which was calculated during drying time only ! For correct interpretation of the figures, it is important to understand the operating modes of the system, which are very different from those applicable in most collector-storage combinations given in the book.*

2nd Drying Cycle (July 1st-8th)
- 9.42 tonnes of forage at 0.65 initial m.c. were dried to 20% m.c. (water removal 5.3 tons)
- Due to regular low insolation, the process could not be run under optimum conditions.
- Storage temperature was 81°C initially.
- The drying air was blown through the hay during 125.5 hours.

1/7	2/7	3/7	4/7	5/7	6/7	7/7	8/7		Date
8-19	11-19	8-19	8-16	11-17	8-19	9-16.30	8-19	hour of the day	Time of operation of the collectors
3240	1800	2416	2538	1879	2560	2502	4421	MJ	Incident solar radiation on the collectors over given collector operation time (Q_{ci})
59	38	38	40	48	42	40	44	°C	Mean operating temperature of collectors
778	1008	965	871	745	842	1195	2297	MJ	Thermal energy (Q_{co}) supplied by the collectors [1]
24	56	40	34	40	33	48	52	%	Mean collector efficiency (η_c)
778	90	965	706	745	842	151	835	MJ	Part of Q_{co} transferred directly to the drying air (cf. operating mode stand # 3)
-	918	-	166	-	-	792	1674	MJ	Part of Q_{co} to storage (cf. operating mode stand # 1)
2261	961	-	-	144	-	1044	1426	MJ	Energy output from storage (cf. operating mode stand # 5) [2]

(1), (2) cf. footnotes, given under first drying cycle.

3rd Drying Cycle (August 3rd-5th)
The drying cycle took place under optimum weather conditions.
- 5 tonnes of forage, at 0.55 initial m.c. were processed (water removal 2.2 tons)
- The storage tank had been previously heated to 75°C.
- Ventilation occurred during 46.5 hours in total.

3/8	4/8	5/8		Date
13.30-15.30	8-15.30	12-19	hour of the day	Time of operation of the collectors
1102	4244	2650	MJ	Incident solar radiation on the collectors over given collector operation time (Q_{ci})
70	64	51	°C	Mean operating temperature of collectors
245	1253	1199	MJ	Thermal energy (Q_{co}) supplied by the collectors [1]
22	30	45	%	Mean collector efficiency (η_c)
245	792	1199	MJ	Part of Q_{co} transferred directly to the drying air (cf. operating mode stand # 3)
-	461	-	MJ	Part of Q_{co} to storage (cf. operating mode stand # 1)
1087	1465	-	MJ	Energy output from storage (cf. operating mode stand # 5) [2]

(1), (2) cf. footnotes, given under first drying cycle.

4th Drying Cycle (September 2nd-4th)
7 tonnes of forage at 0.55 initial M.C. were dried (3 tons water removal)
- Weather conditions were optimum.
- Initial storage temperature 51°C.
- Ventilation occurred during 59.5 hours in total.

2/9	3/9	4/9		Date
8.30-19	8-19	8-19	Hour of the day	Time of operation of the collectors
4180	3704	3694	MJ	Incident solar radiation on the collectors over given collector operation time (Q_{ci})
60	40	52	°C	Mean operating temperature of collectors
1278	2066	1246	MJ	Thermal energy (Q_{co}) supplied by the collectors [1]
31	56	34	%	Mean collector efficiency (η_c)
1278	810	1246	MJ	Part of Q_{co} transferred directly to the drying air (cf. operating mode stand # 3)
-	1256	-	MJ	Part of Q_{co} to storage (cf. operating mode stand # 1)
1170	1256	-	MJ	Energy output from storage (cf. operating mode stand # 5) [2]

(1), (2) cf. footnotes, given under first drying cycle.

5.5. Drying results

1.	3/6	1/7	3/8	2/9		from:	PERIOD	
2.	6/6	8/7	5/8	4/9		to:	1980	
3.	20 000	20 000	20 000	20 000	m^3/h	flowrate		
4.	3.2	3.0	4.1	3.9	K	Δt (1) INLET AIR		
6a.	55	65	55	55	%	moisture a. initial		
6b.	20	20	20	20	%	b. final PRODUCTS		
7.	2 700	3 150	3 250	3 150	kg	useful quantity (dry matter)		
10.	2.8	2.1	2.2	1.9	$MJ/kg\ H_2O$	Thermal energy supplied by the collectors per kg evaporated water		
	0.31	0.26	0.23	0.22	$MJ/kg\ H_2O$	Energy input into ventilation per kg evaporated water		

(1) Temperature difference of the drying air with ambient air. Calculated over the entire time of operation of the drier (global average).

5.8. Considerations concerning the development of simple liquid collectors adapted to the situation of medium size farms in Northern Italy.

I. The need to lower the cost of solar collectors led the Institute of Agricultural Engineering, Milan University, to experiment with solutions other than the costly metal collectors. The research was carried out, keeping in mind the characteristics of the agricultural sector.
Specifically :
- Modular components were designed and built, with which solar collectors of the required size can be assembled by unskilled labour.
- Solar collectors for roof retrofitting were designed to cut down on overall construction costs (building + solar collectors).
As to the materials, preference was given to plastic transparent covers and absorbers and to metal frames.
A. The following absorbers were successfully tested :
- A synthetic (EPDM) rubber module.
- A polypropylene module, suitable for high pressure.
These materials were selected for the following reasons :
- Farming requires simple, reliable facilities both for farm and domestic applications.
- A feature of the plastic absorber is its simplicity : it requires fewer and simpler hydraulic connections and can be built to form large, easy to move modules. (60 m^2 – able to be rolled up).
- No anti freeze liquid is necessary, since water can freeze in the chosen plastic pipes without harmful effects. This does away with the need for a heat exchanger, since the water that is to provide the heat demand can flow through the collector directly *(1)*. This has a twofold advantage : lower cost and higher efficiency.
- The lifetime of these materials is of the same order of magnitude of the metals used for such applications (aluminium, copper, stainless steel).

(1) As can be found in the conclusions of this report, the problem with this point of view is that frost danger of the metal pipes of the collector circuit remains, so in fact antifreeze liquids remain essential. (N.O.T.E.)

B. Two alternatives were considered for the transparent covers :
  - Corrugated GRP plates, made of glass fibre or polyamide fibre reinforced polyester (ave. weight 1.3 kg/m^2), coated with a polyvinylfluoride film on the side exposed to the sun.
  - Polycarbonate alveolar plates, 6 mm thick, weighing 1.3 kg/m^2. (Glass has excellent optimal properties, but was not considered because of its weight and fragility and because it requires a complicated carrying framework).

  GRP - plates (1st alternative) are supplied at a standard width of 1.6 m, and can have any length, up to 14m. Placement is easy : sealing can be done simply by overlapping the plates. The mechanical properties are very good, while optical and thermal properties are fair (transmissivity = 70-75% ; k = 6 W/(K m^2)). The life time of the coated version is excellent. If the collector has to be assembled on an 8-10 m high building and the plates have to be manually carried to the roof, an unskilled labourer can place up to 60 m^2 in one day (5 screws per m^2 are needed for safe fastening).

  The Polycarbonate plates (second alternative) are normally used for small collectors (20 m^2 max.), which operate at high temperatures even in winter. Installation is however more time consuming and complicated than that of GRP plates. Good mechanical and fair optical characteristics, as well as good thermal characteristics (transmissivity = 65-70% ; k = 4 W/(K m^2)). For the user, however, the cost is 60% higher than that of a coated GRP unit.

  Altogether the most suitable transparent cover material for farming use is coated GRP, mainly because of its favourable cost/performance ratio.

C. Among insulation materials, tests found 2 types suitable :
  - 20 to 40 mm thick polyurethane foam, coated with a glass fibre sheet (for roofs).
  - a self-supporting sandwich structure composed of two metal sheets (preferably aluminium) with an inserted polyurethane core (for freestanding collectors).

D. Placement procedure :
  - For properly oriented roofs, the roofing material (asbestos cement or tiles) is removed and a suitable underlayer (wood planks or brickwork slab) is laid, if this does not yet exist. Afterwards, the insulation material, absorbers and transparent cover are installed. A 0.08 to 0.1 m deep box of the required surface (mostly 80 to 200 m^2) is built to house the plastic absorber. When no properly oriented roofs exist, self-supporting collectors are built with metal sandwich components.

II. As mentioned before, heat exchangers and storage account for a sizeable part of the cost of a solar plant. One could decrease this cost by :
  - Doing away with the exchangers, given the synthetic material of the absorbers [1].
  - Using open storage tanks made of asbestos cement or polyethylene (8 mm thick, max. capacity 4 000 litres). When open tanks are used, it is better that they be drained when idle.

  The Institute of Agricultural Engineering, Milan University, used the above materials and technologies to build some plants on farm premises. Total manufacturing costs for those solar systems range

(1) See footnote on previous page.

from 130 ECU to 200 ECU per m^2, depending on surface size and complexity. Under northern Italian weather conditions (45°N), and with high utilization ratios (0.7-0.8), the expected payback period is of the order of 6 to 10 years.

6. Remarks

1. Even though the energy input into the milking parlour has not been assessed, one can get a rough idea of its importance by means of a simple calculation.
   Assuming
   - the cold water is available at about 12°C (= average annual ambient temperature)
   - the cows are dried off at the same moment of the year. (this means hot water is only needed 300 days/year)
   - the efficiency of the hot water preparation (1) is comparable to the % of the total energy need that can be provided by the solar system ($\eta2$).

   Then the energy, collected by the solar system over a whole year for hot water preparation (Q) is :

   $$Q = \eta2 \left(\frac{1\ 000\ kg/day\ .\ 300\ days/year\ .\ (40-12)°C\ .\ 4\ 180\ J/(kg\ K}{\eta1} \right)$$

   ≈ 32 GJ/year.

   This amounts to a value comparable to the total energy ($\pm$ 35 GJ) collected during the drying tests described above, which run over a period of 18 days.
   If the average energy collection over the 18 test drying days would be representative for the total of 75 full capacity drying days, the extra solar heat gain for the hot water provision would only be $\pm$ 20% of that for the drying activity.
2. This project description is based on 2 reports, received from the study center in question. Unfortunately, several inconsistencies and contradictions were spotted. Even though some of those are still present in the given description, for most of them, we chose the more likely value of interpretation. It should be clear then, that due care needs to be taken with the interpretation of the given information.
   We went ahead with this description because of following reasons :
   - Where contradictions occurred in given results, etc... the different net resulting values still remain in the same range.
   - The goal of this book, as mentioned in its introduction, is not so much to provide final conclusions, but to present a wealth of materials, with conclusions of an often provisional nature, in order to best illustrate the direction in which research on the subject is going.
   - This project has a few interesting aspects connected to it : e.g. Its assessment of the feasibility of <u>water</u> collectors for drying purposes, as well as its way of revealing the fact that effective design of a multi-purpose solar system is not an easy task.
   An essential element in feasibility of such a system is finding suitable adjustment of the demand of the different applications to the supply, and this over the major part of the year.

6. The use of solar energy for bulb conditioning and soil heating

Project location

ANNA PAULOWNA, NETHERLANDS

Study Center :
- Project leader : Ing. D.E. Brethouwer
- Institution : Technisch Physische Dienst TNO-TH
  (Institute of Applied Physics)
  Stieltjesweg 1
  2628 CK DELFT
  or P.O. Box 155
  2600 AD DELFT
  Netherlands

Running since 1981

Project description

A common practice in hyacinth bulb growing firms is the application of soil heating, in order to enable growth of hyacinths from April to June, rather than during summer. (This practice allows for early flowering -around Christmas time- of the plants grown from the produced bulbs).
After harvesting, the bulbs are conditioned during summer-time by means of ventilating air at a temperature of 20° - 30°C.
Given the amounts and distribution throughout the year of the heat demand, the potential of integrating a solar system is worth assessing. Therefore

2 practically identical roofs, covered with a total surface of 700 m² of air collectors, in combination with an auxiliary heater were installed on two new halls of a bulb-farm.

The intercepted solar heat is transferred via an air to water heat exchanger to 2 storage tanks in parallel of 6 m³ each.

This heat is being used in 3 different ways :
- From April to June for soil heating.
- From June to November for bulb conditioning.
- From November to April for floor heating of the work area.

A one year monitoring period is being carried out with the intention of gaining experience with the set-up, and also in order to validate a simulation model that is being developed.

This model will make the design of optimized set-ups possible, and will help determine the feasibility of such installations.

Scheme

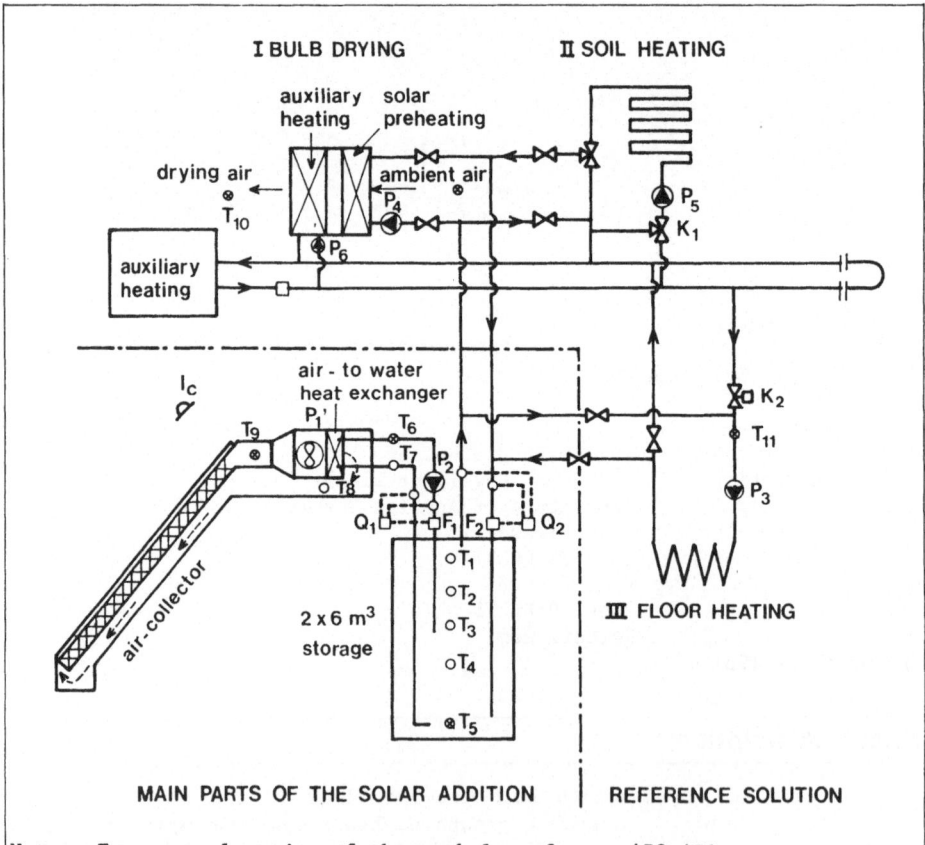

Note : For an explanation of the symbols, cf. pp. 470-471.

Operating modes
I. Primary solar circuit : only if $T_9 > 30°C \rightarrow P_1'$ on.
  Secondary solar circuit : if $T_6 > T_5 + 5°C \rightarrow P_2$ on
  if $T_6 < T_5 + 2°C \rightarrow P_2$ off

II. Load-circuits

(Note : In the following description of the operating modes, only the positioning of K_1 and K_2 are explained. The other valves are obviously also manually adjusted in such a way as to close off the non-operating circuits).

* soil heating
  - April to Mid-May : . K_1 is set in position manually (no auxiliary heating)
    . P_5 is on for 15 minutes out of every 2 hours (stored heat is used in that time)
  - Mid-May to Mid-June : . K_1 is set manually, in such a way that solar and auxiliary heater work in parallel.
    . P_5 is on continually.
* heating of ventilation air.
  - only if $T_{10} > T_{ambient} \rightarrow P_4$ on (= solar heating on)
  - only if $T_{10} < T_{desired} \rightarrow P_6$ on (= auxiliary heating on)
* floor heating
  - K_2, which creates a mix between solar and auxiliary heating, is adjusted by a thermostat in order to achieve a desired temperature in T_{11}. (P_3 is on)

Conclusions

- The 700 m^2 of collectors saved about 305 GJ of energy over the whole year.
- The solar system proved reliable over the 2 years of operation. (No break-downs occurred). Thanks to the design of the collector and duct system, pressure losses are small, and air leaks less significant than for most other collector designs.
- Air collectors were chosen because of financial and constructional considerations. From the efficiency viewpoint, the use of water collectors would be preferable.
- Of the 3 loads, the soil heating load proved to be very compatible with the solar supply : low temperature (18°C-60°), high load at a time when sun is available. Unfortunately the type of heating only occurs over a short period of the year, and the number of firms applying this practice is small.
  The air heating proved to be rather inefficient, since, given the low temperature-demand (20-30°C), little air heating needed to occur over the period of the day when sun was available. The need for storage of the collected heat strongly reduced the efficiency of the solar collection.
  When no storage is needed, all the heat exchangers could be bypassed. Such a type of application would be very promising.
  As to the floor-heating, the winter sun did not contribute any heat to this load, given the low solar supply, as well as the small angle of the collector surface (20° from horizontal).
- Some problems were experienced with the operating modes :
  + The operating mode of the primary collector circuit required high energy input for little gain, since the fan would remain on, even if

no heat was being collected in the storage tank ($T_6 < T_5 + 2°C$).

Therefore the operating modes of both collector circuits were changed into :
if $T_{absorber} > T_5 + 12°C \rightarrow P'_1$ and P_2 on

if $T_{absorber} < T_5 + 2°C \rightarrow P'_1$ and P_2 off
+ The operating mode for the solar heat supply to the drying air contains the possibility that the solar supply exceeds the demand, while P_4 remains on.
Thus the temperature of the drying air could exceed the desired value. In another instance P_4 could be on while no heat is available in the storage.
In order to solve these 2 problems, the operating mode for P_4 was changed.

PROJECT ANALYSIS

1. Site and climate

1.1. Site

Latitude :	52°50' N
Longitude :	4°45' E
Altitude:	0 m
Nearest main city :	Amsterdam
Distance from main city :	60 km
Direction from main city :	North
Obstructions :	None

1.2. Site Micro-Climate

The long term averages are derived from the meteorological station at De Kooy, only 10 km from the project. So there are no local variations from the described conditions.

1.3. Annual long-term averages

			Project	Country in general
1. Prevailing wind direction	W		S	SW
	S		SW	SW
2. Average wind speed	W	m/s	6.8	4.2
	S	m/s	6.1	3.5
3. Total precipitation		mm/yr	795	775
4. Absolute humidity	W	g/kg	4.8	4.5
	S	g/kg	7.9	7.7
5. Global irradiation on horizontal plane		MJ/(m^2.yr)	3 867	3 525
6. Diffuse proportion		%	$\simeq$ 60%	$\simeq$ 60%
8. Hours sunshine		hrs/yr	1 550	1 505
9. Ambient temperature		C	9.0	9.2
10. Average max. temp. (Jul/Aug)		C	19	21
11. Average min. temp. (Jan)		C	0.5	− 0.7

12. Meteorological station   de Kooy
13. Latitude   52°55'
14. Longitude   4°45'
15. Altitude   0 m
16. Distance from site   10 km N

Note : W = Winter
S = Summer

2. Plant

2.2. Drying-place

1. Drying system : Heating of ventilation air.
   The supplied fresh air is heated by solar energy and post-heated by auxiliary energy.
2. Material : 1st Hall type : equipped with a greenhouse roof (part of old installation)
   2nd Hall type : equipped with collector roof
3. Dimensions : 1st type : 12 x 41 m height : ca. 6 m
   2nd type : + 16 x 41 m height: ca.6 m
5. Ventilation   − Power : Hall type 1 :   80 x 10^{3}W
                       Hall type 2 : 110 x 10^{3}W
       − Flow rate : Hall type 1 : 2.8 m^3/s
                      Hall type 2 : 4.2 m^3/s
6. Conventional heating source : gas-fired heater
   Calorific power : 600 x 10^3 W
7. Diagram : cf. figure 2.2.7.

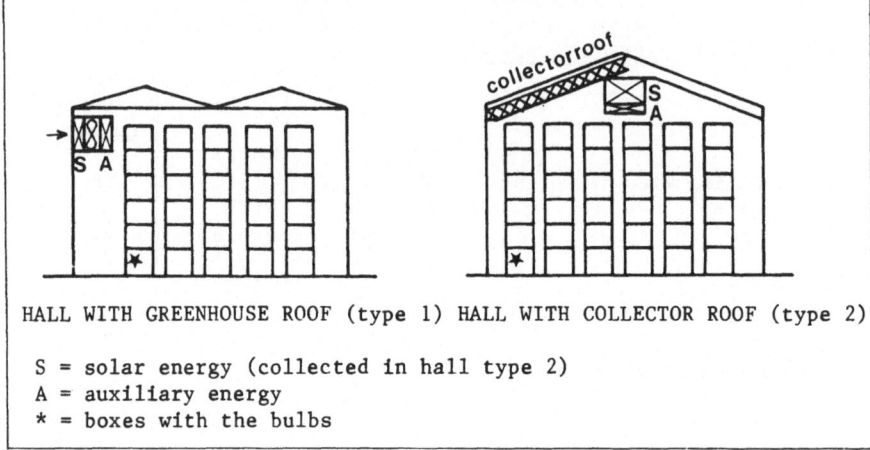

HALL WITH GREENHOUSE ROOF (type 1) HALL WITH COLLECTOR ROOF (type 2)

S = solar energy (collected in hall type 2)
A = auxiliary energy
* = boxes with the bulbs

Figure 2.2.7. : Diagram of bulb conditioning halls.

2.3. Collector (flat-plate collector)

1. Absorber
   - Coolant :   Air
   - Material :   Aluminium
   - Aperture area :   700 m²
   - Absorber area :   700 (320 + 380 m²) m²
   - Collector fluid content :   29 1/m²
   - Working pressure :   0 Pa
   - Flow rate : $3.5 \times 10^{-3} \cdot m^3/sm^2$
   - Emittance of plate :   0.95
   - Coef. of absorption of
     visible light :   0.95
2. Glazing :
   - Emittance of glass :   0.85
   - Number of glass covers :   1
   - Material :   glass
   - Special characteristics :   none
3. Insulation : Poly-Urethane min. 5-10 cm
4. Mounting :
   - Orientation :   201.5° (SSW)
   - Tilt :   20°
5. Collector performance (cf. fig. 2.3.5)

2.4. Heat storage

1. Medium of heat storage : water in 2 tanks of 6 m³ each.
2. Description, location in the system : The heat storage is situated in one of the halls on which the collectors are placed. It is intended as a short term storage.
3. Insulation : The storage is placed in a square sheet piling which is filled with polyurethane foam. The thickness of the insulation is at least 10 cm.
4. Dimensions : Height of the storage tank : 3.10 m
                Diameter   1.55 m
                Sheetpiling : height   3.50 m
                              length   1.80 m
                              width   1.80 m

5. Thermal capacity : 25.0 MJ/K
6. Overall heat loss coefficient : 140 W/K
7. Diagram : cf. figure 2.4.7.

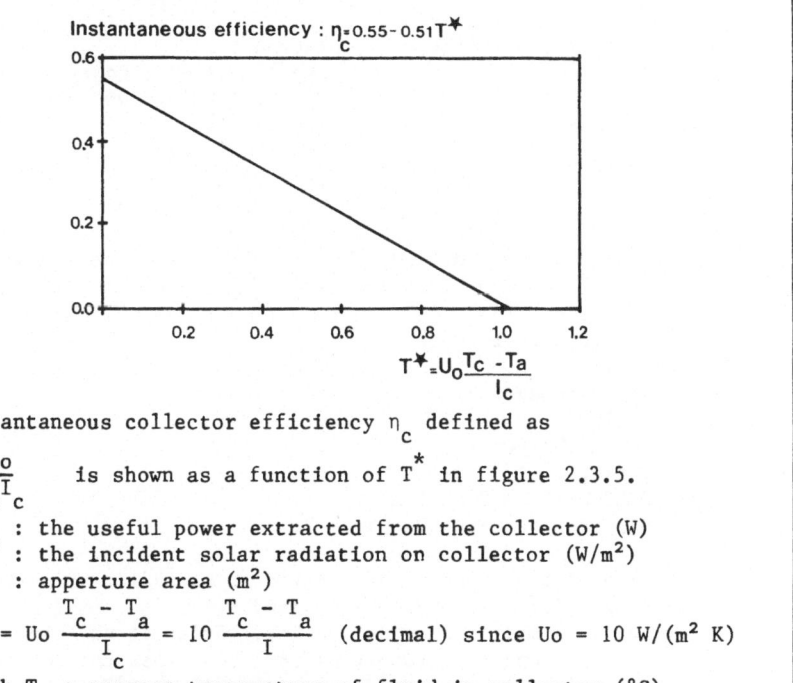

Instantaneous efficiency : $\eta_c = 0.55 - 0.51 T^*$

$$T^* = U_0 \frac{T_c - T_a}{I_c}$$

The instantaneous collector efficiency η_c defined as

$$\eta_c = \frac{E_{co}}{S_c I_c} \quad \text{is shown as a function of } T^* \text{ in figure 2.3.5.}$$

with E_{co} : the useful power extracted from the collector (W)
I_c : the incident solar radiation on collector (W/m²)
S_c : apperture area (m²)

$$T^* = U_o \frac{T_c - T_a}{I_c} = 10 \frac{T_c - T_a}{I} \quad \text{(decimal) since } U_o = 10 \text{ W/(m² K)}$$

with T_c : average temperature of fluid in collector (°C)
T_a : ambient temperature around collector (°C)

Figure 2.3.5. : Instantaneous collector efficiency as a function of solar radiation and temperatures.

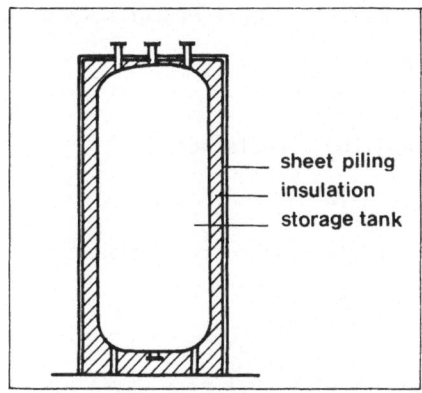

sheet piling
insulation
storage tank

Fig. 2.4.7. : Diagram of heat storage tank.

2.5. Installation parts and measuring devices

1. Technical description of the installation parts :
- Solar collector system
  (per collector roof)
  - Fan at collector outlet :   Thermo Air
    Type HAT 18-18, Power 4.0 kW,
    Flow : 7 000-15 000 m^3/h
  - Heat exchanger :   Thermo Air (2x)
    Type WWHB 18/3
    Capacity 110 kW
  - Pump   WILO RS 50 0.54 A
- Soil heating :
  Soil heating is part of the old installation. The heated water is
  fed through a plastic tube situated 20 cm deep in the ground. The
  flow is about 9 m^3/h.
- Heating of ventilation air :
  Fan 1st hall :   2.78 m^3/s
  Solar heat exchanger :   80 * 10^{3}W
  Fan 2nd hall :   4.17 m^3/s
  Solar heat exchanger :   110 * 10^{3}W
- Floor heating :
  Pump : WILO RS 25 - 80 0.4 A
      Distance between the pipes 30 cm

2. Description of the sensors (cf. scheme on p. 290) :
   - heatmeters (incl. flowsensor) (Q_1 and Q_2)
     principle : multiplication and integration of the flow through
             and temperature difference across the section
     precision : all-over precision 1°C < ΔT < 5°C : 8%
                               5°C < ΔT < 12°C : 4%
                            12°C < ΔT < 80°C : 2%
     measuring range : ΔT : 80°C
     flow F_1 : 1 - 7 m^3/h   flow F_2 : 0.12 - 12 m^3/h
   - pyranometer (including integrator) (I_c)
     principle : temperature difference measurement using a thermopile
     precision : 5%
     measuring range : 0 - 1 200 W/m^2
   - temperature sensors
     a. T_6 & T_7
         principle : platinum 100 Ω resistance
         precision : 0.5%
         measuring range : 0 - 100°C
     b. $T_1 - T_5$, T_8 & T_9
         principle : thermocouple copper-constantan
         precision : 1.0°C
         measuring range : 0 - 200°C

4. Data-collection
   The ambient temperature, wind velocity and wind direction are
   extracted from the records of the meteorological station of De Kooy
   (10 km NW of the project location).

On site measurements	Measuring frequency	Storage	Units
Temperatures	Continuous reading with 2 multipoint-recorders	paper	°C
Heatmeters	reading every day	manual on paper	kWh/day
Pyranometer (incl. integrator)	reading every day	manual on paper	kWh/(m².day)
Flowsensors	reading every day	manual on paper	m³/day

- multipoint recorder 1 :	Honeywell electronic 15 - 12 points thermocouple (CuCon) - 28 - + 100°C - 4 points mV 0 - 1 mV
- multipoint recorder 2 :	Chessel model 301 accuracy 0.25% of fullspan 3 points (PT 100)
- integrator	Kipp & Zn. type CCI.

3. Products

3.2. Drying-products

1. Designed for special products : bulbs
2. Required conditions of inlet air :
   The ambient air will be heated in summer between 22 - 30°C.
   The flow rate is 2.78 m³/s for the 1st hall and 4.17 m³/s for the 2nd hall. There is no control for the humidity.
3. Properties of the products :
   During the summer, the bulbs have to be conditioned at a temperature level rising from 22°C in June to 30°C in October.
4. Other (related) products which could be dried : None
5. Any special treatment of the crop ? No

4. Economic aspects

4.1. Economic environment

1. Energy price of gas	0.20 or 0.16 ECU/m³ *(1)*
2. Energy price of electricity	0.10 ECU/kWh
3. Energy prices evolution prospects (inflation rate : 2 - 3%)	5%
5. Interest rate of loans for farm equipment (rate without subsidy)	9%/yr
6. Subsidies available for the solar and competitor farm equipment (as capital grant or interest rate reduction)	none

(1) 0.20 ECU/m³ for 0-30 000 m³ per year
0.16 ECU/m³ for > 30 000 m³ per year

4.2. Investment expenses

Prices in January 1981

	Solar only
1. Total investment budget	64 350 ECU
2. Cost of equipment :	
- cost of collectors per m²	20 ECU *(1)*
- cost of regulation equipment	650 ECU
- cost of other materials	13 700 ECU
3. - Time of personnel for the conception, installation and start-up of equipment	2 000 man-hours
- Cost of personnel for the conception, installation and start-up of equipment	36 000 ECU
4. Lifetime of the equipment (evaluation)	30 years

Note : Because the solar energy is integrated in a partly existing installation, it was not possible to provide all the data concerning expenses, because the owner does not know these figures.
(1) Most likely material costs only. The labour costs are then included under point 4.2.3. (N.O.T.E.)

4.3. Operating expenses

	Solar only
1. Energy cost	747 ECU/yr
5. Insurance cost	429 ECU/yr

---

5. Results

5.1. Method of calculation

- Points 5.2.3. and 5.2.4. are data from the meteorological station nearby.
- The other results are based on measurements made on the performance of the 320 m² roof (one of two practically identical parts).

5.2. Weather conditions and energy

1.	1/11	1/12	1/1	1/2	1/3	1/4	1/5	1/6	1/7	1/8	1/9	1/10		from: 1/11/1982 PERIOD
2.	30/11	30/12	31/1	28/2	31/3	30/4	31/5	30/6	31/7	31/8	30/9	31/10		until: 31/8/1983
3.	8.4	4.2	6.2	1.1	5.1	7.7	10.2	14.9	17.9	17.6	14.2	11.8	°C	average ambient TEMP.
4.	7.5	7.0	9.5	6.5	6.5	5.5	5.5	5.0	4.0	4.6	6.2	6.8	m/s	average ambient WIND
5.	32.3	18.7	28.6	66.8	115.4	137.4	152.9	224.0	200.2	192.5	117.5	95.0	GJ	SOLAR RAD on collector
7.	0	0	0	6.20	14.79	27.32	23.50	35.24	27.96	18.06	5.87	4.12	GJ	input energy STORAGE
8.	0	0	0	6.08	14.50	27.24	22.62	31.57	24.92	13.71	1.63	1.07	GJ	output energy STORAGE
9.	(19)	(15)	(12)	9.4	10.2	13.9	19.3	33.7	34.9	41.0	(32)	24.3	°C	average TEMP. storage
10.	0	0	0	6.08	14.50	27.24	22.62	31.57	24.92	13.71	1.63	1.07	GJ	USEFUL energy
11.	(0)	(0)	(0)	(0)	(0)	(90)	(330)	(95)	(65)	(40)	(0)	(0)	GJ	AUXILIARY HEATING energy (non-solar)
12.	0.11	0	0	0.83	1.73	2.94	2.18	3.21	3.02	2.45	0.8	0.65	GJ	energy for OPERATION

Note : Values between brackets were estimated, because no measurements have been carried out.
These results were obtained for only 1 of 2 roofs (320 m² collectors) cf. section 5.1.

5.5. Drying results

1.	1/11	1/12	1/1	1/2	1/3	1/4	1/5	1/6	1/7	1/8		from	PERIOD
2.	30/11	31/12	31/1	28/2	31/3	30/4	31/5	30/6	31/7	31/8		until	
3.							6.94	5.11	5.0	3.33	m^3/s	flowrate	
4.	No drying from 1/11 to 1/5						24	26	28	30	°C	temperature	INLET AIR
5.							42	39	41	40	%	humidity	

Note : given values are estimations.

5.7. An example of some measured data

1. Measured values over a day when the soil is being heated
   (system in continuous working mode). Cf. figure 5.7.1.

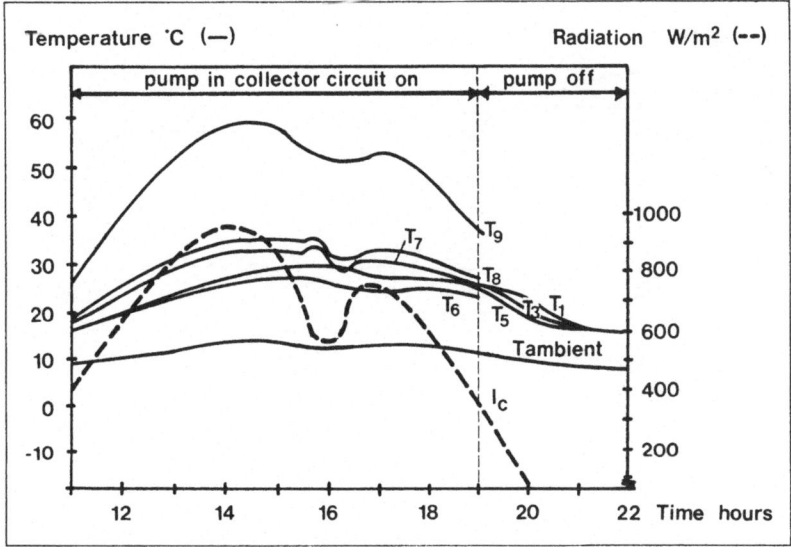

Figure 5.7.1. Measured values over a day when soil is heated.

2. Measured values over a day when air is being heated.
   (cf. Figure 5.7.2)

6. Remarks

This project, which strictly speaking does not fit as a drying project,
still has some main characteristics in common with drying.
E.g. : Some of the problems one has to face in trying to apply a solar
system for many purposes - typical for solar drying - are clearly
represented.
- Quite a few applications cannot be included because of their distance
  from the heat storage and the losses associated with heat transport.
  (Certain applications were not included in this project because of this
  reason).
- It is very hard to combine different applications in which the character
  of the load is compatible with the solar supply.

(cf. the incompatibility noticed in ventilation air and floor heating). The first factor could in some cases be influenced by building design, or by the use of portable collectors with all the difficulties associated with them.
Nevertheless the problems raised are of a prohibitive nature in quite a few cases.

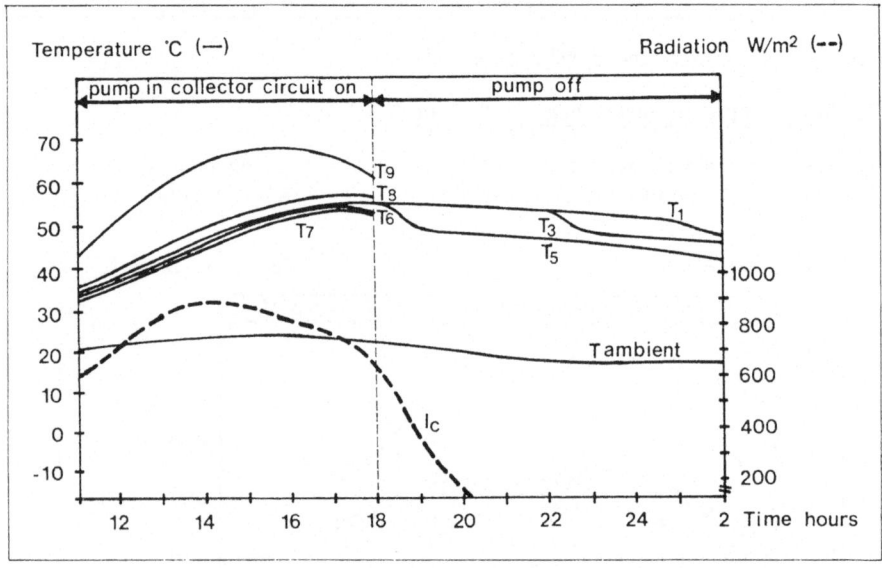

Figure 5.7.2 : Measured values over a day when air is heated.

7. Solar drying of grains

Project location

BAZIEGE, *(1)* FRANCE

Study Center :
    - Project leader :   J.C. Lasseran
                              Institut Technique des Cereales et Fourrages
                              (Station expérimentale)
                              Boigneville
                              F-91720 Maisse
                              FRANCE
    - Institution :   G.I.E. Secograins
                              3, allée des Sylphes
                              F-31520 Ramonville Saint Agne
                              FRANCE
Running since 1981

Project description

A new technique for bin grain driers (sunflower-seed; soja-bean; sorghum; maize) is being developed in order to transfer the dependence on fossil fuels to solar energy combined with a rather small amount of electrical power that can be provided by nuclear plants.

(1) 20 km SE of Toulouse.

- For 3 identical bin driers (diameter 5.34 m and height of grain 5.4 m) different cheap collector types were developed and compared.
  c_1 : a vertical collector of 80 m², wrapped around 2/3 of the bin drier.
  c_2 : a 90 m² flat plate collector set on a south facing roof.
  c_3 : a "greenhouse collector" of 160 m² ground surface with a black plastic absorber sheet covering the soil.
- In all 3 cases a 10 kW fan blows air through the collector at a rate of 3.2-3.8 m³/s. During the day the air is heated in the collector. At night an auxiliary heater of 14 kW heats up the incoming air by 4 K (the fan itself adds a temperature increase of 2 K).
- The bins are equipped with augers.
  Mixing augers for c_1 (batch drying only) ; discharge augers for $c_2 + c_3$ (continuous drying).
- Different parameters of the low temperature drying process were assessed
  - batch versus continuous drying
  - optimal automatisation of the ventilation as a function of the humidity of the air and the solar radiation.
- The resulting grain quality was tested (commercial value, nutritional value, hygienic value)
Goals of the experiment :
- Next to creating an independence from fossil fuels, the aim is to reduce the cost of drying in comparison to the conventional hot air drying methods.
New models are developed that can easily be added in existing situations.
- Finally the limits of the system are assessed (initial moisture content versus weather conditions ...)

Scheme

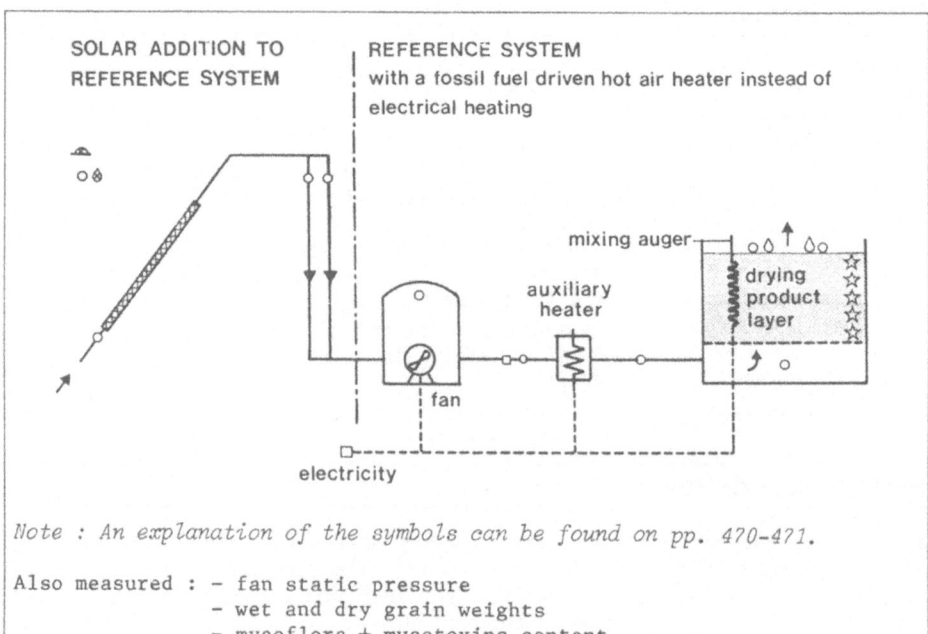

SOLAR ADDITION TO REFERENCE SYSTEM

REFERENCE SYSTEM with a fossil fuel driven hot air heater instead of electrical heating

mixing auger

drying product layer

auxiliary heater

fan

electricity

Note : An explanation of the symbols can be found on pp. 470-471.

Also measured : - fan static pressure
                - wet and dry grain weights
                - mycoflora + mycotoxins content

Operating modes :
+ Between 6 AM - 10 PM (Auxiliary Heater off)
  - If light intensity > 4 lux *(1)* → Fan on !
  - If r.h. < 80%   → Fan on !
  - If light intensity < 4 lux *(1)* and r.h. > 80% → Fan off
+ Between 10 PM - 6 AM
  Auxiliary heater on and Fan on
+ Mixing augers in c_1 break down the final moisture gradient when the average moisture content of 15.5% is reached.
  Discharge augers in c_2 and c_3 are operated based on the moisture content measured at 0.5 and 1 m height.

Conclusions

- Some tests were executed in November and December 1982 on the driers c_1 and c_2; c_3 was tested in 1983 (for encouraging first drying results of c_3, cf. section 6).
  The drying time of maize in c_1, from 21%-15.5% m.c., was 22 days. The evaporation rate was 236 kg of water/day.
  This required 5.38 MJ of electricity per kg of evaporated water.
  To dry sorghum from 19.7% - 15.2% m.c. it took 29 days in c_2 ; evapor. rate was 147.6 kg water/day ;
  specific electricity consumption 7.15 MJ/kg of water.
  Given the late time of the season, and the poor weather conditions during the measuring period, these results are a serious understatement of what can normally be attained with the described systems.
- The quality test results of the grain were positive.
- The average collector efficiency measured for an air temperature rise of less than 10 K was 50-55% for c_1, 40% for c_2, and 30% for c_3. (Some problems were experienced with accurately measuring collector air inlet temperatures).
- A simulation model was run over a 95 day-period of September to December and predicted a payback period of 4 - 5.3 years and specific electricity consumption values of 2.2 MJ/kg of evaporated water.
- Even though the results seem promising both from technical, economic and grain quality viewpoint, this drying technique, which implies a transfer from fossil fuels to nuclear or hydro-electricity, is suitable only in specific regions like the South of France and seems to be convenient for sunflower-seeds and sorghum (both expanding crops and with low temperature drying requirements). Only maize hybrids with low initial moisture content are adapted for solar drying.

- Further references :
  + "Sechage solaire ou bi-énergie du sorgho et du maïs ; Bazièege 1982" La Station expérimentale du G.I.E. secograins.
  + Technical and economic report of the 1983 drying season, available in April, 1984.
  + After the 1984 drying season, publication of a booklet on solar or dual energy drying of grain.

(1) "Lux" is a measure of light intensity as the human eye perceives it.

PROJECT ANALYSIS

1. Site and climate

1.1. Site

Latitude :	43.18° N
Longitude :	1.49° E
Altitude :	160 m
Nearest main city :	Toulouse
Distance from main city :	20 km
Direction from main city :	SE
Obstructions :	No solar obstructions

1.2. Site micro-climate

Mediterranean type climate, with oceanic influence. Very often windy.

1.3. Annual long-term averages

(30 years, 1951–1980)			Project	Project drying season only (1)
1. Prevailing wind direction	W		NW	NW
	S		SE	
2. Wind speed > 16 m/s	W	days/yr	14	
	S	days/yr	6	11
3. Total precipitation		mm/yr	671.3	163.3
4. Absolute humidity	W	g/kg	4.8	
	S	g/kg	10.5	7.0
5. Global irradiation on horizontal plane			4 935	1 044
			MJ/(m².yr)	MJ/(m².season)
6. Diffuse proportion		%	42.6	46.2
8. Hours sunshine		hrs/yr	2 038	529
9. Ambient temperature		C	12.7	11.6
10. Average max. temp. (Jul/Aug)		C	26.5	16.0
11. Average min. temp. (Jan)		C	1.5	7.1

12. Meteorological station :	TOULOUSE-BLAGNAC Airport
13. Latitude :	43.38° N
14. Longitude :	1.22° E
15. Altitude :	152 m
16. Distance from site : 25 km	

Note : W = Winter (Jan. Feb. Mar.)
*   S = Summer (Jul. Aug. Sept.)*
(1) Sept. Oct. Nov. Dec. (4 months)

2. Plant

<u>2.2. Drying-place</u> (identical for c_1, c_2, c_3)

1. Drying system :
   Solar and dual energy low temperature in-bin drying
2. Material : round metal bin
3. Dimensions : ϕ = 5.34 m
                    False bottom : 0.4 m
                    Height of grain (maximum) : 5.4 m
4. Capacity : maximum 90 tonnes
5. Ventilation - Power : 10 kW
               - Flow rate : 3.20 to 3.76 m³/s
               (for bin c_2 : 2.5 m³/s).
   6. Auxiliary heating source : electricity
               heating power : 14 kW
   7. Diagram : cf. figures 2.2.7.1 and 2.2.7.2.

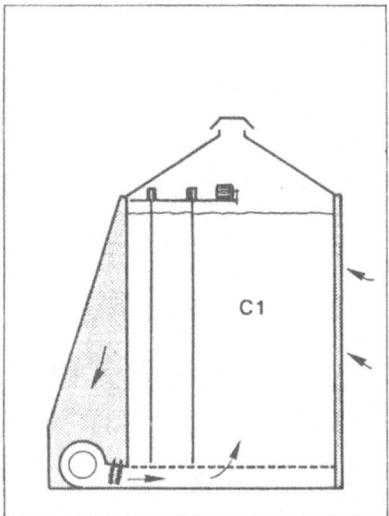
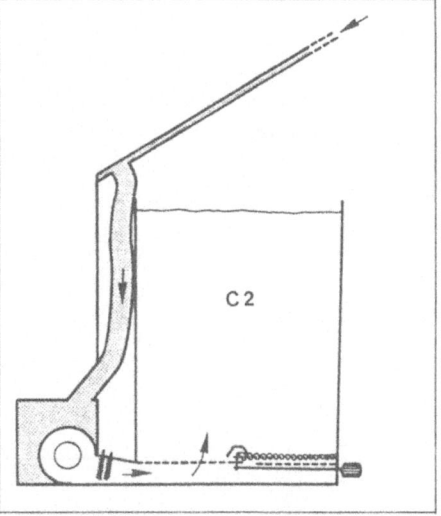

Figure 2.2.7.1 : Scheme of bin-drier C1 with a vertical air solar collector attached on bin walls. C1 is equipped with stirring augers.

Figure 2.2.7.2. : Scheme of bin-drier C2 with a 30° inclined roof solar collector. C2 is equipped with a floor discharge auger. (except for the solar part, the drier set-up is the same for C2 and C3).

<u>2.3. Collector (three types)</u>

Type I : Vertical cylindrical collector for bin-drier C1.
   1. Material :
      - Glazing : 6 mm twin-wall polycarbonate plates ("AKYVER", BEGHIN-SAY
                comp.) coefficient of transmission : 0.85 (when new),
                0.76 (when 10 years old).
      - Absorber : coolant : air
                   material : black painted iron sheets

2. Dimensions :
   Solar collector : enveloped area : 10.8 m (length) x 5.3 m (height) = 57.24 m^2
   Fan's glasshouse : 20 m^2
3. Flow rate :
   3.32 to 3.56 m^3/s, i.e. 58.0 x 10^{-3} m^3/(s.m^2 collector surface), depending on grain species and height in the bin.
4. Diagram of collector : cf. figures 2.3.4.1-3.

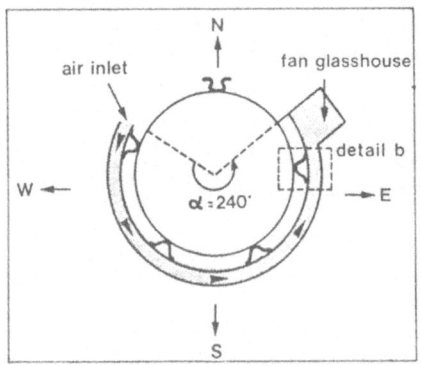

Fig. 2.3.4.1 : Collector and bin-drier horizontal cross-section

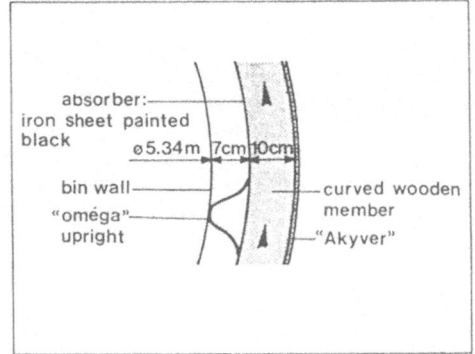

Fig. 2.3.4.2 : Collector : mounting details

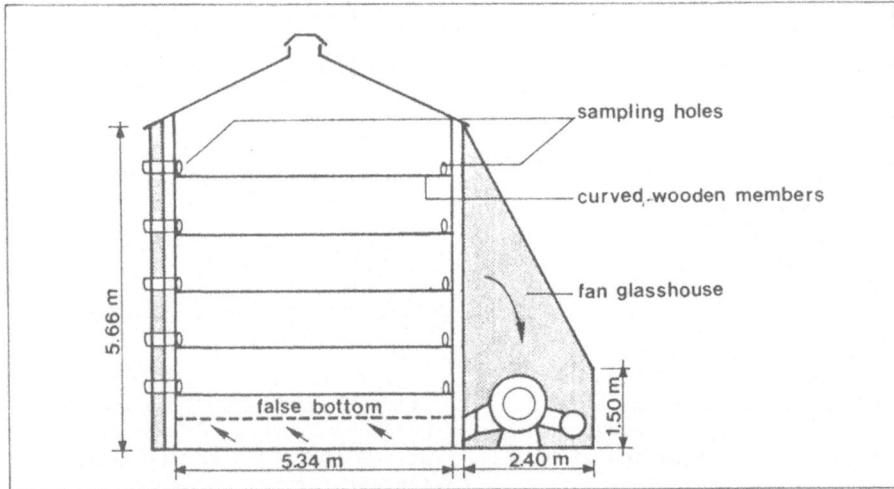

Figure 2.3.4.3 : Vertical cross-section of bin-drier C1 : solar collector, fan glasshouse and drying bin.

5. Some average daily efficiency values

Date	Day-light time (hours)	Efficiency (η_c)
27.11.82	9.0	0.512
04.12.82	8.9	0.519
05.12.82	8.85	0.555

The average daily efficiency ranges from 0.50 to 0.55

Type II : Roof collector for bin-drier C2.
1. Absorber
   - Coolant : air
   - Material : black painted corrugated asbestos cement plates
   - Aperture area : 0.14 x 6.36 = 0.89 m² (air inlet cross-section)
   - Absorber area : 89.04 m²
   - Collector fluid content : 140 1/m²
   - Working pressure drop : 100 Pa
   - Flow rate : (3.2 to 3.76 m³/s) 35.9 x 10⁻³ to
     42.2 x 10⁻³ m³/(sm² collector surface) (1)
   - Emittance of plate : $\varepsilon = 0.95$
   - Coef. of absorption of visible light : α : 0.95
2. Glazing
   - Emittance of glass : not measured by manufacturer
   - Number of glass covers : 1
   - Material : corrugated clear transparant polyester plates
               "CLAIRFLEX" from ETERNIT
   - special characteristics : - thickness : 1.3 mm
                               - sine waves : 177 x 51 mm
   - Coef. of transmission : 85% (when new), 68% (when 10 years old)
3. Insulation
   5 mm thick asbestos cement plate only
4. Mounting
   - Orientation : South
   - Tilt : 30°
   - Distance absorber-glazing : 14 cm
5. Over a long period (Nov. 23 to Dec. 21), the mean of the average
   daily efficiency is about 0.40. (Cf. fig. 2.3.5.1) But as the tests
   were run very late in season, the efficiency rates are probably
   higher in Sept. and Oct.

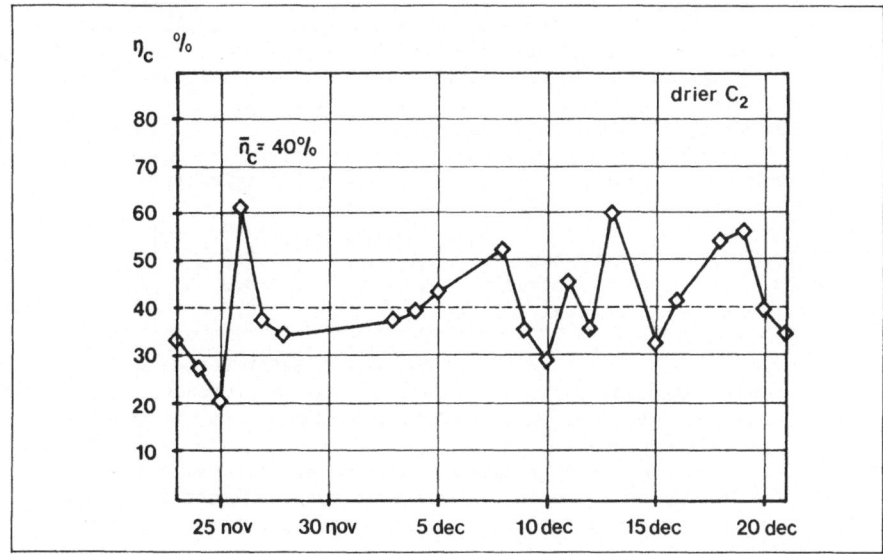

Fig. 2.3.5.1. : Collector efficiency measurement data for collector
               type II.

(1) *depending on grain species and height in the bin.*

Type III : Greenhouse collector for bin-drier C3
   1. Material :
      - glazing : 200 µm of transparent polyethylene film
                  (anti Ultra Violet treated)
      - absorber : 200 µm black polyethylene film on the ground
   2. Dimensions :
      - 20 m long
      - 8 m wide
      - 3 m high in middle
      - 160 m² groundsurface
   3. Flow rate : 2.5 m³/s (air speed ≃ 0.2 m/s)
   4. Diagram of collector : cf. figure 2.3.4.

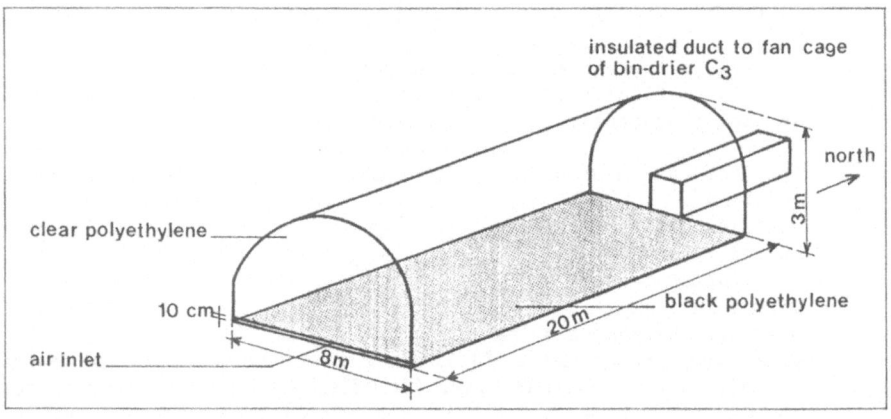

Fig. 2.3.4 : Diagram of collector-type III.

Note : for reasons of space limitations the experimental greenhouse was
       oriented parallel to the South-North axis. For operational
       set-ups an East-West orientation would obviously be better.

5. Some average daily efficiency values

Date	Weather conditions	Temp.rise (ΔT)	Efficiency (η_c)
3/10/83	Bright sunny day	6.6 K	0.3
8/10/83	Partly cloudy	5.0 K	0.3
21/10/83	Completely cloudy	1.5 K	0.36

Note : The very low cost of this type of collector compared to the other
       ones compensates for the low collector efficiency. However, for
       the next year, a fex improvements will be made in order to raise
       the efficiency :
       - a 3 m wide black film will be hung up vertically in the axis of
         the greenhouse over the whole greenhouse length.
       - the north side will be made in black plastic also.

2.5. Installation parts and measuring devices

1. Technical description of the installation parts :

<u>Fan</u>

The 10 kW fans for C1 and C2 are centrifugal fans : the DS 62 type of SOLYVENT. The running speed is about 2 000 R.P.M.. The air flow-total pressure curve is shown in figure 2.5. The total pressure drop through a 5 m bed of grains and through the other parts of the drier (solar collector, ducts) is about 2 200 Pa, so the air flow is about 3.2.-3.4. m^3/s when the bin is filled up.

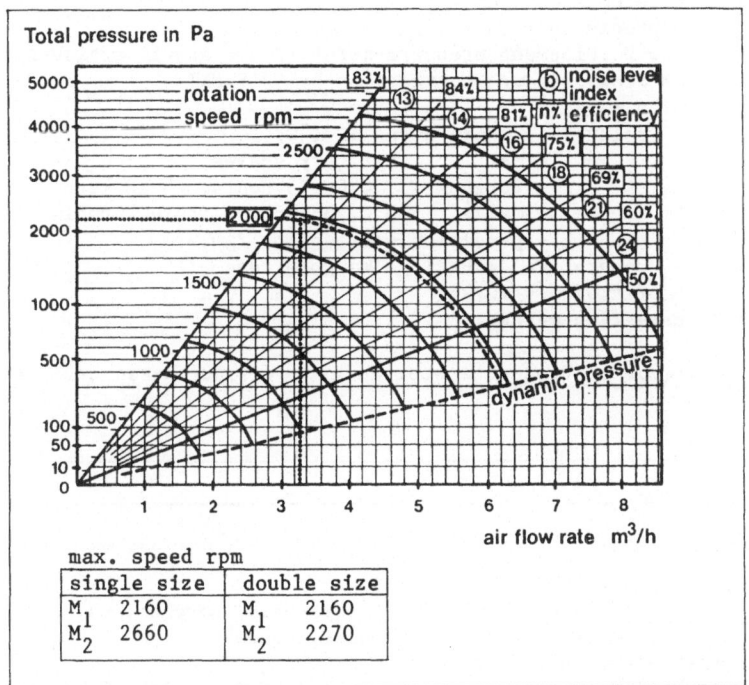

Fig. 2.5 : Fan characteristics.

There are no pumps, radiators or heat exchangers.

<u>Bin-drier C1</u> (around vertical collector)
. Energy levels
  – solar collector : 0 to 18 kW (clear day average)
  – fan's constant heating : 8 kW
  – auxiliary heating source at night
    (electric resistance) : 14 kW
. Flow rates : 3.32 to 3.56 m^3/s

<u>Bin-drier C2</u> (roof collector)
. Energy levels (same as bin-drier C1)
. Flow rates : 3.20 to 3.76 m^3/s.

<u>Bin-drier C3</u> (greenhouse collector)
. Flow rate : $\pm$ 2.5 m^3/s

2. Description of the sensors
- Temperatures : - copper - constantan thermocouples
                 - 2 recorders 16 channels (MECILEC)
                 - precision + 0.5 °C
                 - measuring range - 10 ... + 50°C
- Relative humidity of air - laboratory made psychrometers
  Absolute humidity of air - precision 2% for both relative and absolute
                              humidity.
- Solar radiation : - Kipp and Zonen thermopile
                    - recorder 0 - 900 W/m^2
                    - integrator J/cm^2
                    - precision + 1%
- Air flow : - Air Flow Development pressure-meter (fan's static pressure
               and air flow-pressure curve given by SOLYVENT)
             - precision + 5%
- Moisture content of grains : - Chopin ovens (+ 1°C)
                               - Mettler laboratory scales
                               - Sorghum : 130°C, 23 hours, whole kernels
                               - Maize : 130°C, 38 hours, whole kernels
                               - precision + 0.5 point (%w.b.)
- Masses of grain : - Coopérative de Bazlège Weigh-bridge
                    - precision + 10 kg
- Mycotoxins : multidetection method.

No data logger, cassette or disquette storage (not yet purchased because
of lack of money).

---

3. Products

3.2. Drying-products

1. Designed for special products : all grains
   In the South-West of France : sunflower, soya-bean, sorghum, maize;
   Drying season : September, October, November, December.
2. Required conditions of inlet air :
   - specific air flow rate : 140 m^3/h and per tonne of grain
     (110 m^3/h and per m^3 of grain)
   - temperature : 2 to 8 K above ambient
   - relative humidity : below 70%
3. Properties of the products :
   - maximum quantity : 90 tonnes
   - temperature : harvest temp. (10 to 30°C)
   - maximum allowable moisture content at harvest : 28% (w.b.)
4. Other (related) products which could be dried :
   Tropical products in tropical areas : coffee, cocoa, rice, etc...
5. Any special treatment of the crop ? :
   Eventually, precleaning of foreign materials (leaves, husks, cobs,
   etc.)

4. Economic aspects

Basis
- the economic data were extrapolated to actual on-farm set-ups, based on the knowledge of the described experimental ones, and referring to a simulated 92 day drying season of sorghum and maize
- prices are 1982 values, TVA excluded

4.1. Economic environment

	Solar + support energy	Reference solution HOT AIR DRYING
1. Energy price of oil (ECU/1)		0.4
Sept.	D = 0.028	
	N = 0.017	
2. Energy price of electricity		
(ECU/kWh)(1) Oct. Nov.	D = 0.062	
Dec.	N = 0.031	
3. Energy prices evolution prospects		
- of electricity (%)	8	
5. Interest rate of loans for farm equipment (rate without subsidy) (%)	11(2)	13.5
	6(3)	
6. Subsidies available for the solar and competitor farm equipment (capital grant) (%)	20	10

(1) *Middle voltage (2 500 volts)*
   D = Day 6 : 00 - 22 : 00
   N = Night 22 : 00 - 6 : 00
   Prices of electricity will be modified in 1984. Night kWh will be cheaper and Oct. will become a summer month.
(2) *Normal interest rate of loans for farm equipment*
(3) *Special interest rate of loans for young farmers, or in the case the equipment is used jointly by several farmers (i.e. foundation of a cooperative store).*

4.2. Investment expenses (ECU)

(Estimated values, cf. "Basis" p. 311)

		BIN-DRIER		
		C1	C2	C3 *(1)*
7. Cost of collector and man-power (total) (per m2)	a)	3 803 47.6	4 375 48.7	2 141 13.4
8. Storage bin only		7 840	4 030	4 030
9. Share of the storage bin in drying cost (1/3) *(2)*	b)	2 613	1 343	1 343
10. Cost of drying equipment (augers,fan, false-bottom, ducts)	c)	7 389	9 213	9 213
11. Total actual cost without subsidies (a + b + c)		13 806	14 931	12 697
12. Total actual cost including subsidies=investment for farmer		11 045	11 945	10 157
13. Annual instalment (annuity) if storage bin has to be bought *(3)*	A	1 535	1 660	1 412
14. Total actual cost without subsidies if storage bin is already existing (a + c)		11 192	13 588	11 354
15. Total actual cost including subsidies when storage bin is already existing = investment for farmer		8 854	10 870	9 083
16. Annual instalment (annuity) if storage bin is already existing *(3)*	B	1 245	1 512	1 263

(1) Bin-drier with a 160 m² greenhouse as solar collector, installed for the 1983 drying season.

(2) The bin used for drying functions as a storage bin after the drying season. So, amortization is 1/3 for drying, and 2/3 for storage the rest of the year.

(3) Writing-off costs calculated over the lifetime of the equipment, taking into account a subsidy of 20% and interest rate of loans of 11%.

    Estimated life-time - of collector (mainly glazing) : 10 years
                        - of the driers   : 20 years

4.3. Operating expenses

(estimated values, cf. "Basis" p. 311)

	BIN-DRIER		
	C1	C2	C3
1. Energy cost (electricity)(ECU/yr)	1 791	1 791	1 791
2. Maintenance cost ± 0.5% of the investment per year (ECU/yr)	95	102.5	91
4. Manpower cost *(1)* 46 h (ECU/yr)	(644)	(644)	(644)
5. Insurance cost 0.5% inv./year (ECU/yr)	95	102.5	91
7. Total operating expenses (ECU/yr)	1 981	1 996	1 973

(1) The manpower cost is indicated but not taken into account, because it is largely compensated by the queuing cost at the country elevator (reference solution)

4.4. Hidden cost and benefits (PROFITABILITY OF SOLAR DRYING)

(1)

(estimated values, cf. "Basis" p. 311)

	BIN-DRIER		
	C1	C2	C3
1. Production obtained (dried substance, tonnes)	SORGHUM : 252 tonnes(15% m.c. ; dried from 21% m.c.) MAIZE : 160 tonnes (15% m.c. ; dried from 28% m.c.)		
2. Annual elevator drying charges*(2)* (ref.) (ECU/yr)	4 231	4 231	4 231
3. Operating expenses (ECU/yr)	1 981	1 996	1 973
4. Annual instalment *(3)*			
- with bin A	1 535	1 660	1 412
- without bin B	1 245	1 512	1 263
5. Solar drying annual cost(ECU/yr)			
- with bin A	3 516	3 656	3 385
- without bin B	3 226	3 508	3 236
6. Annual profitability			
(benefits) (ECU/yr) A	715	575	846
(4.4.2. minus 4.4.5.) B	1 005	723	995
7. Pay-back period (years) *(4)*			
- bin to buy A	4.9	5.3	4.5
- bin already bought B	4.0	4.9	4.0

8. Conclusions

Bin-drier C3 (bin inside a building) coupled with a greenhouse as solar collector appears to be the most profitable system. If an open-air bin already exists, then a vertical cylindrical collector, attached on bin walls may be an interesting solution.

9. Energy consumption		CONVEN
- Electricity consumption for obtaining 252 t of dried sorghum and 160 t of maize implying an evaporation of	SOLAR DRYING	TIONAL DRYING
48.03 t of water (MJ)	104 585	6 915
- Oil consumption (MJ)	0	241 225
- Total energy consumption (MJ)*(5)*	104 585	248 170
- Difference (%)	57.8	

(1) *PROFITABILITY is based on the expenses to build the driers in the experimental station, and on certain assessments about grain drying, interest rate of loans (11%), subsidies (20%) and residual value of investment (= 0).*

(2) *Based on elevator drying tariffing. Sorghum : 8.60 ECU/tonne resulting dried product + hidden profit due to dry matter losses reduction : 1.2 ECU/tonne resulting dried product Maize : 14.5 ECU/tonne resulting dried product.*

(3) *Cf. footnote nr. 3 on previous page.*

(4) *Calculation method*

$$(years) = \frac{\text{Total Investment for farmer (= section 4.2.12 or 4.2.15)}}{\text{Net Annual Cash Flow (N.A.C.F.)}}$$

N.A.C.F. = savings in operating expenses, i.e. difference in cost between reference solution (drying at country elevator) and operating expenses in solar drying (section 4.4.2. minus 4.4.3.).

(5) *The researcher justifies the straight comparison of electrical energy figures with primary fossil energy figures, by stating that nuclear*

or hydro-power are assumed as the electrical power source. These specific sources cannot possibly be applied in a more "energy profitable" way. Hence experts in the field would consider nuclear and hydro-electricity equivalent to any other form of primary energy. (N.O.T.E.)

---

5. Results

5.1. Calculation method

Data are processed at the Laboratoire de Biotechnologie Solaire de l'Ecole Nationale Supérieure d'Agronomie de Toulouse.
<u>Data and computer programs used</u> (APPLE 2)
The values of temperatures, checked manually each hour for every MECILEC channel, are noted, each day in two tables.
One is a matrix 16 x 24 (16 channels, 24 hours) concerning bin-drier C1 and ambient air ; the other one, a matrix 9 x 24 for bin-drier C2.

<u>Computer program I</u> "Resultats BAZIEGE"
The quoted values are inserted in the first computer program in order to create a data-processing file.
The other inserted data are :
  - fan running time (hours)
  - solar energy (J/cm^2)
  - electric energy (kWh)
  - air flow (m^3/h)
The printer lists daily averages of temperature, r.h., theoretical and potential water loss, as well as various energy inputs (collector, fan, resistance).

<u>Computer program II</u> "CYLSOL"
For each day the <u>Cl collector</u> efficiency is calculated.
Knowing the global solar irradiation on one horizontal m^2, the computer program first calculates the amount of solar energy which strikes the cylindrical surface over one day (Q_{ci}). Next, the collector efficiency η_c, is calculated as follows :

$$\eta_c = \frac{\rho . \mathring{V} . Cg . \Delta T_c . t_d}{Q_{ci}}$$

where ρ : specific mass of air (kg/m^3)
  $\mathring{V}$: air flow (m^3/s)
  Cg : specific heat of air under atmospheric pressure $(J/(kg.K))$
  ΔT_c : temp. rise in collector (K)
  t_d : daylight duration (seconds/day)
  Q_{ci} : solar energy striking the cylindrical collector (J/day)

The accuracy of the program becomes poor when the diffuse proportion is high (cloudy weather). Furthermore, to allow for reliable program calculation results for days with good or medium type weather, the aeration has to be kept on all day without stopping.
Taking away both cloudy days as well as days when grain samplings were taken, only 3 measuring days were left (Results cf. section 2.3. type I point 5).

Computer program III

The collector efficiency is calculated in the same way, but now for collector C2. The first step in this calculation – calculation of the global solar energy striking the 30° tilted roof Q_{ci} – is done as follows:

$$Q_{ci} = \frac{<I'_c>S_c \cdot I'_h}{<I'_h>}$$

with $<I'_h>$ annual average of the daily solar radiation on a horizontal plane of 1 m² (based on 30 years of measurements) (J/m² and per day)

I'_h solar radiation on a horizontal plane of 1 m² on the day for which Q_{ci} is calculated (J/m² and per day)

$<I'_c>$ solar energy striking a plane of 1 m² tilted at an angle of 30° on a day with annual average solar radiation (obtained, using correlation formulae of values in combination with $<I'_h>$ – AFEDES tables) (J/m² and per day)

S_c collector surface (m²)

This formula is somewhat approximate, as is also the numerator of the fraction in which η is calculated. (cf. formula under computer program II). In the numerator of this formula ΔT_c is underestimated, since the thermocouples at the inlet of the collector were less protected from solar rays than the ones at the outlet (leading to a slight rise in the temperature registration of the inlet air, mainly at noon time).

Computer program IV

A drying simulation model (BARRE et al. modified by S. SOPONRONNARIT) calculates the simulated average drying kinetic and allows for comparison with the observed drying kinetic.

5.2. Weather conditions and energy (actually measured)

	BIN DRIER C1 MAIZE	BIN DRIER C2 SORGHUM			
1.	22/11/1982	24/11/1982		from PERIOD	
2.	14/12/1982 22 days	21/12/1982 29 days		until	
3.	8.0(86.6% R.H.)	8.1 (86.6% R.H.)	°C	average ambient TEMPERATURE	
4.	2.6	2.6	m/s	average WIND	
5.	7093	26225	MJ	average SOLAR RAD on collector over the drying period	
10.	3752	10490	MJ	useful energy over the drying period	DUAL ENERGY HEATING Solar + Electricity
11.				AUXILIARY HEATING energy (non-solar)	
a.	15486	20142	MJ	a. fan's temperature rise (1)	
b.	8638	11675	MJ	b. electricity at night (1)	
14.	24124	31817	MJ	ELECTRICITY consumption (fan + elecric heating) (1)	

(1) *The values in lines 11 a), 11 b) and 14 are calculated from the measured net heat input into the air.*
Those values were found by measuring the electricity consumption.

5.5. Drying results (actually measured)

	BIN DRIER C1 MAIZE	BIN DRIER C2 SORGHUM				
1.	22/11/1982	24/11/1982		from	PERIOD	
2.	14/12/1982	21/12/1982		until		
3.	3.44	3.60	m^3/s	flowrate	⎫	
4.	12.6	12.9	°C	temperature	⎬ INLET AIR	
5.	60	60	%	relative humidity	⎭	
6.	21.0 to 15.5	19.7 to 15.2	%	moisture content (initial and final)	⎫	
7.	70.9	74.9	ton	useful quantity		
8.	good	good		quality	⎬ PRODUCTS	
9.	< 13°C	< 13°C	°C	temperature	⎭	
17.	5196	5064	kg	evaporated water		
18.	4.64	6.28	MJ/kg	energy efficiency (i.e. electricity consumption/mass of evaporated water) for dual energy drying		
19.	5.0 to	6.3 (1)	MJ/kg	energy efficiency for oil heated air drying reference		

(rightmost vertical heading: DUAL ENERGY DRYING Solar + Electricity)

Quality : MAIZE
- good marketable quality - very good wet milling ability
- dry matter loss : less than 1%
- mycotoxins : only some zearalenone (120 ppb) (2), both on check and drying samples (so not induced by drying technique)
        : SORGHUM
- good marketable quality
- mycotoxins : only some zearalenone (25 to 40 ppb) (2)
- dry matter loss : 0.86%

(1) These values were obtained in an unfavourable period of the drying season. A simulation over a whole representative drying season yielded energy efficiency values of ± 2.2 MJ/kg evaporated water. (cf. section 4.4.9.). Recent experimental results with C3 yielded an energy efficiency of 2.5 MJ/kg evaporated water (cf. section 6) (N.O.T.E.).
(2) Minimum content in animal feed : 500 to 1 000 ppb, according to species.

5.7. Example of some measured data (cf. figures 5.7.1-2

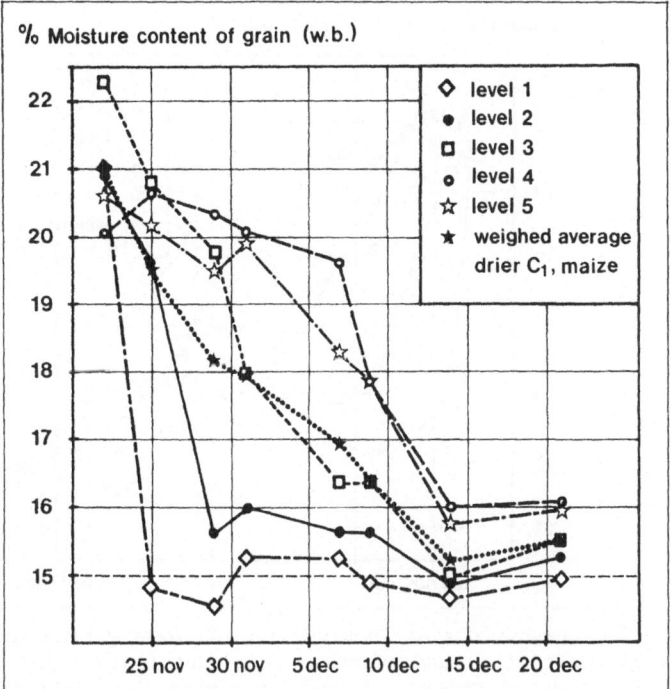

Figure 5.7.1 : Maize drying in bin-drier C1; drying kinetics of 5 layers
(bottom to top) and average, as a function of time. Initial wet
mass : 76.03 tonnes. Final dried mass : 70.86 tonnes.

6. Remarks

- A few data have recently become available of the greenhouse collector
  drying performance :
  A test was done with 125.98 tonnes of sorghum at initial moisture
  content 19.8%. This was dried to 13.9% m.c. (116.57 tonnes) over 26
  very sunny and warm days (30/09-26/10/1983).
  (5-10 tonnes of wet product per 2 days continuous drying via discharge
  auger.)
- Electricity consumption :
  - fan 13 740 MJ (19.5 hrs/day - 11.5 hrs solar collector/8 hrs
    resistances)
  - aux. heater 8 080 MJ
  - discharge auger 76 MJ
- Evaporation of water 8 714 kg (dry matter loss 0.6%)
- Energy efficiency 2.5 MJ/kg evap. water
- Using experimental data a profitability over the traditional elevator
  costs can be predicted of about 1.5 ECU/tonne of product to be dried.

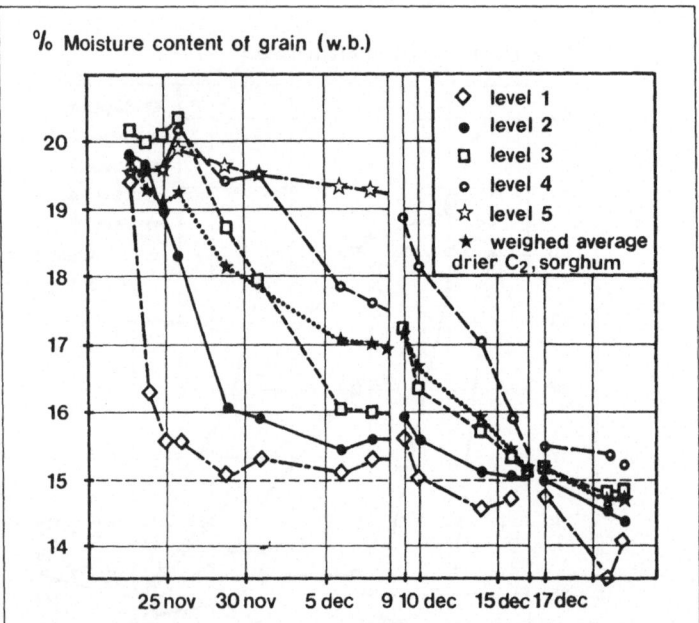

Figure 5.7.2 : Sorghum drying in bin-drier C2; drying kinetics of 5 layers (bottom to top) and average, as a function of time. Initial wet mass : 80.48 tonnes. Final dried mass : 74.86 tonnes. On Dec. 10 and 17, a discharge of nearly 10 tonnes with the special auger was carried out, allowing continuous drying.

PART III

Applications of solar energy for hot water production

CHAPTER I :
Process requirements - Technological options

This chapter gives a theoretical background of five studied applications of solar hot water production :
- biogas digester heating
- waterheating for intensive aquaculture
- floor heating for pigs
- hot water preparation for hygienic purposes in dairy farms
- milk preparation on calf-rearing farms.

The different processes in question are described, drawing on elements such as economic importance, process description, hot water requirements (energy needs and temperature levels), techniques of hot water preparation, etc. ...

I. BIOGAS DIGESTER HEATING BY MEANS OF HOT WATER

1. Motives of the recent interest in biogas production

Biogas production has recently received a lot of attention, and much research is being done in the area.
Contrary to what might be presumed in light of this recent attention, it is a process that has been practised for many years past ... In India ten thousand, and in China several millions of primitive small low yield biogas producing units exist, providing energy mainly for heating, lighting and/or cooking purposes. In fact, outside of the realm of man's intervention, biogas is constantly being produced in the anaerobic conditions of marshes, covered by stagnant waters.

Nevertheless, actual research in the area of biogas production is fairly new. In Europe, it is only in the second part of the 20th century that a particular interest for this process originated, in the context of two study domains of increasing importance:
- the problem of waste water treatment;
- the problem of manure disposal.

At that time, the growing presence of highly polluted waste waters, and the need for their treatment led to the development of different techniques. Given the high investment and/or energy costs of most of the other systems, the option of anaerobic fermentation, allowing among others for significant energy cost reductions - by using the biogas produced during the waste water treatment process itself - attracted much attention, and found several applications on town or inter-town scales.

At the same time in agriculture, the continual upscaling, specialization, and vertical differentiation led to the existence of very densely populated and intensive animal farm systems, in which manure disposal became a real problem.

Indeed, in regions where this transition towards industrial stock-breeding systematically occurred, the common means of manure disposal, namely its distribution over the arable land, could not keep up with the total manure production : spreading all the manure over the limited available land would cause excessive fertilization rates, resulting in lower soil production capacities, accumulation of toxic and harmful components in the soil, pollution of the groundwater with bacteria from the animal faeces, as well as pollution of underground and surface waters with organic materials and inorganic salts... It is unnecessary to evoke the additional smell nuisance that accompanies the above mentioned manure abundance, just at a time when stricter rules were being set as to the quality of the environment.

Also in this context of manure disposal, a particular interest originated in anaerobic fermentation, since this process can transform the organic carbon fraction of the manure into (useful) biogas. At the same time, it reduces the smell of the resulting manure residue as well as the toxic nature of certain of its harmful elements. Furthermore, the manure residue remains a valuable fertilizer.

Given the close link between biogas production on the one hand and the potential for energy recovery and resource recycling on the other hand, -centres of focus during the "energy crisis stricken seventies"- the interest in anaerobic fermentation was even further increased.

Theoretically, this biogas production seems very attractive in the agricultural realm.

By way of illustration, the theoretically possible energy production per animal per day is given in table 1.

Table 1: Some representative figures of manure production, manure contents and biogas production (3)

animal type	kg of fresh manure production (urine included) per animal per day	% of dry matter	% of organic matter in the dry substance	m³ gas production per kg organic matter	% of methane in the biogas	Net Energy recuperation potential (E) in biogas-form (*) MJ/animal/day
cow	35	15	80	0.262	61	16
pig	5	7	85	0.308	65	1.4
chicken	0.18	25	65	0.469	64	0.2

* *Assuming 1/3 of the biogas is reinvested in the biogas producing system itself for digester temperature regulation purposes (cf. below), and given the combustion heat of methane of 35 MJ/m³.*

However, the technological problems to be dealt with are far more complex than this - in principle - simple biogas production process would suggest. These problems are expressed - among others - as a consistent lack of reliability of the currently existing agricultural biogas plants. E.g.: A CEC-study on 96 recently built biogas production plants identified the occurrence of an average of two major problems (e.g. temporary shutdown of the plant or non-reversible souring of the methane digestion mixed liquor) and four minor problems (without major operation disruption) per plant, and this over their short lifetime of a few years at most (2). This factor (system complexity) might very well prove to be the major obstacle to a larger breakthrough of this otherwise promising technology.

2. Description of the biogas production process

The biogas production process consists of three main steps, which take place in an anaerobic environment.

1) A primary enzymatic hydrolysis of insoluble organic compounds (e.g. cellulose) into smaller soluble molecules (carbohydrates, lipids, alcohols...)

2) Those soluble molecules are then transformed into organic acids (mainly acetic and propionic acids) by facultative heterotrophs (e.g. bacteroids spp. and Closeridium butyricum for pig manure)

3) Finally those acids are transformed by methanogenic bacteria (methano-bacterium spp. Methanobacillus spp.) into dissolved methane and Carbon dioxide, which eventually transfer to the gaseous phase: Biogas.

A scheme of a possible installation is given in figure 1.

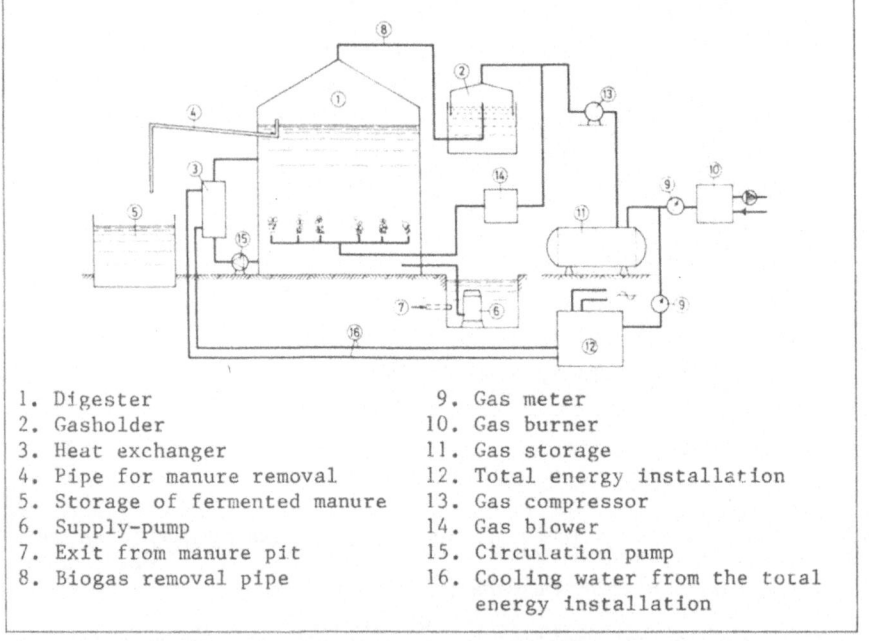

1. Digester
2. Gasholder
3. Heat exchanger
4. Pipe for manure removal
5. Storage of fermented manure
6. Supply-pump
7. Exit from manure pit
8. Biogas removal pipe
9. Gas meter
10. Gas burner
11. Gas storage
12. Total energy installation
13. Gas compressor
14. Gas blower
15. Circulation pump
16. Cooling water from the total energy installation

Fig. 1. Biogas installation in Duiven, the Netherlands (3)

The effluent (1 - 10 % dry matter in suspension) of an animal farm is pumped from the manure pit -7- into the digester -1-, the content of which is being homogenized by means of biogas that is recirculated through the manure. The produced gas is temporarily stored under a low pressure gasholder -2-, and then pumped into a higher pressure container -11-. The gas is finally routed, when needed, to a gasburner -10- and/or to a total energy installation -12-, the cooling of which is done by water, delivering its waste heat to the digester by means of an external heat exchanger -3-. The old manure is discharged gradually from the top of the digester, as fresh manure is pumped in underneath.

3. The digester temperature, parameter in the biogas production process

Since the heart of the biogas production is to be found in the action of micro-organisms, the main factors of influence in the performance of a biogas installation will be factors that influence their life and growth potential : the type of waste product, the pH, the presence of toxic chemicals, and last but not least, the temperature. The fermentation process is workable within the range of 5 - 75°C. Mesophyl bacteria can take temperatures of 5 - 40°C (with optimum temperature 30 - 35°C), while thermophyl bacteria function at temperatures between 40 - 75°C (optima 50 - 55°C). As shown in figure 2, a higher environment temperature not only means a significantly reduced required fermentation time, but also a greater gas volume output per kg organic material.

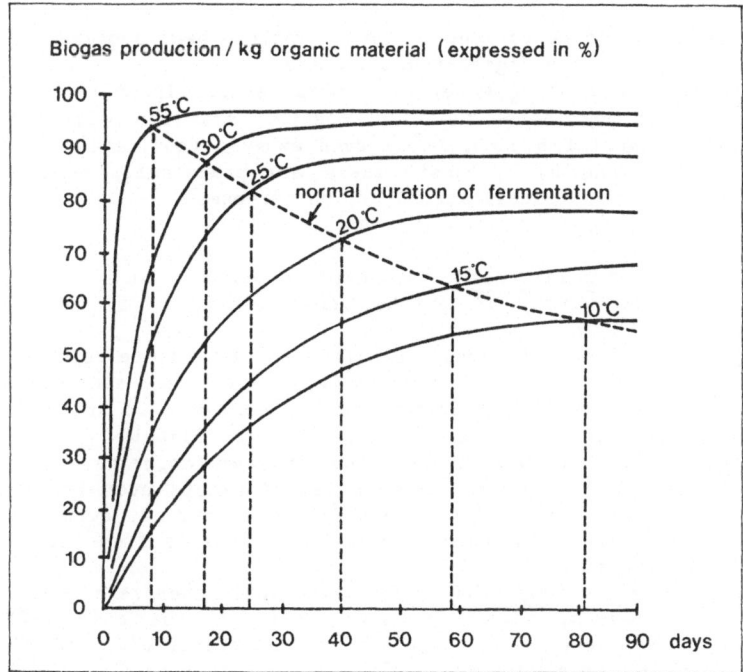

Fig. 2. Biogas production as a function of temperature. (3)

Even given this fact, the process at 30 - 35°C has in practice often been preferred to that at 50 - 55°C, since the latter requires quite an extensive heat input as well as a more sophisticated installation, in order to take due care of the more sensitive thermophyl bacteria.

4. Digester temperature regulation techniques

The heat necessary in order to maintain the desired digester temperatures is most often provided by using part (30 %) of the self-made biogas.
This heat is usually applied by means of a heat exchanger, built into the digester, carrying warm water.

Different alternatives are being considered and/or tested, in order to minimize the biogas consumption for this purpose - and thus to maximize net biogas production efficiency - while still maintaining the desired temperature.

* A first logical option is the <u>minimization of the heat demand</u>:
  - by means of <u>insulation of the digester</u>; for instance, using humidity resistant insulation materials, (e.g.: polystyrene). In this context, it is advisable to reduce the heat transmission coefficient to a value below 0.6 W/m^2K.
    Note: In some cases the reactor is buried underground, as a means of protection from cold.
  - by preheating the inflowing manure suspension with the outflowing counterpart, using <u>heat exchange devices</u>.

* Another option would be to look for other <u>waste - heat sources</u>, which could contribute to the digester heat demand.
  For instance, the cooling water of a total energy installation or of other combustion engines on the farm. Possibly also thermally polluted water from industrial plants; maybe even exhaust ventilation air from the live stock housing. (Fossil fuels are practically never used, except sometimes for installation start up purposes).

* <u>SOLAR ENERGY</u>
  The application of <u>solar energy</u> into this process has been suggested by some as a topic worthy of further consideration (cf. Fundamental Criteria, p. 1).
  . <u>Energy demand</u>: The heat demand for digester temperature regulation on a particular farm is not negligible , and, in case the wider introduction of the biogas production process would be successful in the future, the total heat demand could become sizeable.
  . <u>Solar supply</u>: If one wants to keep the digester temperature at the optimal level all the time, a heat demand exists over the greater part of the year, (and in the northern EC countries, even all year long). Thus, one could theoretically also benefit from the periods of high solar supply.
  . <u>Solar system efficiency</u>: The solar system would operate at low temperature levels, given the temperature of the cold inflowing manure on the one hand, and the still reasonably low desired temperature of the manure (30 - 35°C) on the other hand.
  . <u>Socio-economic factor</u>: Whether the farmer would be able and willing to deal with the added complexity of the solar installation, to the already complex biogas production system, is not so evident ... Very little practical experience is available on this application. (cf. discussion in chapter 2 on p. 383)

* <u>The use of an activator</u> :
  In order to avoid the complications necessary to keep the digester at the optimal temperature for the biogas producing bacteria, an activator - consisting of certain enzymes and bacteria - is being developed, which is expected to double the biogas production rate for a given digester temperature (1).

References

(1) Y. Bertrand: Biomethane S.C., Brussels – Personal communication.

(2) M. Demuynck; E.J. Nyns; W. Palz: "Biogas plants in Europe – A practical handbook" Solar Energy R & D in the European Community. Series E, Volume 6, D. Reidel Publishing Company, (Dordrecht, Boston, Lancaster) for the CEC 1984, p. XIX.

(3) W. Verstraete: "Biogas", 1st edition; Antwerpen; Stichting Leefmilieu VZW – dossier nr. 6, 1981, pp. 207.

II. WATER HEATING FOR INTENSIVE AQUACULTURE

1. The evolution of fish farming

Since 1970, the world yield of fisheries has stagnated at an annual level of 70 million tons (7) and fears exist that this level is close to the maximum value of what this (rather primitive) way of "food collection" has to offer.

One of the characteristics of fish, which in this world of scarcity of resources makes it an interesting study object, is its extremely high efficiency in protein production.
Table 2 should help establish the importance of this factor.

Table 2. Representative energetic efficiency values in protein production for different animals (7).

Production system	$\dfrac{\text{protein-production}}{\text{energy consumption}}$ (grams/MJ)
Beef	0.55
Pig	1.53
Chicken - eggs	2.42
- meat	3.80
Fish	9.57

(From : Rumsey, 1978)

Indeed, the poikilothermic fish don't need to spend such vast amounts of energy on thermo-regulation as do their homoiothermic counterparts. Furthermore, the energy required for their movement is lower, and the way they dispose of their metabolic waste products more economical.

This, among other factors, has led to the development of aquaculture systems; a step comparable with that of the transition from hunting to agriculture on land. As is the case of agriculture, here too a strong trend can be noticed towards production rationalization, to such a degree that commercial fish production systems are developing, which are comparable with the current large scale, specialized, capital intensive industrial forms of agriculture.

Schematically, the evolution from extensive to intensive aquaculture can be outlined as follows:
- In an early stage are to be situated the extensive forms of aquaculture, where a limited amount of fish is kept in a pond. The maximum production level is determined by the capacity of the ecological system in the pond, to provide food for the raised fish.
- Extra provision of food for the fish (either indirectly, by adding manure, which stimulates the ecological system; or in the form of foods, directly consumed by the fish), allows for a significant addition to the fish production potential.

- The production level increase is however limited, since at a certain
  density of pond population all the oxygen in the water - essential
  element for fish survival - is consumed. Several means of bypassing
  this limitation are applied.
    - Simply by regularly renewing the water. For instance so called
      "Raceways", with a constant through flow of fresh water. In some
      other instances, the fish is kept in densely populated cages, in a
      running stream, so fresh water is constantly being provided.
      However, this process requires large amounts of water. By way of
      example, a 100 ton per year production unit of carp would require an
      average water flow rate of 38 200 litres per minute! (1) (2)
    - In order to reduce the fresh water requirement, the water is often
      artificially aerated.
- Even if aeration is duly provided for, at a certain point, further
  production intensivation runs into yet another obstacle: the build-up
  of metabolites and other water pollutants. In fact, ammonia - one of
  the metabolites produced by the fish - quickly reaches levels of
  toxicity (generally speaking, 0.025 mg NH_3/l is considered as the
  maximum safe concentration. (6)
  Considering the limitation of ammonia concentration, the minimum
  required average water flow rate for the above mentioned 100 ton per
  year carp-production unit would be in the area of 2 000 litres per
  minute. (1) (2)
- To allow for even higher levels of aquaculture intensivation, one can
  make use of water recirculation units, which in the extreme case
  completely renew the water, and this, following three steps:
  1. Decantation of solid wastes by sedimentation.
  2. Elimination of ammonia and nitrites by means of an aerobic bacterium
     filter.
  3. Artificial aeration of the water.

Table 3 and figure 3 give an idea of the production intensity levels that
are being obtained in some of the above mentioned production systems.

Table 3. Representative annual carp production levels for different
         aquaculture systems. (7)

Aquaculture system	Yearly production (tons/ha)
Extensive pond culture, without artificial feeding (Europe)	0.025 - 0.4
More intensive pond culture, with artificial feeding (Southern Europe)	1.5
More intensive pond culture, with artificial feeding (Japan)	5
Intensive aquaculture in running water (Japan)	400 - 2 000
Intensive aquaculture in recirculated water	4 000

(1) These are representative values for a water temperature of 23°C.

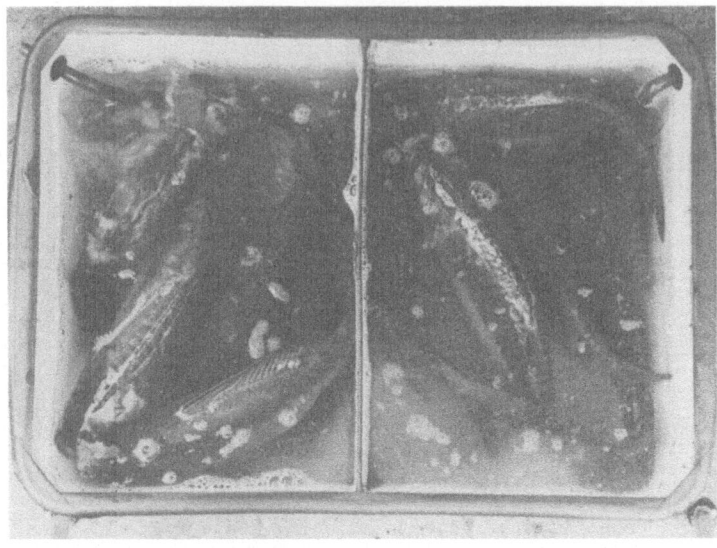

Fig. 3 : An example of the extremes to which intensivation can go: Ten
         fish (mean weight of 913 g) reared in a 40 litre aquarium with
         water recirculation (Max Planck Institute, Hamburg, W. Germany)
         (5)

Even though the more extreme forms of intensivation allow for a better
control of the parameters by the culturist, and so too, for a higher
production rate, there are evidently some related aspects which restrict
practical consideration to certain fish species.
- Not all fish species can successfully deal with high population density
  rates. Certain territorial fish species can't even adapt to it at all
  (e.g.: Some of the centrarchids). Their growth rate is reduced, among
  others reasens, because of their tremendous aggressivity, and their
  health is jeopardized by physiological stress. As can be seen in the
  above picture, this is not a problem for the common carp. For tilapia,
  crowding problems occur at densities of 500 fish/m^3. Here, an optimal
  population density seems to be 300 fish/m^3. (7)
- Given the high investment involved in such intensive aquaculture
  systems, a high return on these investments is imperative.
  This limits its applicability to quickly growing, easily reproducing,
  high-price luxury fish species, for which a sizeable market exists, or
  could be created.
- Finally, it is clear that, in order to minimize the non-negligible
  risk-factor in such an operation, the chosen fish species should be
  both capable of dealing with unexpected parameter deviations and
  resistant to diseases the spreading of which could be extremely rapid
  and disastrous in such crowded conditions.

Some data on production values of commercial intensive fish cultures are
given in table 4.

Table 4 : Production values of commercial intensive fish cultures *(1)* in the five major intensive aquaculture producing countries within the EC. (1)

Country	Fish species	Production in 1981 (tons)	Production in 1985 (tons)
Italy	Trout	25 000	25 000
	Eel	2 000	3 000
	Mullet	1 000	2 000
	Catfish	1 000	2 000
France	Trout	25 000	30 000
	Turbot	5	70
	Eel	90	150
	Salmon	100	120
	Carp	2 000	2 500
W. Germany	Trout	12 000	11 000
	Carp	5 000	5 000
Ireland & U.K.	Trout	8 000	8 000
	Salmon	1 000	4 000
	Eel	150	400
	Turbot	20	200

2. Water temperature, an important parameter of the fish production process

If environmental temperature is an important factor in the efficient production of the homoiothermic pigs (cf. section 3 in "Floorheating for pigs", p. 341) it is even more so for the poikilothermic fish !
Figure 4 compares the effect of environment temperature on the growth, weight gain, or production rate of some homoiothermic animals (pigs, cows, hens, broilers) with that of a few homoiothermic ones (catfish, shrimp, sole).

Some illustrations might help clarify this idea :
- A Dover sole, kept on the East coast of England, with water temperatures between 5°C in winter and 17°C in summer, takes between two and three years to grow from a length of 5 cm to the minimum marketable size of 24 cm. Kept at 19°C (temperature corresponding to maximum feed efficiency) this period is reduced to less than a year. (cf. figure 5)
- In the more northern regions of Europe, in natural circumstances, the eel takes six to seven years to grow from 10 to 200 grams. Over this period, it gains weight only during the summer months. For the rest of the time, its weight is stationary or even decreasing. In intensive heated and aerated water cultures (at $\pm$ 25°C), this growth period can be reduced to one year only ! (6)
- Carp is another of the more marked examples of sensitivity to the environment temperature. In figure 6, two sibling one-year-old carp are pictured : the upper one was raised in a "wild" pond, while the

(1) In the table, the term "intensive fish culture" includes any culture where artificial feeding (directly in the form of fish food) occurs.

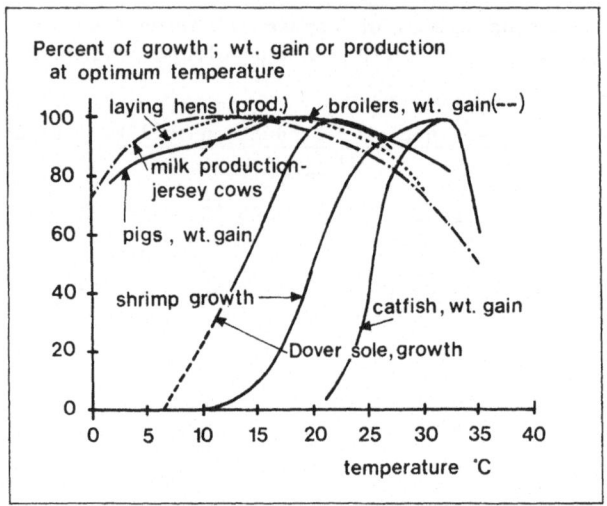

Fig. 4 : Effect of temperature on the growth, weight gain or production rate (result of food assimilation and feed efficiency) for a few animal types. (3) (9)

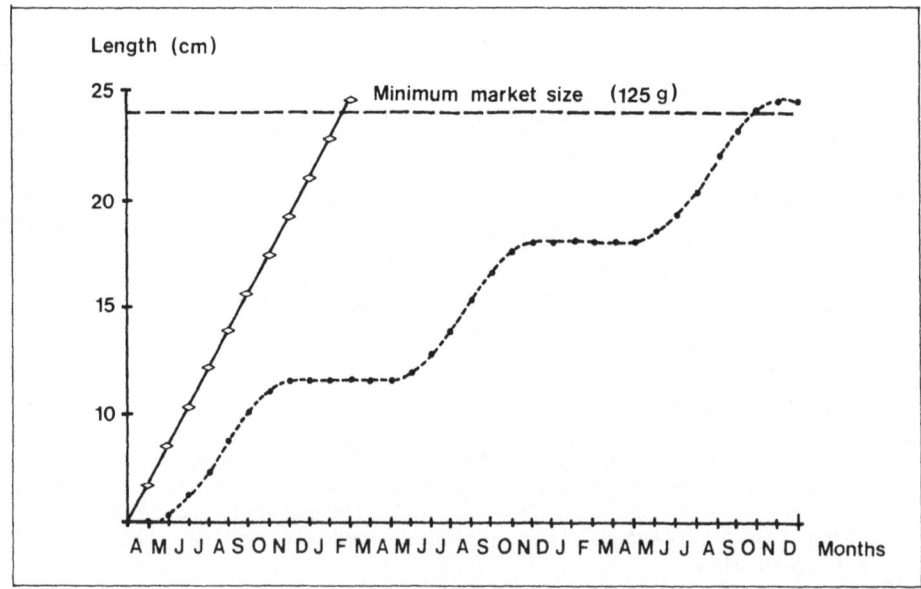

Fig. 5 : Predicted growth in length of Dover sole at ambient temperatures off the east coast of England (solid circles), versus a constant temperature of 19°C (open triangles); (3)

lower one was kept in an intensively cultured, temperature controlled aquarium (optimal temperature ± 24°C).

Furthermore, it is important to note that temperature optimization permits not only higher growth rates, but also provides a means of control of the reproductive cycle (in combination with the so called "hypophysation technique": a hormone injection).

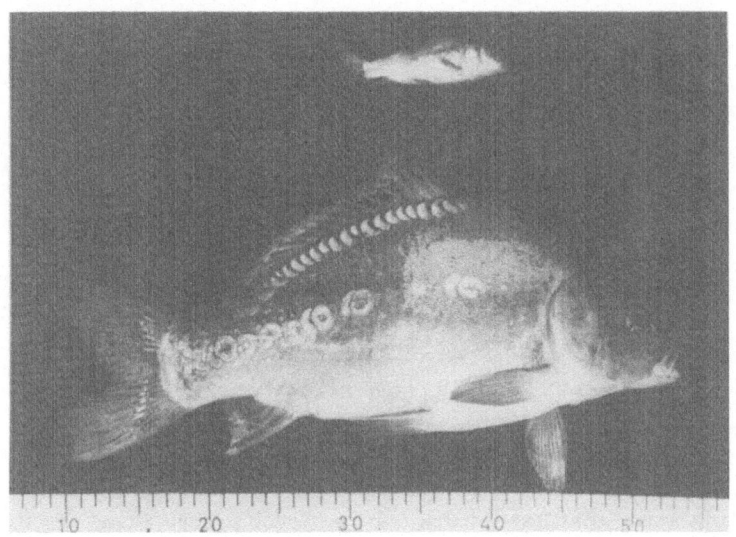

Fig. 6 : Two sibling one-year-old carp. The upper fish, raised in a
         "wild" pond, weights 40 g. The lower, intensively cultured in an
         aquarium with temperature control, weights 1 750 g (Courtesy Dr.
         Christoph Meske). (5)

This enables one to regulate the fish production so as to be able to
supply fish at periods of high demand, and hence obtain high sales
prices.

The tremendous response of the fish to the temperature of the environment
has drawn a lot of attention in the last decennia, and in this context
several freshwater fish, marine fish, oysters and shrimp are being
studied in Europe.

3. Methods of hot water production

An important question, which has in many cases not yet been addressed
sufficiently, is the effect of this increased growth rate on the
production costs of the fish, and consequently on the economically
justifiable cost of the provision of heated water. A study on this
subject by means of a simulation model yielded the following results for
the production of carp. (cf. figure 7)

By means of this diagram, the author of the model has calculated the
maximum economically acceptable cost per unit of energy required to heat
the water to the near optimal temperature of 23°C, assuming 15°C as the
natural water temperature. The results of this calculation are given in
table 5.
In fact, the calculated values of the allowable heating costs could be
considered as somewhat representative for situations with natural water
temperatures varying between - say - 13 and 23°C. This generalization
would correspond to a linear response of the per-unit-production-cost to

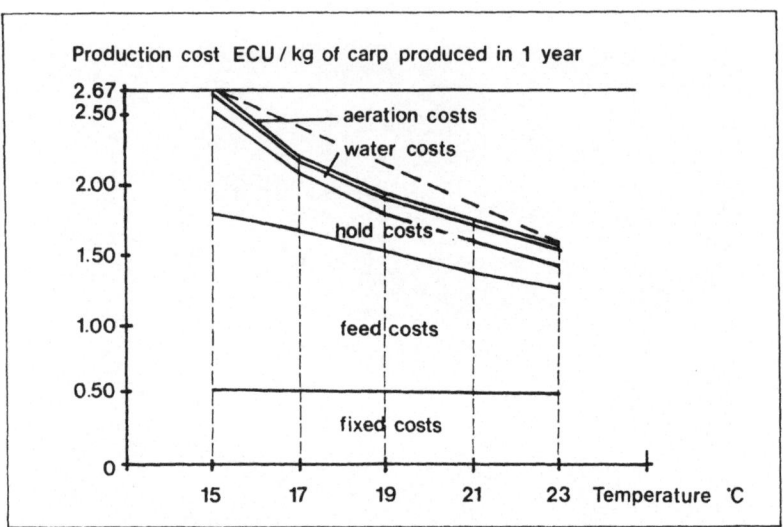

Fig. 7. Simulated cost structure of a model carp farm at different
temperatures. (2)

Table 5. Estimated maximum economically acceptable energy costs for carp
water heating from 15°C to 23°C for the case where no
recirculation occurs. (2)

	aeration level		
	0 %	50 %	100 % (1)
Average cost savings per K temperature increase and per kg fish produced over the year (ECU)	0.13	0.13	0.13
Average water renewal rate, required per kg carp produced over the year (l/min)	0.382	0.202	0.020
Max. allowable heating cost per K and per litre/min. water flow (ECU)	0.35	0.65	6.5
Max. allowable heating cost per kJoule (ECU)	1.6×10^{-7}	3.0×10^{-7}	3.0×10^{-6}

the water temperature, as shown in the dotted line of the cost structure
diagram above, (rather than the simulated curve of the total production
cost).

(1) At the 100 % aeration level it is not the lack of oxygen in the water,
but the build-up of toxic concentration of pollutants which becomes
the limiting factor. Further reduction of the throughflow of fresh
water will require the application of recirculation units.

A quick analysis of the quoted maximum allowable heating energy costs for this very temperature sensitive fish would lead to the conclusion that, certainly for systems without water recirculation, the use of conventional fossil fuels is not economically feasible in this case. Recirculation systems on the other hand are expensive, and – while reducing the heat demand – would significantly add to the investment ...

Consequently the utilisation of waste heat has been an important study object in the domain of heated aquaculture. The most important waste heat providing candidates so far have been the power plants. Indeed, the fossil fuel or nuclear power plants produce their electricity at an efficiency level of only 33 – 42 %. The main part of the other 58 – 67 % goes to waste in cooling waters. By way of example, the nuclear power plant, Tihange I in Belgium, with electricity generation capacity of 870 MW, uses 33 m³/s of cooling water (nearly 1 000 times as much as the required water-flow rate of the above mentioned 100 ton per year carp production unit with a 100 % aeration level), which it both aerates and heats up (by an average of 12 K). (cf. figure 8)

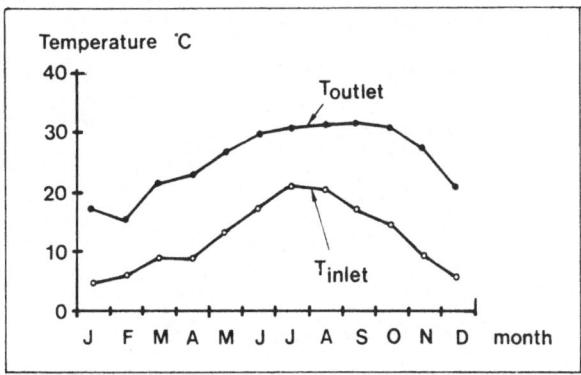

Fig. 8 : Average monthly water temperature at the condenser inlet and at the condenser outlet of the nuclear power plant Tihange I in Belgium. (8)

Several aquaculture systems heated in this way exist, most of them still in a research stage, some of them however already commercial operations (1). A few important installations of this kind within the EC are listed in table 6.

An example of the organization of a fish farm is schematically depicted in figure 9.

(1) Commercial heated fish farms are reported to exist at least for eel and Tilapia. Heated water use for larvae or very young fish - as opposed to use for their whole life-time - is more widely spread (e.g. in carp and salmon hatcheries).

Table 6 : Some important heated aquacultures associated with a power
plant. (6) (8)

Country	Power plant	Fish species
Belgium	Tihange	Tilapia, Carp, Eel, Trout
	Doel	Eel, Seabass, Gilt-Head
France	Cadarache	Eel
	St.-Laurent-des-Eaux	Eel, Pike
	Pierrelatte	Eel
the Netherlands	Flevocentrale (Lelystad)	Carp, Trout, Grass-carp
U.K.	Ratcliffe-on-Soar	Carp, Eel
	Trawsfynydd	Trout
	Ironbridge	Carp
	Drax	Eel
	Hinckley point	Eel
W. Germany	Bergheim	Carp, Eel, Tilapia, Catfish
	Kleinenberg	Carp, Eel, Trout

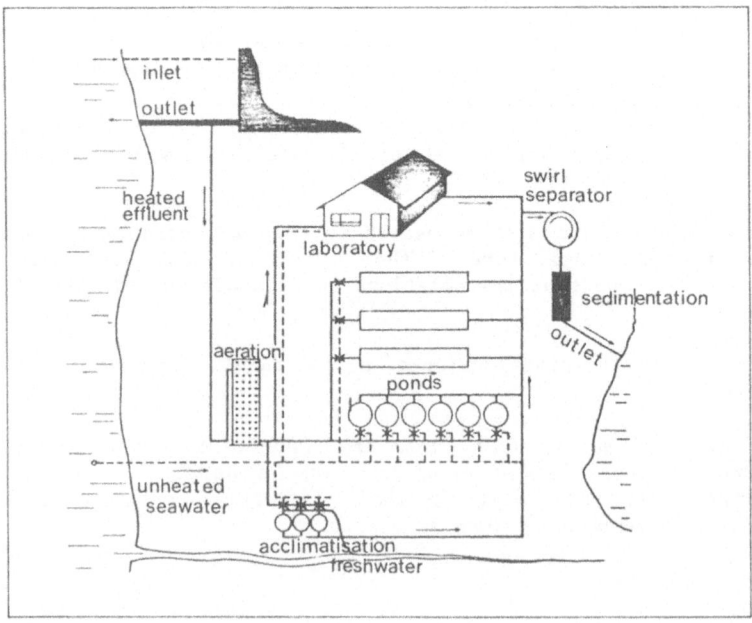

Fig. 9 : Diagram of a fish farm planned for the Ensted power plant in
Denmark. (4)

Other methods

Some examples of other energy sources under consideration or even under development are listed below.

+ Freezer-plants, washing plants, drying plants, etc. ... Given the low temperature level of the heat demand for heated fish farming, less sophisticated equipment would be required for the recuperation of the waste heat for the fish, than for its - at first sight more logical - use in filling the diverse heating needs of the waste heat producing plants themselves.
+ In colder regions and at lower temperature levels:
  - freshwater heating during wintertime by running it through a pipeline (= heat exchanger) in the seawater (which is warmer than the river-water at that time)
    (Such an installation exists in the fjords in Scandinavia) (4)
  - geothermal water usage
  - heat transfer via heat-exchangers from sewage waste water, which is usually significantly higher in temperature. (cf. study in Kiel, W.-Germany, where winter sewage waste water temperatures of about 10°C were measured (4))
+ One could conceive of developing different combinations, like the combination of a fish farm with a greenhouse: Heating of the water to be applied to the plants, and keeping fish in that water would not only affect fish growth due to the higher temperature, but also plant growth, due to both higher temperature as well as fertilization effect of the produced fish faeces.

Note:
1. As the above examples indicate, heated aquaculture today is nearly always planned in combination with some other activity, and plays a role of waste recycler. Combination and recycling systems of this kind are certainly worthy of attention for agriculture in general, in its efforts to deal with the increasing problem of scarceness of resources, and of energy in particular.
2. Whichever of the above mentioned options is considered, some crucial questions need to be dealt with:
   - the quality of the hot waste water: the presence and concentration of biocides, leached metals, radioactive or other harmful elements in the waste water will determine whether or not indirect heating by means of heat exchangers is required. (Heat exchangers involve a considerable investment expense, especially when the water is not recirculated, and/or when brackish or seawater are involved (cf. corrosion)).
   - the aeration level of the water: if the waste water is well aerated, the additional investment in aeration devices can be avoided.
   - The quantity of the on-site available water: the smallest viable production capacity for an intensive fish farm seems to be around 50 tons of fish per year. If the water is not being recycled, this means flow rates of 500 - 1 000 m³/h should be constantly available. (7)
   - The temperature of the water: Other than the hot water source, a cold source should be available, in order to regulate the temperature. In addition, the hot water should be constantly present, unless the fish can withstand periods of lower temperature, obviously at lower growth rates.

In this context, it is important to realize that certain power plants only function during winter time to meet the extra energy demand at that time.
- The <u>reliability of the heat and water providing system</u> is obviously a crucial factor in the degree of risk involved.

SOLAR ENERGY

<u>The application of solar energy</u> in heated aquaculture has been proposed by some as a subject worthy of research (cf. Fundamental Criteria, p. 1).

<u>Energy demand</u>: As can be concluded from the above discussion, the heat demand on an individual fish farm scale can be tremendous, and if the presently limited heated fish culture gains in importance, the heat demand on the EC level could also become sizeable.

<u>Solar supply</u>: The heat demand occurs during a good part of the year (at least eight months per year in Northern regions).

<u>Solar system efficiency</u>: Given the very low required temperature level, at first sight the efficiency to be expected is very high. (comparable with the efficiency of solar systems for swimming pools)

<u>Solar system cost</u>: Whether this high efficiency rate will allow for cost rates of the solar collected energy falling within the extremely tight limits of what is economically feasible (cf. above) is not at all clear.

<u>Socio-economic factor</u>: Should this application ever become economically feasible, it seems likely that the fish farmer could easily be motivated to use solar energy. Indeed, if fish production would benefit from a solar system <u>without</u> auxiliary heating, this technology could possibly free him from his dependence on another production system (like power generation), which, in case of conflicts of interest, will most likely be the dominating partner, with all the problems associated with that situation.

Conclusions about this application based on practical experience are given in chapter 2, p. 384.

References

(1) Dekker: "Scientific research foundation of TROUW Fish feeds". Putten, The Netherlands. Personal communication. 1986.

(2) Hambrey J.: "Technical and economic consequences of using waste heat for aquaculture" in "Energy Conservation and use of Renewable Energies in the Bio-industries". Editor F. Vogt, Pergamon Press, 1981, pp. 217 - 229.

(3) Howell B.R.: "The use of heat in fish farming" in "Energy conservation and use of renewable energies in the bio-industries". Editor F. Vogt, Pergamon Press, 1981, pp. 275 - 283.

(4) Klaus Tiews editor: "Aquaculture in heated effluents and recirculation systems". Proceeding of EIFAC meeting in Stravanger, 28-30/5/1980. Berlin, Volume I and II, 1981.

(5) Meske C.H.: "Breeding carp for reduced number of intermuscular bones, and growth of carp in aquaria". Bamidgeh, 20, H.4, pp. 105 - 119 (1968)
and
"Karpfenaufzucht in Aquarien". Fischwirt, 16, H.12, pp. 309 - 316 (1966).

(6) Ollevier F., Zoological Institute, Catholic University of Leuven; Personal communication and documents, 1984.

(7) Philippart J.C.: "L'élevage intensif des poissons dans les eaux chaudes industrielles" in "Demain" (Société d'Etudes et d'Expansion), nr. 290 : pp. 641 - 653, 1981.

(8) Philippart J.C., Melard Ch.: "Utilisation des rejets thermiques industriels pour la pisciculture (Bilan d'une première recherche en Belgique)". Institut de Zoologie de l'Université de Liège, 1979, p. 14.

(9) Rimberg D.: "Utilization of waste heat from power plants". Noyes Data Corporation London, England, 1974, p. 88.

III. FLOOR HEATING FOR PIGS (FARROWING HOUSES)

1. Introduction

In order to help establish the importance of the pig production sector for the different EC countries, the production values for 1984 are given in table 7.

Table 7 : Production and consumption of pigs in the EC for 1984.
(Eurostat 1/1985)

Geographical area	x 1 000 tons
Total EC	10 553
West-Germany	3 184
France	1 712
The Netherlands	1 543
Italy	1 166
Denmark	1 039
U.K.	964
Belgium / Luxembourg	677
Ireland	119
Greece	148

2. Recent evolution

The recent shifts and developments in the agricultural field (specialization and scaling up ...) are very clearly reflected in the pig-production sector.
Over the period of 1973 to 1979 for instance the size of the pig production within the EC increased by 6 %, while the number of pig producing farms decreased by 25 %!

This evolution is especially noticeable in the more industrialized regions and countries, where it has in some cases reached extreme forms ...
In the more conservative or poorer regions, less change might be visible at present, but the chances are great that here too the same course will be followed in the future.

Labour savings and rationalization on specialized farms can be phenomenal, especially in the area of pig fattening, enabling one person to care for nearly 4 000 fattening pigs at a time, on a quasi eight hour per day basis.
(Given the higher need for control and specific care, this figure is lower for the breeding part. One person can usually take care of 100 – 150 sows, which each yield about 14 – 24 piglets per year).

The other important consequence which can certainly be enhanced by specialization is maximization of the per-unit production. The more limited nature of the activity to be handled (characteristic of specialization), allows the farmer to build up an extensive know-how on the subject, which will enable him to obtain high quality selection and production.
Furthermore, the larger scale makes it possible to put to use differentiated housing facilities, equipment and food, which are

maximally adapted to the specific needs of the pigs at the different stages of their lives. Thus different housing facilities exist: for the new-born, unweaned piglets with their mothers (farrowing house); for the young weaned piglets; for the older piglets (and even those are separated, depending on whether they are kept for breeding or fattening purposes).
The boars are kept in separate facilities, as also are the dry sows, ready to mate, and the pregnant sows.

On the other hand, specialization and labour rationalization has created several typical new problems for which adapted solutions are being developed :
  e.g. - local excesses in manure
       - increased risks of agricultural holdings
       - abondonment of straw provision as a factor in environmental control (cf. section 3).

3. Ambient temperature and microclimate manipulation as a means of ecological optimization

Ambient temperature is one of the factors that has an important influence on production, and hence, is worth controlling at different levels in different housing environments.
The optimal temperature for a pig varies according to its age, production and feed-level. It is around 33°C for a newborn piglet, and decreases with time to about 16°C for an adult.
Figure 10 illustrates how the ambient temperature influences growth of young pigs of different weight classes.

Given the fact that a pig lies on the floor about 80 % of the time, which means 20 % of its surface is in direct contact with the floor (1), the total picture of the heat loss from the pig can be changed quite drastically by changing the temperature and thermal properties of the floor (cf. figure 11).

The influence of the floor characteristics on critical temperatures of the pigs (= temperatures which delimit the thermal neutrality zone, where the metabolic heat production is minimal) is shown by an experiment in which pigs of 40 kg were kept in groups of 9 on floors of straw, asphalt and concrete slats. The calculated effective critical temperatures were 11-13°C for straw, 14-15°C for asphalt and 19-20°C for concrete slats.
The above figures would suggest that the difference in conductive heat loss between straw and concrete slats has an effect on the heat balance of the group of pigs equivalent to that which would be caused by a 7-8 K decrease in ambient temperature for pigs kept on straw bedding (1).

The influence of the heat conducting characteristics of the floor on the weight gain of piglets is shown in Fig. 12.

In modern farm buildings, the provision of straw for insulation purposes has become problematic. On the one hand, the straw provision is quite labour-demanding, while on the other hand, it would immediately plug up the slatted drainage system for automatic manure removal. Here, a possible solution could be the application of floor heating, which can strongly contribute to the establishment of an optimal micro-climate, while avoiding the above-mentioned objections.

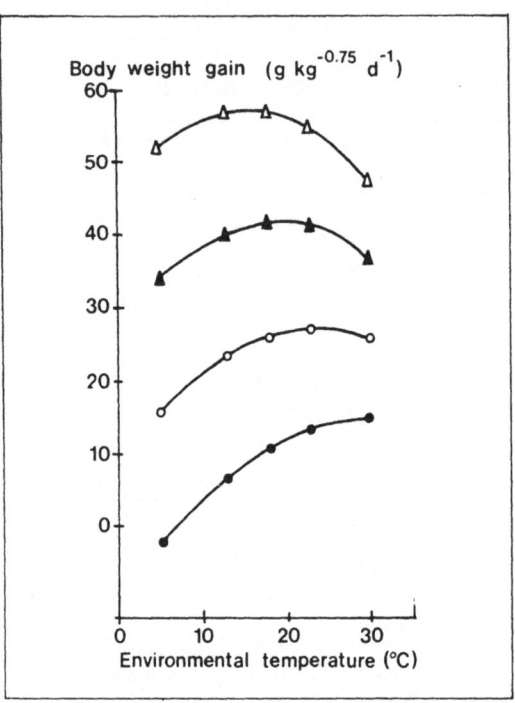

Fig. 10. Body-weight gain of pigs, 20-105 kg body weight, in relation to
environmental temperature (°C) and level of feed intake. Both
body-weight gain and feed intake are expressed relative to
'metabolic body size', that is, g kg$^{-0.75}$ d^{-1}. In these limits
the lines correspond to : (●) 40; (o) 80; (▲) 120; (△) 160 g
feed kg$^{-0.75}$ d^{-1}. (The following sources of information were
used : Heitman and Hughes, 1949; Hicks and Webster, 1968;
MacGrath et al., 1968; Jensen et al., 1969; Sugahara et al.,
1970; Fuller and Boyne, 1971; Morrison and Mount, 1971;
Verstegen et al., 1973; Close, Mount and Brown, 1978; Verstegen,
Brascamp and van der Hel, 1978; Phillips et al., 1979 and
Stahly, Cromwell and Aviotti, 1979). The general equation
relating body weight gain (ΔW) to environmental temperature (T)
and feed intake (I) was : ΔW = 26.26 (± 0.32) + 1.48 (± 0.17) T
− 0.015 (± 0.005) T^2 + 0.45 (± 0.01) I − 0.00020 (± 0.00003)
I x T^2 (r = 0.88)
(The SEs of the coefficients are given in parentheses) (1)

Given the instinct of pigs to dispose of their wastes in the colder part
of their pens and to leave the heated part clean, floorheating can be a
valid solution.

The problem of micro-climate optimization is most pronounced in the
farrowing houses, where the sow (with optimal temperature of 16 - 18°C)
is kept together with the piglets (with optimal temperatures of 25 -
33°C, depending on their age) (cf. figure 13).

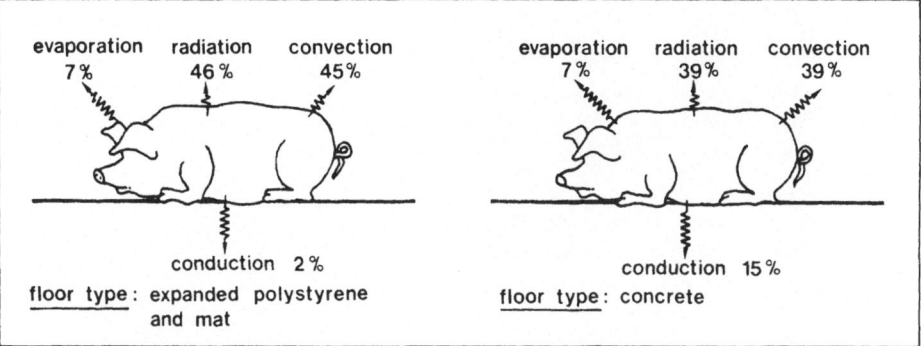

Fig. 11 : Distribution of heat losses from a piglet of 2 kg in a relaxed
posture on different floor types, given an air temperature of
20°C (4).

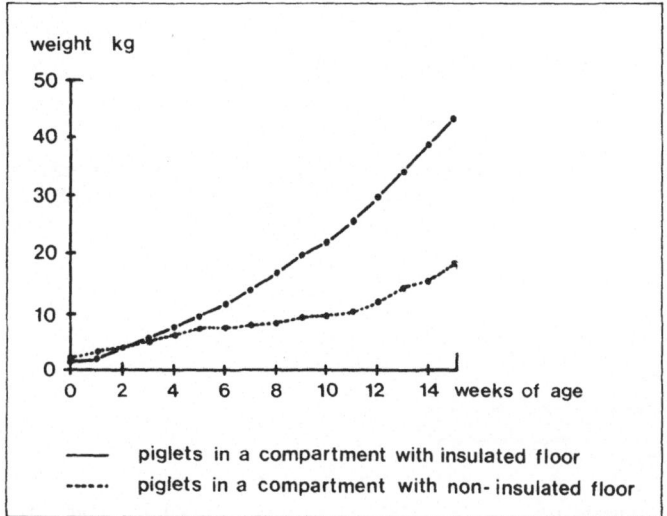

Fig. 12 : Influence of floor characteristics on weight gain of piglets
over time. (4)

It should be emphasized that the newly born piglets in the farrowing house
are most sensitive to the climatic conditions, because among other
reasons, their body thermoregulation systems are still developing.
Thus, the establishment of an optimal micro-climate for these piglets is
essential for their body development as well as for low sickness- and
death rates.

The only practical solution to this specific problem is the application
of local heating in the section which is not accessible to the sow, and
where the piglets can stay when not sucking milk. Here again,
floorheating is one of the more logical alternatives and therefore also,
one of the currently used applications.

Fig. 13 : A modern farrowing sow stand.

4. Methods of floor heating

Floor heating can be provided by electrical resistances, the heat being accumulated in the floor during the night (night tariff electricity) and given off during the day (cf. figure 14).

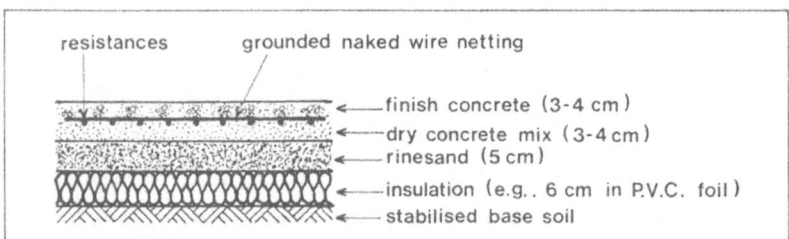

Fig. 14. Scheme of a floor heating plate, heated by electricity.

A floor heating system, more effective in the area of thermal control, uses water tubes, wound through the concrete.
An installation of this kind is shown in Figure 15.

In order to reach the desired floor temperatures (<40°C), water temperatures between 40 - 55°C are applied. Average rates of floor heating energy use are 100 Watt/nursing sow, which for a particular house for growing pigs was found to correspond to more than 40 % of the total heat demand. (2)

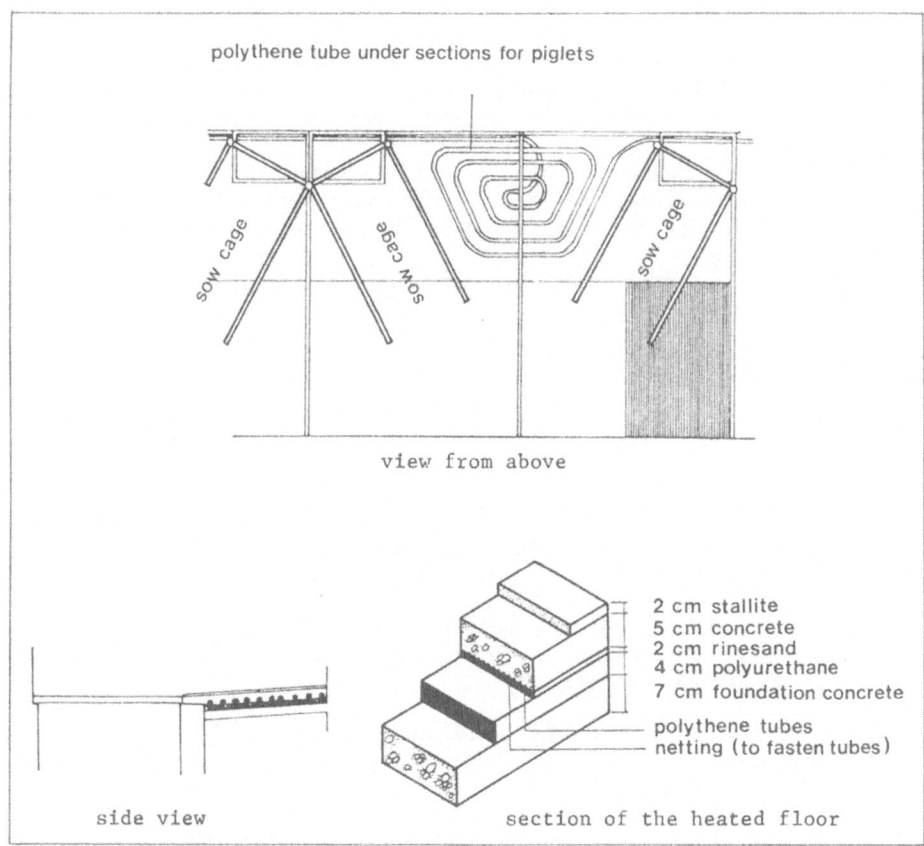

polythene tube under sections for piglets

sow cage sow cage sow cage

view from above

2 cm stallite
5 cm concrete
2 cm rinesand
4 cm polyurethane
7 cm foundation concrete

polythene tubes
netting (to fasten tubes)

side view   section of the heated floor

Fig. 15. A possible water-heated floor set-up. (5)

The water for the floor heating system is traditionally heated by means
of the conventional central heating system, burning any of the fossil
fuels.
Research is being done on the applicability of different alternative
energy sources :
- Use of (air-water) heat exchangers and/or heat pumps, extracting heat
  from the exhaust ventilation air which has a high energy content. (3)
- The integration of solar energy collecting devices.
With a view to improving the efficiency of both these applications,
research is being done to assess the feasibility of using heated mats
with a lower inertia, in which the floor water temperature can be
reduced to below 40°C. (3)

SOLAR ENERGY
As regards the applicability of solar energy, theoretically speaking
this process seems worthy of further investigation : (cf. Fundamental
Criteria, p. 1)
- Energy demand : The potential heat demand is sizeable. (As stated
  above, + 100 Watt/nursing sow, corresponding in some cases to more
  than 40 % of the total heat demand of a house for growing pigs).
- Solar supply : Since a heat demand exists even in summer time, one
  could expect that a reasonable amount of the solar supply could be

put to use.
- Solar system efficiency : The operating temperatures 40 - 55°C (or possibly lower, cf. the above mentioned research) are within the range of what could be efficiently obtained by means of solar collectors.
- Socio-economic factor : On modern industrial type farms, the technical skill level of the farmer will most likely not be a limiting factor for the integration of the new solar technology.

For conclusions based on practical experience, cf. p. 385.

References

(1) Clark J.A. Editor : "Environmental aspects of housing for animal production". First edition; London, U.K.; Butterworths; 1981; pp. 511.

(2) Ghys M. : "Onderzoek naar de integratiemogelijkheden van zonneënergie in kraamstallen". Engineering student dissertation. Catholic University of Leuven, Belgium, 1981.

(3) Laboratory of Agricultural Building Research (Prof. V. Goedseels). Catholic University of Leuven, Belgium, in collaboration with AKZO, Arnhem (Netherlands).

(4) Mount L.E.: "The climatic physiology of the pig". 1st Edition, Monographs of the physiological society, London, Great Britain, Edward Arnold Ltd., 1968, p. 271.

(5) Van der Stuyft E.: "Rekenmodel voor energiebesparingsberekeningen bij vlakke plaat-watercollectoren in kraamstallen". Engineering student dissertation. Catholic University of Leuven, Belgium, 1982.

IV. HOT WATER PRODUCTION FOR CLEANING PURPOSES IN DAIRY FARMS

1. Introduction

Within the EC, a total of 25 million dairy cows produce around 104 million tons of milk per year. (Eurostat, 1983 values)

A tremendous variety of dairy farm types exists within the EC :
Both variety in degree of specialization, as well as variety in exploitation size.

- At the lower end of the specialization scale one can find mixed farms, combining a host of different activities, for instance the cultivation of several crops (grains, sugarbeets, ...), the production of pigs as well as cattle (including green crop cultivation, meadow management, etc ...).
  In some such cases the cows are even kept for the dual purpose of both milk and meat production.
  At the other end of the scale are to be found the highly specialized farms for dairy cows only, obviously with the necessary grass and fodder production capacity (on some extremely large farms, even the task of green food production is delegated to others).
- As far as sizes go, most European farms are family run exploitations. In this context, an exploitation herd with 70-100 cows would be considered large, while farms with 20 cows or less are considered small.

Given the extreme diversity in the production system characteristics, very different milk yields are obviously obtained. Some of the highly specialized dairy farms produce more than 6 000 kg per cow per year, while others have a yield that is less than 50 % of this value ... (E.g. : the average production value on a small Italian farm with less than 20 dairy cows is about 3 000 kg per cow and per year). (1)

The above described wide variety in milk producing systems and yields is a result of different factors, mainly climatological, social, geographical and pedological.
The impact that - say - geography can have on farm structure can be exemplified by the existence of many small labour intensive specialized milk producing farms, in certain mountainous areas in Italy, France, U.K., etc ... where crops other than grass (meadows) can only be cultivated with difficulty.
Table 8 gives an idea of the above mentioned diversity between the different countries within the EC.

2. Recent evolution

In the past few decades there has been a strong trend, which is still continuing, towards increasing size and specialization of the farms.
For example, a study - conducted over a geographical area comprising 60 % of Belgian dairy cows - showed that over the period 1971 to 1982, the number of dairy farms decreased by 6 % per year, while the total number of dairy cows remained the same. (2)
This trend is an unavoidable result of the economic objective of obtaining wages for the farmers that are comparable with those of

Table 8 : Number and average size of dairy farms in the EC countries in 1983.
(Eurostat data bank , 1985)

	Numbers of farms (x 1 000)	Average number of cows per farm/herd
United Kingdom	58	58
The Netherlands	64	40
Denmark	35	28
Luxembourg	3	27
Belgium	49	20
France	420	17
Ireland	91	18
W. Germany	397	14
Italy	332	8
Greece	66	3
EC total	1 515	17

equally skilled workers outside the agricultural industry. Indeed, increasing size and specialization make mechanization possible, and thus allow for significant labour savings and increased income. (In the same study, the income of a worker on a dairy farm with more than 90 cows was found to be 367 % of that earned by a worker on a farm with less than 30 cows.) (2)
This economic objective of labour wage increases to match those paid in the non-agricultural area has among other things been the cause of a complete readaptation of the milk production chain, including techniques for feed-production and provision, for animal waste disposal, for milking and milk processing, etc ...

As regards milking and milk processing, a gradual shift towards mechanization can be noticed over the range of the different dairy farm types mentioned above.
In the first stage are to be found the portable milking-machines, where the milking itself is done mechanically, but the transportation of the milk to the tank, as well as the equipment from one cow to another, stall to stall, is still done by the farmer himself. (This system allows 1 person to take care of 20-35 cows.)
In the next stage, the milk is automatically transferred to the milk-storage tank by means of pipelines installed in the cowshed. (This system allows 1 person to take care of maximum 45-60 cows).
For larger exploitations, the whole housing concept has been altered, and the milking operation isolated in a separate milking parlour, while the cows are loose housed in a cubicle house (see fig. 16) whence they can move in and out of the milking parlour (see fig. 17) at the appropriate time.
In this way, even the need for transportation of the milking equipment (and as a matter of fact, in some cases also of the food since the animals sometimes have access to it themselves) disappears, while at the same time this extra specialization of the milking facility normally allows for better working conditions for the farmer, more rational set-up of the milking equipment, and more hygienic milking conditions. Depending on the system sophistication and dimensions and the degree of automation, 1 person can handle up to 100 (120) cows.

Fig. 16 : Cubicle house.

Fig. 17 : Inside view of milking parlour.

Not only on the farm itself, but obviously also in the milk-processing factories, labour-savings and work rationalisation are imperative in order to survive economically. In this context is to be found another new development in certain locations : The installation of milk-cooling tanks on the farm itself, enabling the milk processing factory to reduce its collection rounds from at least once a day in the past to every other day at present.

3. Hygienic standards and requirements

Next to mechanization, a second factor which has also significantly influenced the characteristics of modern dairy farms is the consistent effort, exerted by public health departments and organizations, to attach a prestigious "quality label" to this very important element in human nourishment.
By means of sizeable additions or subtractions from the base milk price, depending on the milk quality characteristics, these organizations have provided strong incentives to farmers to adjust their milk production methods in order to reach maximum milk quality.

In fact, the development of milking parlours as units, separated from the live stock housing, the use of milk pipelines and cooling tanks for storage of the produced milk, are not only a result of the quest for labour-savings, but also of the efforts to increase milk quality by means of upgraded hygiene potential.
In this context, and given the rich nutritional value of milk, and its susceptibility to spoilage by microorganisms, it is clear that regular rinsing and cleaning of any equipment that has been in contact with the milk is essential.

4. Hot water requirements

In order to clean milking equipment, and the dairy parlour itself, as well as the cow udders - which get soiled more rapidly because of the current housing methods without straw - significant amounts of hot water are being consumed. Amounts varying from 4 - 19 litres per cow and per day have been reported within the EC. (3)
(Among other things, the degree of mechanization has a role to play in the level of hot water consumption. Hand cleaning of the equipment allows for more intense rubbing of the areas to be cleaned than automatic water-spray cleaners, and thus requires less water for equal performance. A representative average value for the agro-climatic zones II and III seems to be 7 litres of hot water per day and per cow.
(2.5 to 3 l at 80 to 95°C; the rest at 40°C) (3)

If the average cold water temperatures are assumed to correspond to the average yearly ambient temperature (which is usually the case for well water) - say 12°C - an estimation can be made of the overall energy demand of this treatment for different farm sizes, as shown in Table 9.

Table 9 : Daily energy demand for hot cleaning-water production on some typical size farms.

Number of cows/farm	20	40	60	80	100	150
Daily energy demand for hot water on the farm (MJ)	30	60	90	120	150	225

5. Methods of hot water production

Traditionlly, geysers or boilers are used for the production of hot water.
- With geysers, the need for a hot water storage tank is obviated, since the water is heated as it is tapped. This solution is only appropriate for smaller farms, due to the limited water flow (commonly maximum 7 litres per minute, heated to 70°C).
- For the larger farms, a boiler is essential in order to have a sufficient amount of heated water readily available.
  These boilers can be heated directly by an oil or gas burner, or an electric immersion heater; they can also be heated indirectly by means of a central heating unit.
- Electric heating involves a rather long heating-up time, and so requires rather high water storage capacities, but may allow for operation at low night tariffs.

An alternative method which has gained much success in the last few years is the recovery of heat from the milk that is cooled on the farm, and this by means of the already present cooling heat pump. (milk is produced at 38°C, and is mechanically cooled from 30°C to 4°C within 3 hours. In this process about 140 kJ of heat is dissipated per litre of milk.) (4)
A possible design of such an installation is given schematically in fig. 18.

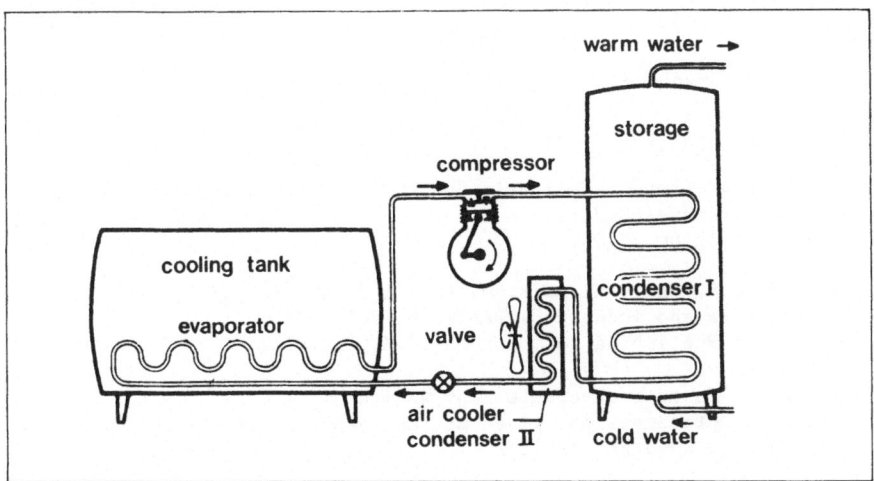

Fig. 18 : Schematic representation of heat recovery from milk cooling.

The heat extracted from the cooling tank is discharged to the water in the storage and heats it to 60°C. When that temperature level is exceeded, the heat is discharged through the air-cooled condenser II instead.
The water at 60°C is then further heated to the desired temperature of 80°C by a conventional heat source. In this way the total required amount of conventional heating energy can be more than halved. (5)

Based on the already acquired experience, in the case of certain heat pump set-ups payback periods are expected of 2 years for farms with + 40 cows. Payback periods increase exponentially with decreasing farm size, and reach economically unfeasible levels for farms with less than 15 cows. (5)

On the basis of detergent composition, as well as optimization of the cleaning process and its energy efficiency, there is a trend to lower the required water temperature from 80°C to closer to 55°C. (6)

SOLAR ENERGY

The criteria for selection of the water heating process as an application of solar energy seem promising at first sight : (cf. Fundamental criteria, p. 1)

- Energy demand : As explained above, the energy demand is important.

- Solar supply : The level of hot-water consumption is rather constant and spread over the whole year. The corresponding heat demand pattern allows for making good use of the periods of high solar supply.

- Solar system efficiency : If solar preheating to e.g. 40-50°C of the water is the aim, reasonable collector efficiency values can be obtained.

Conclusions about this application based on practical experience are given in chapter 2, p. 386.

References

(1) Belgische Boerenbond : "De 'kleine' melkveehouder in de Europese Gemeenschap".
     De boer : 8/4/1983   14, p. 31.

(2) Belgische Boerenbond : Nationale Boerenbonddagen 1983, jaarverslag.
     Sector : Rundveehouderij.
     Leuven, Belgium, 1983.

(3) G. Schepens, D. Mahy, H. Buis, W. Palz : "Solar Energy applications in Agriculture and Industry", Series H, Volume 1.
     D. Reidel Publishing Co., For the C.E.C., 1986.

(4) J. Thysen : "Warmterecuperatie door warmtepompen."
     Report on presentation, given at the Catholic University of Leuven, Belgium, on November 1981.

(5) J. Ubbels, A.P. Meulman, C.P. Verhey, J. Brouwer : "The saving of energy when cooling milk and heating water on farms".
     CEC-EUR 6915, p. 65, 1980.

(6) G. van der Gaast : "Energiebesparing bij Melkwinning"
     Landbouwmechanisatie 32, 11 p. 1043 - 1047, November 1981.

V. HOT WATER FOR MILK PREPARATION ON CALF REARING FARMS FOR VEAL PRODUCTION

1. Introduction

The 1984 production of high quality white veal meat in the EC was more than 900 000 tonnes. More than 80 % of this value was accounted for by France, the Netherlands and Italy.
Table 10 gives an idea of the relative importance of the production in different countries.

Table 10 : White veal meat production in 1984. (Eurostat 1/1985)

Country	Production	
	x 1 000 tons	%
France	418.4	46
the Netherlands	175.1	19
Italy	160.1	18
W. Germany	84.4	9
Belgium	40.6	5
UK	12.6	1
Greece	5.6	<1
Denmark	3.5	<1
Ireland	0.9	<1
Luxembourg	0.0	0
EC	901.3	100

2. Recent evolution

The calf-rearing sector is another clear example of the recent shifts and developments in agriculture – especially in the more industrialized countries – involving mechanization, scaling up, specialization, vertical differentiation on the farm level, and reintegration on an intercompany level.

Before the introduction on the market of artificial milk powders around 1955, milk-feeding of calves for veal production only occurred as a marginal activity on cattle farms : periodically, a few calves were kept and fed on a whole milk diet. Given the expensive nature of the resulting luxury meat product, the market was very limited.
The production of cheaper artificial milk powders brought about lower veal production and sales costs on the one hand – resulting in a higher demand of the increasingly wealthy population –, while on the other hand it opened the flood-gate for specialization and vertical differentiation – resulting in farms devoted exclusively to calf rearing.

A few figures for the Netherlands will help illustrate this trend :
- the increase in production and demand :
  The number of calves reared for veal production only in May 1950 was 2 512. Twenty seven years later, this number had increased by a factor of almost 200. (496 000 calves in 1977) (5)
- specialization and scaling up :
  In 1966, the 10 757 farms where calves were fattened had an average

population of 22 calves. By 1977, the number of farms had been
reduced to 3 272, with an average population of 152 calves. (5)
Today the normal population of an industrial type calf-rearing farm is
about 300-500 units per full-time worker.
- Vertical reintegration on an inter-company level :
  90 % of the veal is currently produced under the status of "contract-
  production", an extremely strict form of reintegration. (1)

These examples are taken from one of the countries where the recent
trends are most extreme. It is obvious that this evolution is less
clear in certain other, often less industrialized regions, and that
simpler, more conservative production systems still exist. Since
however the larger part of the veal production occurs under the more
recently developed conditions, we will from now on focus on these
production type settings.

3. The industrial type calf-rearing process.

"White veal meat" is the result of a very intensive feeding process
based exclusively on artificial milk.
Indeed, for the digestion of this nutrient type, only one stomach needs
to function, while the others can remain inactive. In this way, the
calf can maintain its initial capacity to take in extremely large
amounts of food, more than that needed for normal development. At the
same time, this diet is (or can be made to be) low in iron, this being
the prime requirement for the production of white meat. (The milk
powder should contain less than 20 p.p.m. of iron (5); the water less
than 0.2 mg of iron per litre (4).
The calves are reared in this way up to an age of 16-20 weeks, at which
point they already reach the weight of 180-200 kg !

Bought at an age of 1 week or less, the calves are kept in housing,
where they are either placed in individual boxes on slats (1.6 x 0.6 m)
and fed with a bucket, or in collective enclosures for 15-20 calves,
sometimes fed by automatic feeders (see figs. 19 and 20).

The environmental temperature is kept between 15 and 22°C depending on
the age of the calves (by heating the insulated house during winter and
by increased ventilation - providing a so called free cooling effect -
during summer). A forced ventilation of 25-150 $m^3/(h.calf)$ is applied.
Relative humidity values vary between the limits of 65 and 80 %.

These and other carefully set variables, combined with a watchful
farmer's eye should help prevent sickness of the animals, weakened as
they are by the chosen rearing process. Indeed, the extremely large
amount of rich nutritional elements which they have to digest calls for
a constantly high degree of metabolic activity. This is associated with
an increased transpiration rate, and hence a greater sensitivity to the
environmental climate. Furthermore, as mentioned above, the diet -
however rich it may be in other nutritional elements - is kept low in
iron, causing the animals to be in a quasi anaemic, and thus, weakened
state.

Fig. 19 : Calf house : for veal production with two rows of individual
    boxes.

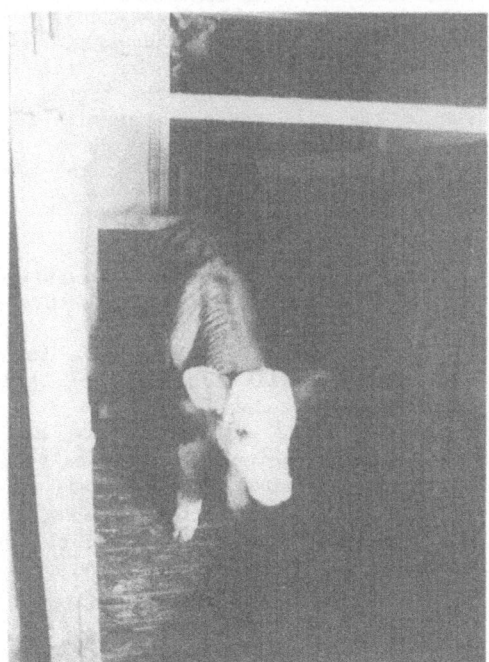

Fig. 20 : Individual calf box.

4. Nutrition

The above discussion has made it clear that the establishment of an optimal "food-program" is the central element of the whole calf-rearing business.
Some of the more crucial factors in such an effective program are discussed below.

1) Contents of the artificial milk-mix.
   There is quite a variety in the mixes used. One that would however be representative for many others is (5) :
   - 45-55 % (up to 60 %) skimmed milk powder, containing 35 % proteins and 50 % carbohydrates.
   - 10-20 % whey
   - 17-20 % fats of animal or plant origin (sheepfat, lard, palm oil ...)
   - 5-10 % carbohydrates (glucose; starches)
   - 1 % minerals, vitamins, antibiotics, ...

2) Concentration and amounts of food as a function of age.
   As the age increases, the total volume of the milk, as well as its concentration are increased, and its composition is changed.

   For example, during the lifetime of the calves their milk intake increases from $\pm$ 2 litres per day (for a 1 week old calf) to $\pm$ 17 litres per day (for a 16-20 week old calf). (5) The calves are generally fed twice a day.

3) Adequate emulsification of the fats in the milk.
   It seems important for digestibility that the fats be well emulsified in the milk. This is often done through melting the fats at temperatures of a least 55-60°C, and then mechanically emulsifying the fine fat globules into the rest of the liquid. In order to reach these temperatures of the milk powder-water mixture, water of 70-80°C needs to be available (no more than 80°C, or vitamins would disintegrate). Cold water is then added to the mixture (at 55-60°C) to bring it to the desired temperature for feeding.

4) Feeding at a constant food temperature.
   In this process, where maximum food intake is required, it is important that the milk be served at a constant temperature of about 35-39°C. This for several reasons.
   1) If colder milk is served, the calves have to produce energy of their own to heat up the milk to their body temperatures (a less efficient and more costly process).
   2) The milk temperature level seems to have an influence on the reflex of an oesophageal sphincter of the calves, which causes the milk to come into the rennet-stomach (instead of the rumen) and to coagulate there. This coagulation allows for the necessary enzymes to do their job. Poor coagulation of the milk will cause it to be indigestible for the calf, and so unavailable for conversion into useful energy through metabolism.

5. Milk preparation and distribution

The milk preparation should be done with all of the above-mentioned factors in mind.
- The <u>appropriate</u> amounts of milk powder with the right composition need to be fed into the mixer.
- Hot water at <u>70-80°C</u> needs to be added, and those 2 components <u>mixed</u> for about 3-5 <u>minutes</u>.
- Cold water needs to be added to the mixture to reach the <u>desired concentration and temperature level</u>.

Installations for milk preparation exist in different sizes and degrees of complexity. They range from very simple devices for smaller farms to more complicated ones for the larger farms.
In order to buffer the differences in installed heating capacities, mixing capacities and feeding capacities, different storage tanks are often needed.

A schematic representation of a possible milk preparation set-up is given in Fig. 21.

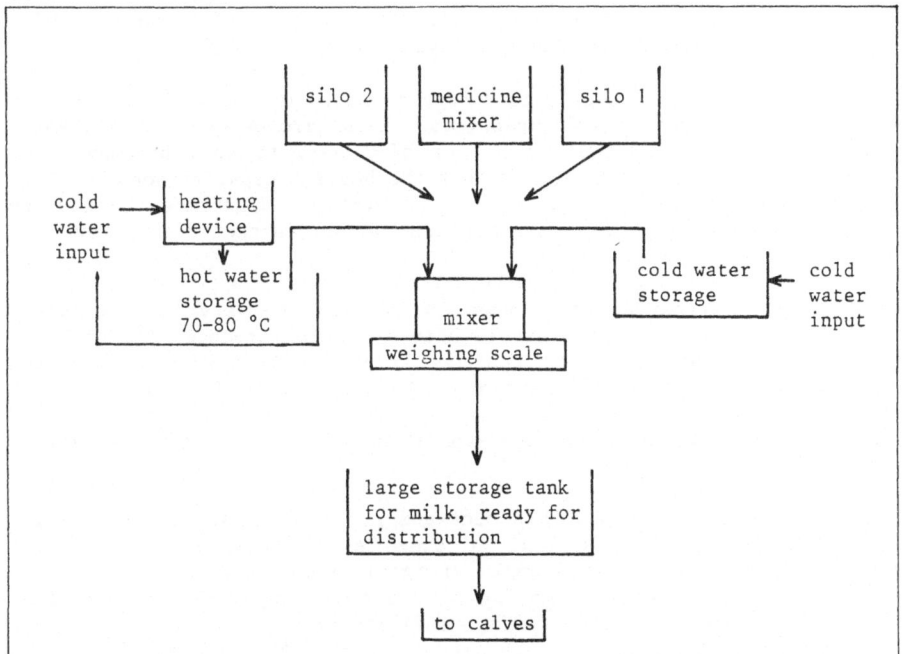

Fig. 21 : Schematic representation of a possible milk preparation set-up.

On farms with 300-500 calves the milk preparation is often completely automated in order to allow <u>1 person</u> to manage the whole farm.

Also the DISTRIBUTION of the milk can take various forms :
- On small farms, the milk is transported in <u>buckets</u> from one or more stationary mixers of 100-300 litres.
- As the size of the farm increases, a transition is often made to the

use of transportable mixers, eliminating the need for laborious and time consuming carrying of buckets. The milk ingredients are added to the mixer in the "food-kitchen" and mixed. The mixer is then rolled to the different compartments, where milk distribution occurs by means of buckets, or by means of a hose.
- For even larger farms, pipelines are often used to distribute the milk from the stationary mixer to the different compartments. In each compartment, a hose is hooked up to the pipe-line to enable further distribution to the individual calves.
- Automatic feeders are sometimes used in collective enclosures. These could theoretically eliminate the labour requirement for milk preparation and distribution, (estimated as 75-85 % of the total labour requirements in the above described situations). (5)
(These feeding systems are often still in an experimental stage, and seem to pose some problems (1). Collective enclosures call for added attentiveness and control from the farmer).

6. The production of hot water.

Hot water production on calf rearing farms requires tremendous amounts of energy. By way of example : In the Belgian industrial-type farm mentioned on p. 387 the energy consumption for hot water production for milk preparation and hygienic purposes averages 3-4 MJ/per calf per day. (1)

Since the periods over which this energy is spent are very short, the use of gas geysers is not usually practical. Even warm water boilers (heated by means of gas or electricity   or oil) need to be supplemented by larger hot water storage tanks, so that the heating capacity does not have to meet the high feeding capacity of the system. Sometimes the storage tank of the central heating system, kept at a constant temperature, is used.

Few, if any, alternative techniques for heating the water are currently used. However, a new "spray dry technique", which produces fat globules of about 2 μm, surrounded by the dry skimmed milk powder, allows for mixing with water at 55-60°C and in some cases (use of vegetable oils) even 45°C ! (3)
Especially in this last case, the use of heat pumps or solar collectors could be worth considering.

As a heat source for a possible heat-pump, the ventilation exhaust-air, at temperatures of 15-20°C (and at times even more), is probably the best candidate, since a considerable energy flux exists in it.
However, the most logical load for such a heat pump would in the first place be the heating of the incoming ventilation air.
(For this application, a heat exchanger would probably suffice).

Let us consider for instance housing for 100 calves (dimensions 22 m long, 15 m wide, 2.2 to 5 m high; walls and roof insulated so as to have a global heat transfer coefficient k of 0.7 $W/(m^2\ K)$; outside temperature - 10°C; inside temperature 15°C; calf weight 150 kg; minimum ventilation rate of 30 $m^3/(h\ calf)$.)

The energy flow for such a situation, and using a heat exchanger with a recovery capacity of 40 % of the perceptible heat flux, is given schematically in figure 22.

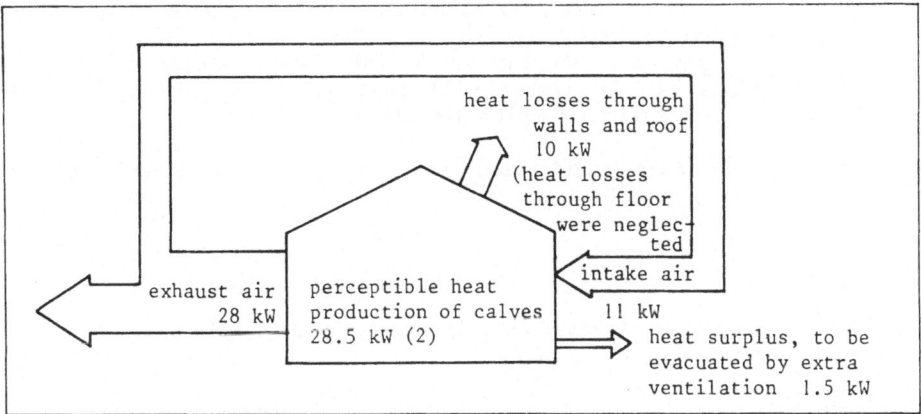

Fig. 22 : Instantaneous scheme of the energy balance of a calf-rearing farm under the conditions described above.

This would mean that, theoretically speaking, even at this low outside temperature, such a heat recovery system could (more than) meet the heat demand of the stable.

As the outside temperature increases above -10°C, in the case of larger farms the excess heat could possibly be put to use to help meet the heat demand in compartments of younger calves, with higher temperature requirements and lower heat production values.

Note :
These heat-flows vary over time (e.g. possible heat excess during the day; heat demand at night) and so heat pump and storage tanks are crucial elements in such a system.

At outside temperature levels where also this load is more than covered, the excess heat would be put to use for water heating via the heat pump.

Even though theoretically possible, questions like extra investment expenses versus returns, or even the question of compatibility of these heat recovery systems with the given farm set-ups (e.g. equipment life-time reduction due to the corrosive nature of the air, dust build-up) are still far from being adequately addressed.

SOLAR ENERGY
Several factors seem to favour the selection of water heating for milk preparation as a subject of research relative to the feasibility of integration of solar energy. (2), (6) (cf. Fundamental Criteria, p. 1)
- Energy demand : The heat demand is high (more than 1 800 MJ per day on a farm with 400 calves)
- Solar supply : A heat demand is present all year long, so the solar supply could be put to use effectively.
- Solar system efficiency : Given the low initial temperature of the well or tap water that is to be heated, the efficiency rate of the collecting system can be expected to be large. This is especially

true if the temperature level which the solar system aims to reach is not too high. For this purpose, following options can be taken :
- Solar preheating only of the water;
- The use of the new "spray dry techniques" with low water temperature level requirements (down to 45°C).
- Socio-economic factor : On the specialized calf rearing farms, the technical skills of the farmer will most likely not be a limiting factor to the integration of solar systems.

Conclusions about this application based on practical experience are given in chapter 2, p. 386.

References

(1) Ghys M. : Walcovit (special animal feeds), Beerse, Belgium. Personal communication.

(2) Goedseels, V. : "Land- en tuinbouw : Toepassingssectoren voor warmtepompen". Report on K.V.I.V.-congress, March 26th, 1982 at the Catholic University of Leuven, Belgium, p. 8.

(3) Grimon, A. : Heat of the laboratory of Walcovit (special animal feeds), Beerse (Belgium); Personal communication and documents.

(4) Mornet, P. : "Le veau"; Paris (France), Maloine S.d., 1977, p. 211.

(5) Van den Hoonaard, L. : "Het houden van Vleeskalveren", publication 79/226/500
Consultentschap voor de Rundveehouderij, Gildemeesterplein 1, Arnhem, the Netherlands. 1979, p. 41.

(6) Van Duyse, N. : "Technico-economische beschouwingen van een zonneinstallatie in een mestkalverij bij middel van een simulatie-model".
Engineering student dissertation
(Catholic University of Leuven), Belgium, 1983; p. 16.

CHAPTER II
Present research within the EC on the application of solar energy for hot water production

In this chapter, an overview is given of some of the research that has been done on solar water heating applications, introduced in the preceding chapter. This material is based on information provided by CEC action 2 and action 3 participants . It will be apparent that this list of research from within the EC, - which includes many projects that are not yet in their final stages, and can thus provide only provisional conclusions - is not complete, and that the individual descriptions do not always contain all the necessary information. We believe however that within the limitations of the material which was available to us, this chapter can give the reader a reliable picture of the state of the art of the current research, as well as help develop a feel for the potential of the applications in question.

The first part of the chapter consists of a list of basic descriptions of research projects, in order of the application. (For an explanation of the format of the basic descriptions, cf. Addendum, p. 467)

Two projects from outside the EC are added to this first part. They are examples of research that is well on its way elsewhere, while still in an early stage within the EC itself.

The general experience gained over the whole range of the technical projects studied is summed up in the conclusions of this chapter.

The conclusions are supplemented by a "Prospects " Section, in which future research activities of interest are suggested.

The "conclusions" and "prospects" sections sum up the core findings and message of this book as far as hot water production is concerned.

1. OVERVIEW OF SOLAR HOT WATER PRODUCTION PROJECTS

1.1. Classification (see table 1)

Table 1 : Overview of the research projects described briefly in this chapter.

Basic description (BD#)	page	Biogas digester heating	Aquaculture heating	Floor heating for piglets	Water heating on dairy farms	Milk preparation on calf rearing farms	Country	title of the project
				application				
1	364	●					Gr	Energy saving by means of solar heating of biogas digesters
2	364	●					I	The use of solar energy to accelerate the fermentation of cattle manure
3	365	●					Nl	The use of solar energy for a biogas installation
4	366		●				I	Bonafous energy farm
5	367		●				Nl	The use of solar energy for heating circulation water for eel- and carpbreeding
6	368		●				UK	Solar heating of fish ponds by active and passive solar energy use
7	370			●			B	The use of solar energy for floor heating in a farrowing house
8	371			●			D	Solar heating of cleaning water and heating of a farrowing house
9	372				●		Nl	The use of solar energy for hot water production for a dairy farm
10	373				●		UK	Appraisal of a solar water heater for Bradmores farm dairy
11	374					●	D	Solar boiler for calf feeding water production
12	375					●	I	Solar heating of water for a calf rearing farm
13	376					●	Nl	The use of solar energy for hot water production on a calf rearing farm
14	377					●	Nl	The use of solar energy for hot water production on a calf rearing farm
Sample 1	380	●					USA	A solar heated anaerobic digester
2	380		●				Canada	Combination of a solar collector with water recirculation units in a fish culture operation

1.2. Location and titles of the projects (see fig. 1)

Fig. 1. Location of the described projects

Note: the circled numbers refer to projects described in more detail in the appendix to part III.

1.3. Basic project descriptions

1. Energy saving by means of solar heating of biogas digesters

+ Project in Athens, Greece.

+ Description (see fig. 1.1) :
  In order to maintain a temperature of + 35 °C on a continually fed experimental manure digester of 5 m³, 20-30 % of the produced biogas needs to be consumed.
  A solar system was designed to provide heat, both to the inflowing manure - 0.25 m³/day - before entering the digester (T_{max} = 50°C), as well as to the manure inside the digester itself (T between 34-37°C). The diagram, given below, shows how this is done for the case of the inflowing fresh manure.

+ Conclusions :
  With the solar heating system, only 5 % of the biogas production need be used to maintain the desired temperature.

+ Study Center : Agricultural College in Athens.

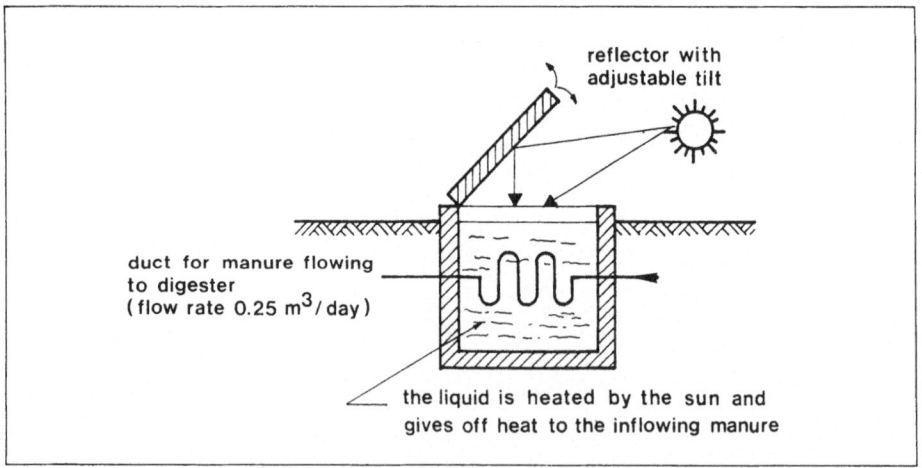

Fig. 1.1 : Scheme of the installation in Athens.

2. The use of solar energy to accelerate cattle manure digestion

+ Project in Pavia, Italy (since 1983).

+ Description (see fig. 2.1.) :
  In order to achieve a higher biogas production rate in a cow manure digester, water collectors are being used to heat up the digester content. The 6 m² collectors are of a very simple kind, and transfer the collected heat via a 100 m long polyethylene tube, embedded in the floor of the 25 m³ digester.
  The experiment aims at assessing the difference in the gas production rate, and at reaching a conclusion about its economic feasibility.

+ Conclusions :
  Experiments are still in an early phase.
  However, simple calculations predict a yearly net energy collection of
  13 000 MJ/year by the system (47 % solar system efficiency).
  During days of fair solar radiation, the expected average temperature
  rise of the manure mass is 0.5 K, which would yield an increase of 7-8 %
  in the biogas production rate.

+ Study Center : G. Castelli
                 Institute of Agricultural Engineering
                 Via Celoria 2
                 Milan
                 Italy

+ For more detailed information, cf. p. 396.

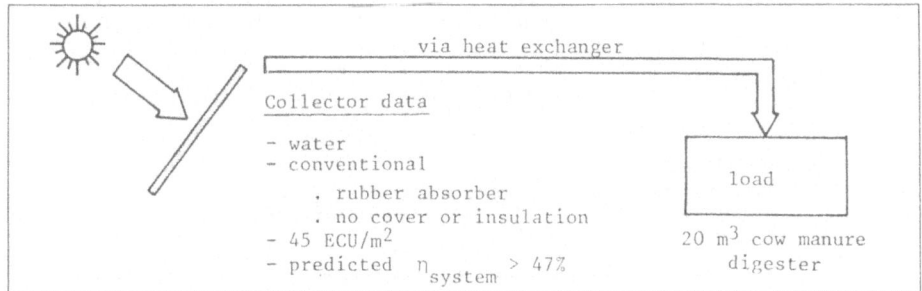

Fig. 2.1 : Data on the installation in Pavia.

3. The use of solar energy for a biogas installation

+ Project in Schijndel, The Netherlands (since 1980).

+ Description (see fig. 3.1) :
  In order to help heat up a biogas digester of 200 m³ to a temperature
  between 27 and 40°C, 240 m² of spectral selective flat plate
  collectors were used. Two heat exchangers transferred the solar heat
  either to the digester, or, if it had reached the maximum desirable
  temperature, to the already sufficiently fermented manure, which was
  dried by the surplus heat. (Resulting in a reduction of manure
  transportation costs).
  The objective of this project is to reduce biogas consumption for
  digester heating purposes from about 30 % to less than 10 %.

+ Conclusions :
  - Over the year 1982 the biogas system was out of operation half of
    the time. The solar installation supplied 103 GJ to the digester,
    which is about 60 % of the energy needed for digester heating, over
    the actual period of operation. The net efficiency of the solar
    system was 8.7 %.
  - The total biogas production, 157 GJ, as opposed to the predicted
    1640 GJ for good operating conditions, was smaller than the applied
    heat.
  - Several technical problems have been experienced with the whole
    set-up, to such a degree that research on this subject has been

discontinued.

+ Study Center : J.M. Lange
                   IMAG (Institute for Mechanization, Labour and Buildings)
                   Mansholtlaan 10-12
                   6708 PA Wageningen
                   The Netherlands

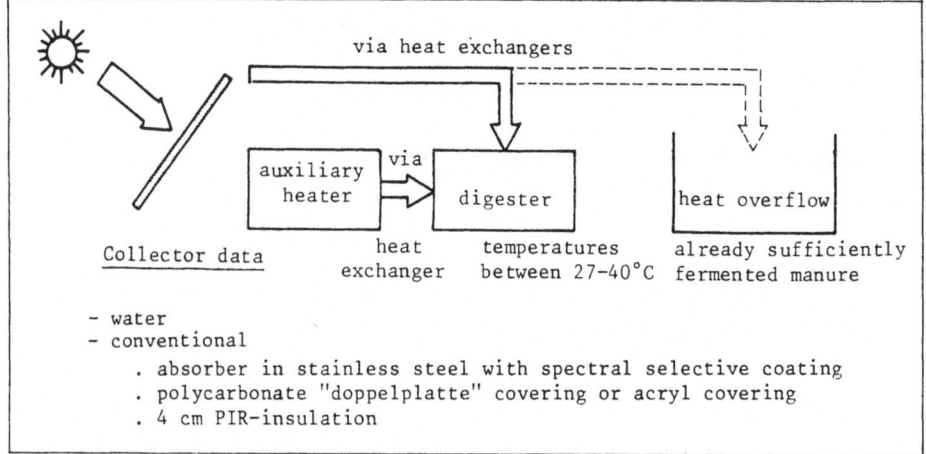

Fig. 3.1 : Data on the installation in Schijndel.

Other similar projects, worth mentioning

In the School for Agriculture in Ath (Belgium) experiments are being done with solar heating of an anaerobic digester.

4. Bonafous energy farm

+ Project in Turin, Italy (since 1983).

+ Description (see fig. 4.1) :
A solar heating system is used to heat a fish-pool.
The auxiliary energy will be provided in the form of biogas, coming from an anaerobic digester.
The solar plant consists of 60 m² of solar panels and a tank for heat storage of 3 m³, as well as several heat exchangers.
Each pool of 25 m³ volume is kept at a temperature of 20°C.

The project aims at testing the efficiency and performance of low cost solar panels for this low temperature application.
The experiment also aims at demonstrating that the use of solar energy and biogas can considerably reduce costs in the agricultural field.

+ Conclusions :
The experiment is still in an early stage, and as such no definite conclusions can yet be made.
The short operational experience would however point to good solar collector performance and efficiency. A projected estimate of

collector performance would predict 4 T.O.E. energy savings per year.

+ Study Center : E. Gautier
S.E.S. S.p. A.
Via Cuneo 20
I- 10152 Torino
Italy

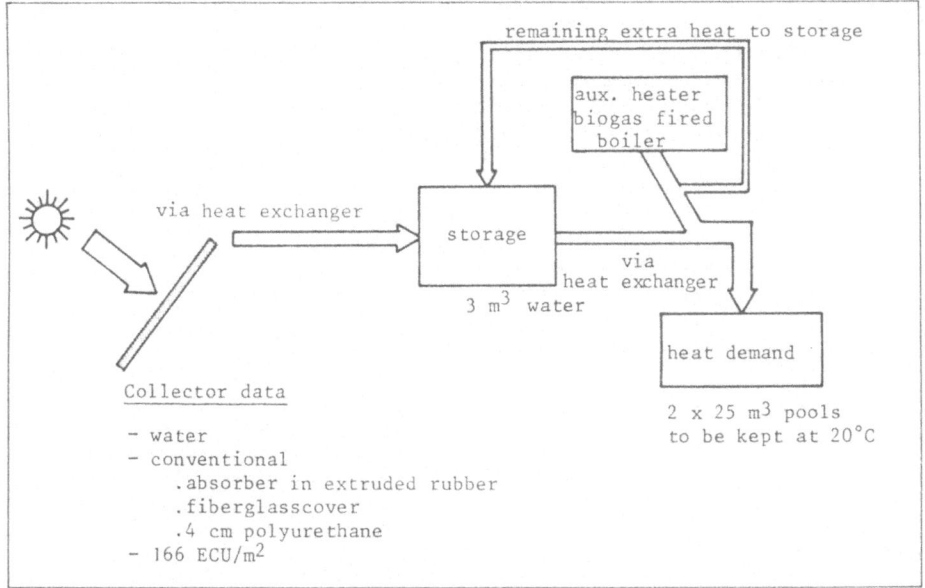

Fig. 4.1 : Data on the installation in Turin.

Note from the editor : It seems that the use of several heat exchangers, as well as the input of heat from the boiler into the solar storage tank is not ideal, since this would raise the collector temperature, and so decrease its efficiency.

---

5. The use of solar energy for heating circulation water for eel- and carp breeding

---

+ Project in Barneveld, The Netherlands (since 1983).

+ Description (see fig. 5.1) :
A low investment solar system with 15 m² of solar collectors was installed to help meet the heat demand in an experimental breeding center of eels and carp with a total water volume of 16 m³ (desired water temperature : 26.5°C).
The goal of the experiment is to gain experience with this technique, as well as to build and validate a simulation model that will allow for optimization of component capacities, and predict system performance for different set-ups.

+ Conclusions :
  - Heating the recirculating water by solar energy poses some
    significant problems :
    . This water should not be in contact with iron, copper or zinc.
    . It is too polluted to flow through the collector.
    Therefore a low performance (plastic-easy to clean) heat exchanger
    has to be inserted in the recirculation system.
    The developed simulation model predicts a production of 500 MJ/m²
    collector per year, which corresponds to a solar system efficiency
    of 15 %.
  - The performance of other set-ups was calculated by the simulation
    model, the best of which yielded a predicted efficiency of 37.5 %.
  - The reliability of the simulation model needs to be supplemented by
    further measurement-calculation comparisons.

+ Study Center : G. Brouwer
                 Van Heugten Consulting Engineers
                 P.O. Box 305
                 6500 AH Nijmegen
                 The Netherlands

+ For more detailed information, cf. p. 404.

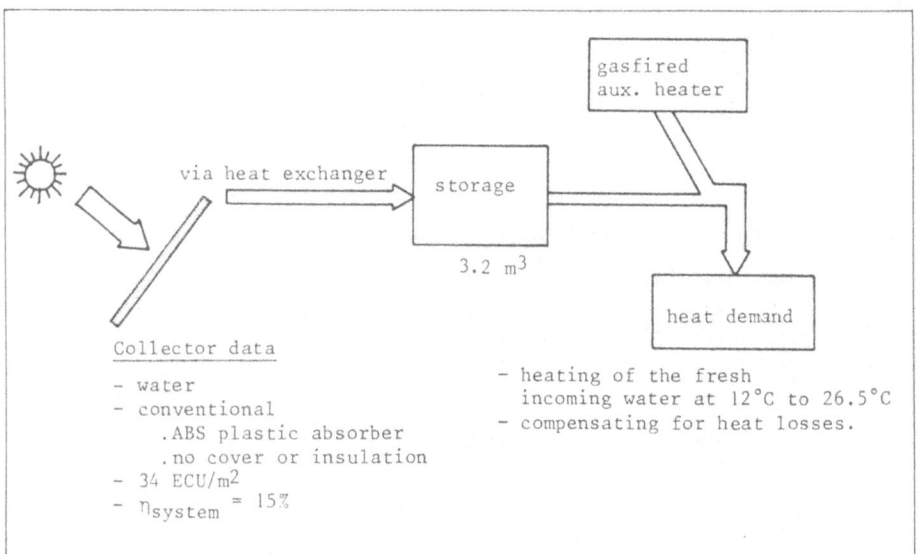

Fig. 5.1 : Data on the installation in Barneveld.

6. Solar heating of fish ponds by active and passive solar energy use

+ Project in Nottingham, U.K. (since 1977).

+ Description (see fig. 6.1) :
  After development and validation of a model, describing the influence
  of water temperature on carp-production, the impact of 2 solar systems
  on water temperature, and consequently on the yield of a fish-farm,

was assessed. For this purpose, 4 fish-ponds of identical size
(100 m² - 50 m³) were compared.
- The first two, naked ponds.
- The third one, a pond covered with a plastic greenhouse.
- The fourth one, identical to the third, but with addition of 25 m²
  of solar collectors, combined with a 100 m³ storage tank.

+ Conclusions :
Monitoring was done over the period of May to December 1978.
- The naked ponds yielded fish of less than 5 grams in December
  (170 days after hatching). The survival rate was 38.5 % for one,
  and 24.3 % for the other pond.
- The greenhouse-covered pond was affected by a fluke infestation,
  which falsified the results. It is estimated that the performance
  of this pond would normally correspond to about 70 % of that of the
  fourth pond.
- The fourth pond yielded fish of 35 grams, 170 days after hatching,
  for a food intake of only 36 % of that in the 2 naked ponds.
  The survival rate was 90.5 %. The higher temperature level of the
  water (Δ9 K on average over the period of June to August compared to
  the first option) caused an extension of the growing season from
  September to December. It is likely that this extension of the season
  would also have occurred during spring-time if the installation had
  functioned at that time.
- There are indications that the third option, covering the pond with
  a plastic greenhouse, would be economically feasible on
  semi-intensive enterprises.

+ Study Center :
F. Vogt   Severn and Trent Water Authority
Built Environmental Research   Calverton Fish Farm
Group   Nottingham
The Polytechnic of Central   United Kingdom
London
London NW1 5 LS
United Kingdom

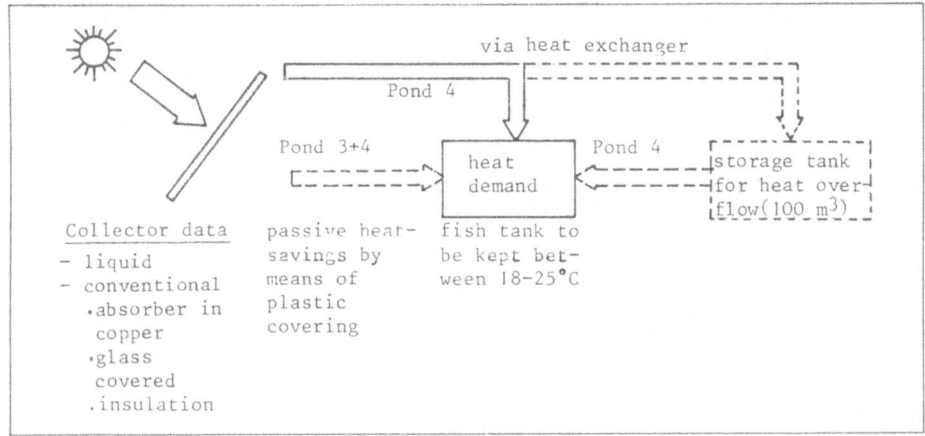

Fig. 6.1 : Data on the installation in Nottingham.

7. The use of solar energy for floor heating in a farrowing house

+ Project in Schriek, Belgium (since 1981).

+ Description (see fig. 7.1) :
A small-scale experimental installation was set up, in which a 4 m^2
solar collector contributed to the heat production for floor heating
in part of a farrowing house. The project was used to evaluate a
simulation model, and to gain practical experience in the process.

+ Conclusions :
- The evaluated simulation model was used to calculate the optimal
  dimensions of the different components for a large full scale farm :
  80 stands for farrowing pigs with piglets; corresponding to an average
  total floor heating primary energy demand (for η heater = 70 %)
  between 8 kW (summer) and 15 kW (winter).
  The calculated economically most desirable dimensions are :
  18 m^2 collectors in combination with a 100 l storage.
- Estimated payback period : $\pm$ 10 years. Given the practical
  experience, which would predict a relatively short lifespan of the
  installation, due to the corrosive and aggressive atmosphere, the
  application would not be economically feasible for the moment,
  unless the operation temperature could be reduced, and hence the
  collector efficiency increased. Research is being done in this
  area.

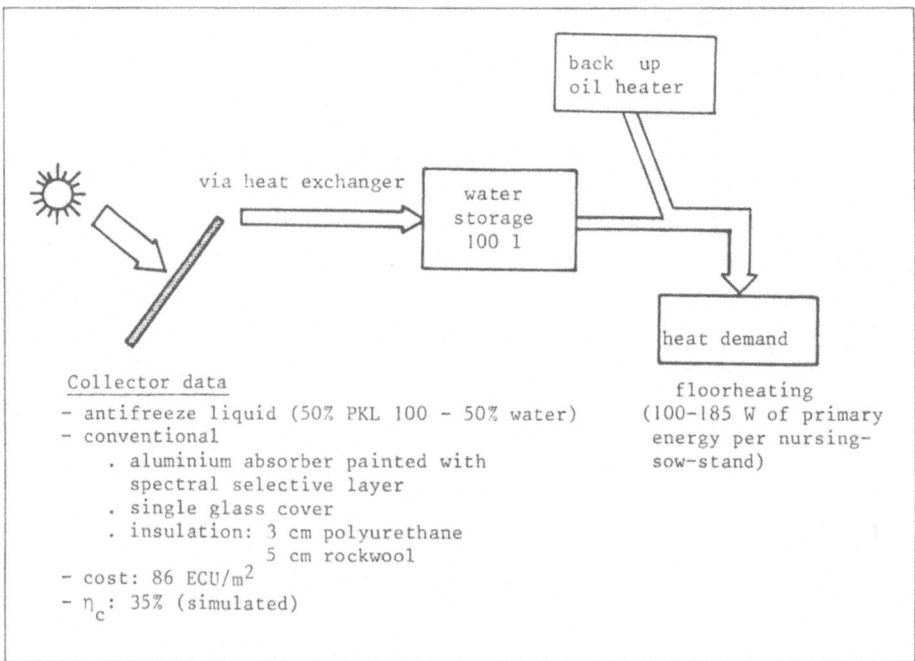

Fig. 7.1 : Data on the installation in Schriek.

+ Study Center : V. Goedseels
                 Laboratory of Agricultural Buildings Research
                 Catholic University Leuven (KUL)
                 Kardinaal Mercierlaan 92
                 3030 Leuven
                 Belgium

+ For more detailed information, cf. p. 414.

---

8. Solar domestic hot water production and solar heating of farrowing
   house

+ Project in Buchhofen (Bayern) West-Germany.

+ Description (see fig. 8.1) :
  124 m² of collectors were installed in order to help heat a farrowing
  house and to help provide for hot tap water.
  The water of the collectors (protected with "drain back" frost
  protection) directly heats up a 6 m³ storage tank.
  If the storage tank temperature is high enough, it is circulated
  through the heating system of the farrowing house and/or through a heat
  exchanger in a boiler for hot water production. In both instances,
  conventional heating devices are used in case the solar system does not
  suffice.
  Finally, cold tapwater is routed through a heat exchanger in the
  storage tank to provide hot water for the pig house.
  The aim of the experiment is to gain experience with the system, and
  to build a simulation model that will make optimization of the system
  possible.

+ Conclusions :
  - Over the one year period of 9/83-8/84 a solar heat production of 68 GJ
    was measured, 26.6 GJ of which are routed to the stable heating, and
    the rest to domestic water heating.
  - Thermosiphonic heat losses were found in the collector. A non-return
    valve took care of that problem.
  - Transparent PVC covers should probably be changed every 6 years.

+ Study Center : K. Meuren
                 Bayerische Landesanstalt für Landtechnik
                 Vöttingerstrasse 36
                 D-8050 Freising
                 W.-Germany

+ For more detailed information, cf. p. 431.

Other similar projects, worth mentioning

An operational system of solar floor heating in farrowing houses was
installed near Dixmuide, Belgium by the firms CONSULT and PORTAL. 20 m²
of collectors help meet the heat demand. No problems were experienced
with the solar installation over the first 2 years of operation.

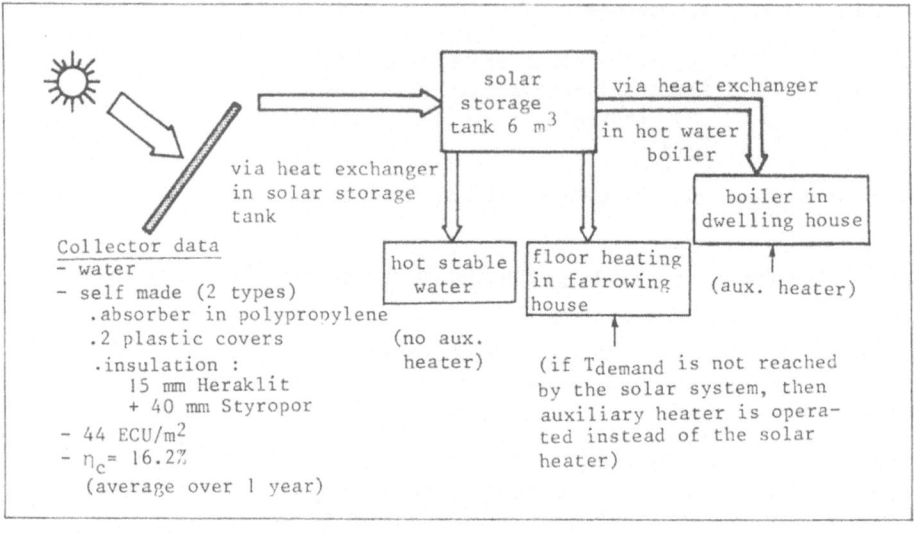

Fig. 8.1 : Data on the installation in Buchhofen.

9. The use of solar energy for hot water production for a dairy farm

+ Project in Duiven, The Netherlands (since 1977).

+ Description (see fig. 9.1) :
  In order to help meet the energy demand in heating up 250 litres of tap water per day to 70°C, 12 m^2 of solar collectors, together with a storage tank of 1 m^3 were installed.
  The experiment mainly aims at estimating the economic feasibility of such a system, based on its energy-gain during a measuring period (1977-1978).

+ Conclusions :
  - Over a one-year period in '77-'78, about 14 GJ were saved (54 % of the total energy need).
  - The spectral selective coating of the absorber made energy collection possible, even in periods of diffuse solar radiation.
  - Compactness of the system - i.e. the need for collector, storage and auxiliary heater to be close together - is essential for reasonable efficiency.
  - One should be aware of the potential of intoxication by anti-freeze liquids.
  - Future research will focus on the feasibility of heat pumps as an alternative system.

+ Study Center : J.M. Lange
                 IMAG   (Institute   for   Mechanization,   Labour   and
                 Buildings)
                 Mansholtlaan 10-12
                 6708 PA Wageningen
                 The Netherlands

+ For a more detailed description, cf. p. 442.

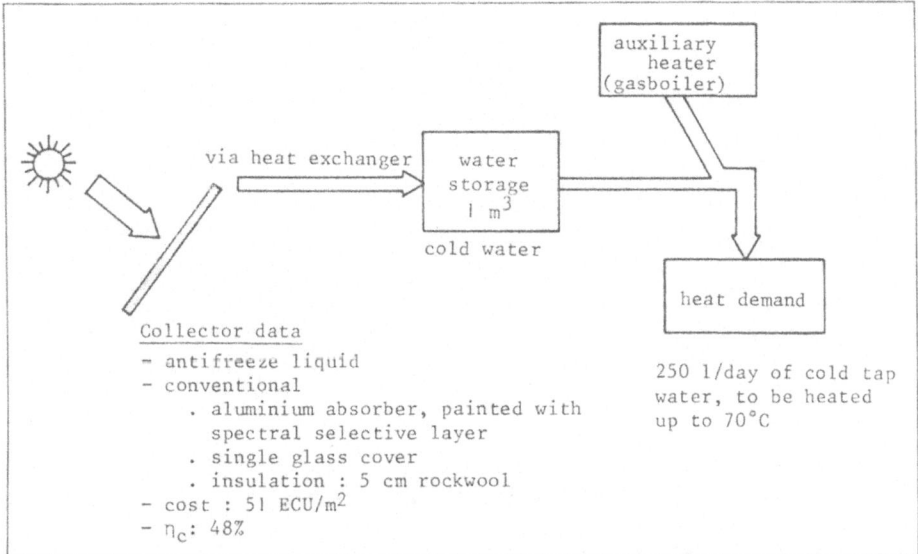

Fig. 9.1 : Data on the installation in Duiven.

10. Appraisal of a solar water heater for Bradmores farm dairy

+ Project in Newton Abbot, United Kingdom (since 1980).

+ Description (see fig. 10.1) :
On a dairy parlour for 100 cows, a solar collector system with adjustable collector (2 m² – 10 m²), and storage capacity (190 –360 l) was installed to assess its performance in hot water production in an "on site" situation, given different simulated heat draw-characteristics (up to 243 l/day, partly at 82°C, partly at 40°C, required at different times of the day).

+ Conclusions :
  - The economic feasibility is disappointing compared to other alternatives, given the low electricity prices at the time.
    (payback periods – based on average expectations of annual solar gain – between 15-20 years depending on different inflation rate values).
  - No mechanical problems were experienced after 3 years of system operation.
  - Installation of a solar collector system in an existing structure can pose significant practical problems.

+ Study Center : J.L. Carpenter
                 Seale – Hayne College
                 Newton Abbot
                 Devon, TQ 126NQ
                 United Kingdom
  (for a list of the supporting U.K. organisations, cf. p. 448)

+ For a more detailed description, cf. p. 448.

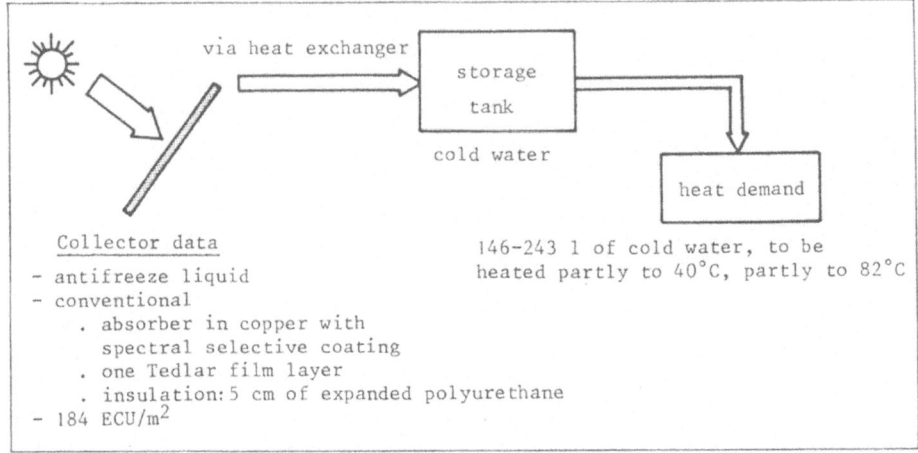

Fig. 10.1 : Data on the installation in Newton Abbot.

Other similar projects worth mentioning

Research on solar preheating of water for use in dairy parlours has been led by M.R.J. Nicholson (Ministry of Agriculture, project in Bristol, United Kingdom).

---

11. Solar boiler for calf feeding water production

+ Project in Siegritz, Bamberg - West-Germany (since 1980).

+ Description (see fig. 11.1) :
  Solar heating of water, used for feeding 80 calves, is being done by means of a low cost do it yourself water collector (33 m^2).
  The absorbed energy is transferred to the storage tank of 3 000 l by a heat exchanger. This water is then post-heated to 50-60°C in an electric boiler, installed in each stable (capacities of respectively 6 and 4 kW).
  The daily water consumption ranged between 900 - 1 400 l/day (15 l per calf per day). The aim of the project is to test the performance of a low-cost roof collector in this application field.

+ Conclusions :
  - The tested system produced 32 GJ/year, which corresponds to about 42 % of the total heat demand.
  - The system efficiency was quite low, due to inadequate insulation, and due to the use of 2 heat-exchangers in series instead of just one : (collector loop - heat exchanger - storage water in the open storage tank - heat exchanger - tap water).
    The storage tank was changed into a pressurized one, where the tapwater plays the role of both storage and load, limiting the need for 2 heat exchangers to just one.
  - The experience with the tested collector led to the development of a new collector of about half the price of the original one.

+ Study Center : H. Schulz
                    Bayerische Landesanstalt für Landtechnik
                    Vöttingerstrasse 36
                    D-8050 Freising
                    W.-Germany

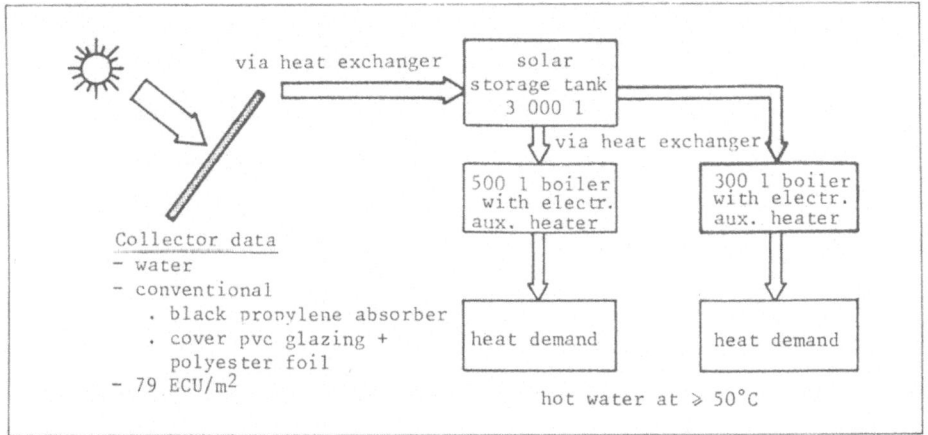

Fig. 11.1 : Data on the installation in Siegritz.

12. Solar heating of water for a calf rearing farm

+ Project in Maccarese, Italy.

+ Description (see fig. 12.1) :
  A solar collector system was designed to help heat up water for food
  preparation and rinsing purposes in a calf rearing farm.
  - The heat demand on this farm with 400 calves is 240 MJ/day (the
    water use is 1 350 litres per day at 60°C).
  - 40 m² of flat plate solar collectors heat up a 2 m³ water storage
    tank by means of a heat exchanger.
  - Preheated water is drawn from the storage tank and, if necessary,
    postheated by a gas-heater.

+ Conclusions :
  - Calculations predict that 71 GJ per year can be saved by using the
    solar system (this is 80 % of the total energy need).
  - Estimated cost of an optimized unit is 13 700 ECU (1983-prices).

+ Study Center : A. Scarpini
                    CTIP Solar S.P.A.
                    Piazza Douhet - Roma
                    Italy

Note : This study center is involved in the set-up and monitoring of
       several such operational solar projects.

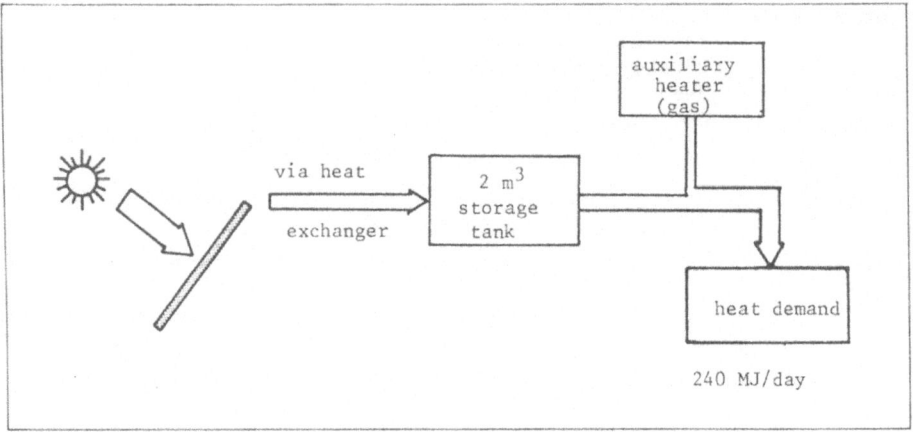

Fig. 12.1 : Data on the installation in Maccarese.

---

13. The use of solar energy for hot water production on a calf rearing farm

+ Project in Een, The Netherlands.

+ Description (see fig. 13.1) :
  The solar energy installation for preheating the water, used in a calf rearing farm (with 800 calves), consists of a 120 m² collector and a 10 m³ storage tank.
  Twice a day, about 2 m³ of water are heated to a temperature of about 75-80°C, and put to use for dissolving milk powders and for administering the milk to the calves at a temperature of 35°C; it is also used for rinsing and cleaning equipment.
  The aims of this project are to determine the possibilities of solar energy in this application field, to measure the heat gain, the temperature levels of operation, to validate the computer simulation model, which will be used to develop other similar installations.

+ Conclusions :
  - Over a monitoring period of 111 days, mainly in the summer season, 60.9 GJ of net useful energy were collected. This corresponds to about 50 % of the total energy need.
    A validated simulation model predicts a yearly energy saving of 143 GJ/yr.
  - Frost protection by draining the collectors when not in use makes direct heat transfer to the storage tank possible.
    This allows for lower working temperatures, and thus for higher collector efficiency values.

+ Study Center : G. Brouwer
                 Van Heugten Consulting Engineers
                 P.O. Box 305
                 6500 AH Nijmegen
                 The Netherlands

+ For a more detailed description, cf. p. 457.

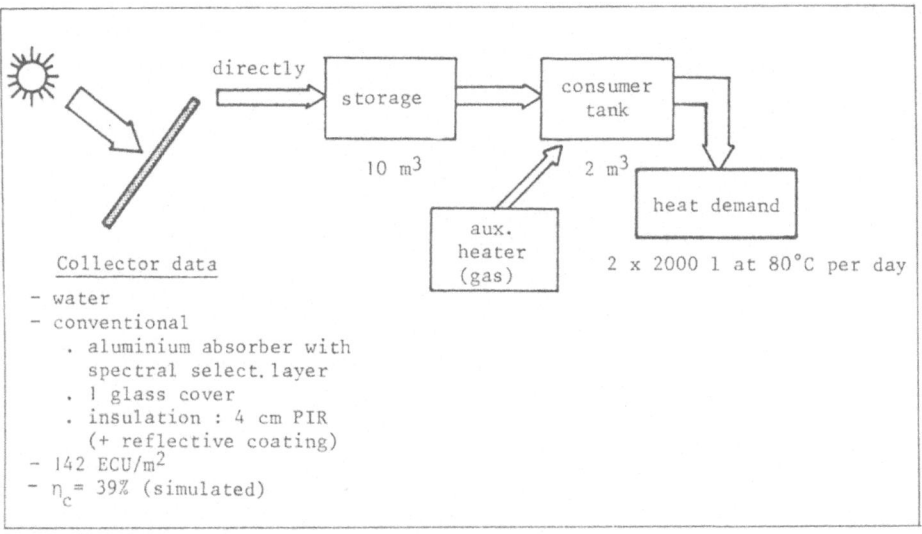

Collector data
- water
- conventional
  . aluminium absorber with
    spectral select. layer
  . 1 glass cover
  . insulation : 4 cm PIR
    (+ reflective coating)
- 142 ECU/m^2
- η_c = 39% (simulated)

Fig. 13.1 : Data on the installation in Een.

14. The use of solar energy for hot water production on a calf rearing farm

+ Project in Lelystad, The Netherlands.

+ Description (see fig. 14.1) :
  With the same intention as in the project at Een (The Netherlands), a quite similar set-up was made on a farm, housing 80 calves, in Lelystad.
  Here, use was made of an 8 m^2 flat plate collector and a 300 l storage tank, with a heat exchanger.
  A total of about 120 l/day of water at 80°C are being used. The set-up and goals of the project, as well as the main conclusions, are practically the same as for the project in Een, The Netherlands.

+ Conclusions :
  In this case, over a monitoring period of 46 days in October and November 1983, 187 MJ of net useful energy were collected by the solar system, which corresponded to 19 % of the heat demand. This monitoring period is however not representative of the normal situation, since due to absence of calves, the heat demand was unusually low.
  Simulation results over a normal year predict an average of 7.4 GJ of annual energy savings, which corresponds to 58 % of the total heat demand, and a solar system efficiency of 27 %.

+ Study Center : G. Brouwer
                Van Heugten Consulting Engineers
                P.O. Box 305
                6500 AH Nijmegen
                The Netherlands

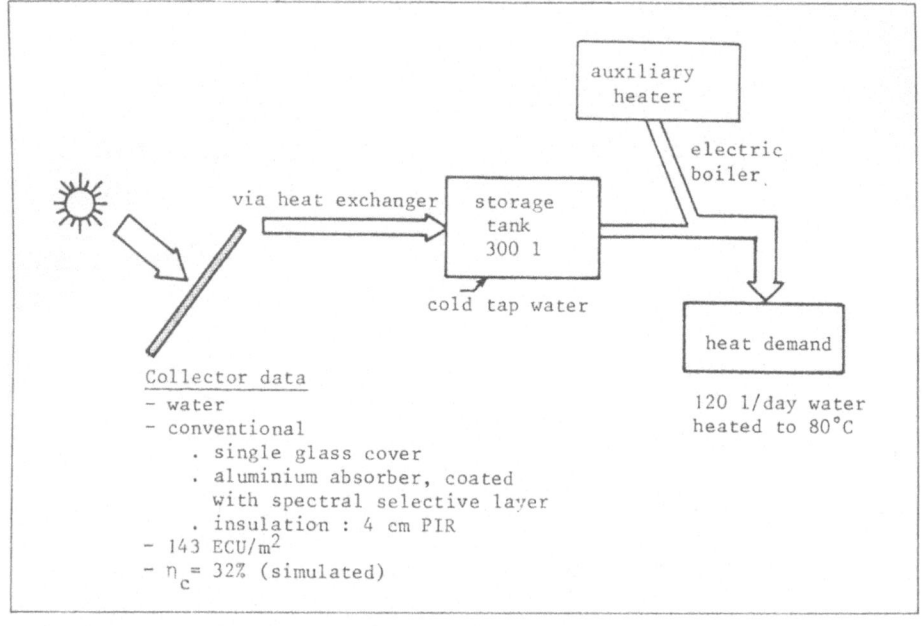

Fig. 14.1 : Data on the installation in Lelystad.

Similar projects worth mentioning :

+ In Kootwijkerbroek, The Netherlands
J.M. Lange
I.M.A.G.
Mansholtlaan 10-12 (P.B. 43)
6700 AA Wageningen
set up a solar heat collector system, consisting of 180 m² of
air-collectors (painted flat black and with double cover), in
combination with a 30 m³ storage tank for a farm with 500 calves, and
a water demand of 2.5 m³/day partly at 40°C – partly at 70°C.
The expected net solar energy output is 216 GJ/yr., which corresponds
to an installation efficiency of 30 %.
Air collectors were chosen, and provide an opportunity to weigh the
advantage of the simplicity of the given collectors against the
disadvantages of a possible loss of system efficiency.

+ In Stolwijk, The Netherlands
The same researcher monitors an installation of 58 m² of collectors
(absorbers coated with a spectral selective layer) in combination with
a 5 m³ storage, for preparation of hot water, needed at a temperature
level of 70°C and for 500 calves.
The expected energy savings are in the range of 80 GJ/yr.

+ In Voorthuizen, The Netherlands
15 m² of collector surface was installed in combination with a storage
tank of 1 m³. The system, set up on a farm of 225 calves, has
functioned well for over 4 years now.

+ Given the fact that more than half of France receives at least
  1 700 h/year of sunshine, a simplified concentrating collector was
  developed by C.N.E.E.M.A. Antony in France (since 1977).
  The mirrors consist of sheet-iron rolled in cylindro-circular, rather
  than cylindro-parabolic shape.
  They remain in a fixed position, while the movement of the receptor is
  very simple.
  Experiments with these collectors in Lot, France, on a calf rearing
  farm, requiring water at 80°C for milk preparation, have produced
  satisfactory results over the first 2 years of their operation.
  (The expected cost for such a collector system, including the other
  installation materials, was 85-140 ECU/m^2 in 1980.)

1.4. Two examples of North American projects

1'. A solar heated anaerobic digester

+ Project in Tucson, Arizona (U.S.A.) (since 1979).

+ Description (see fig. 1'.1) :
The impact has been assessed of the use of solar energy on the biogas production of a plug flow anaerobic digester with a capacity of 0.57 m^3 of a manure - water mixture with 10-12 % solid content, fed with 2.2 kg of fresh manure solids per day.
The digester tank, which was painted black in order to absorb radiation, was set in a housing structure, containing a south-facing window in clear plastic, which could be covered up with an insulated "solar door".
The "solar door" was opened in order to let sunrays through when the sun shone, and the digester could use the solar heat. It was closed when the outside temperature and radiation were low, or when outside temperature exceeded 35°C.

+ Conclusions :
The digester maintained its temperature at an average of 5 K above ambient during spring months.
The variations of the daily ambient temperatures were reflected, although in a somewhat reduced way, in the digester temperature.
Biogas was produced at an average rate of $0.5 \text{ m}^3/\text{day}$.

- Ref. : D.G. Williams, "Operating experiences with a solar heated anaerobic digestor".
New Dimensions of Appropriate Technology
Michigan, LAAATDC pp. 238-246 1980.

2'. Combination of a solar collector with water recirculation units in a fish culture operation

+ Project near Winnipeg, Canada (since 1978).

+ Description (see fig. 2'.1) :
On an experimental indoor fish hatchery, the impact is being tested of using 125 m^2 of solar collectors in helping to heat up the water for rainbow trouts from 6°C (artesian source) to 12°C.
- The solar heat is transferred to the 19.13 m^3 storage tank via a stainless steel shell and tube heat exchanger.
- The storage tank has an electric auxiliary heater.
- The heated water is blended with cold water in order to reach the desired temperature in a blending tank of 10.52 m^3, the volume of which corresponds to one 12 hour day's supply of the usual amount of water renewal (recirculation rate 91.6 % - 1.2 l water renewal per minute and per 1.5 m^3 fishbasin - 12 basins).
- During cloudy days, the recirculation rate is made 100 %, in order to reduce the heat demand. This complete recirculation was kept up for a maximum of 2 consecutive days and nights.

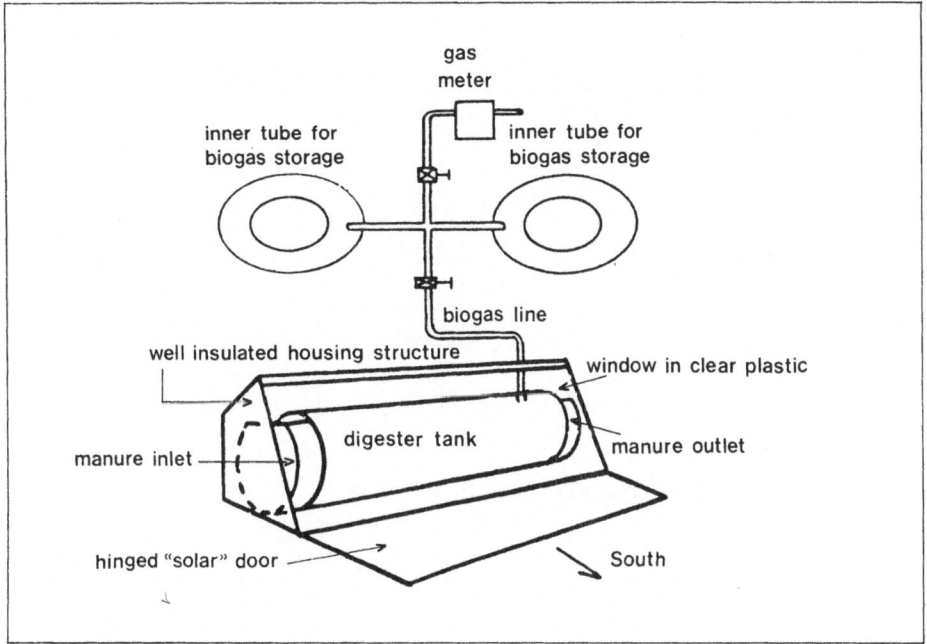

Fig. 1'.1 : Scheme of the installation in Tucson.

+ Conclusions :
  - Measurements over a 6 week period during the worst time of the year (early 1979) yielded a solar capture of 18.6 GJ. Taking into account the total energy need for heating of the pilot scale set-up (energy for operation of the heater included) and accounting for the extra energy required for extra pumps from the solar part etc... this corresponds to a net reduction of the energy need by 72 %.
  - The solar system efficiency was 25 %, due to the fact that a large percentage of the radiation energy (> 50 %) was too low for operation of the system.
  - The fish performed well at an insignificant mortality rate. (Production of 54 kg of fish per 1 500 l basin over the 6 weeks).

+ Study Center : G.B. Ayles
                 Fisheries and oceans Canada Freshwater Institute
                 501 University Crescent
                 Winnipeg, Manitoba
                 Canada R3T 2N6

- Ref. : G.B. Ayles, K.R. Scott, J. Barica + J.G.I. Lark
         "Combination of a solar collector with water recirculation units in a fish culture operation"
         Proceedings of the World Symposium on Aquaculture in Heated Effluents and Recirculation Systems.
         Stravanger 28-30 May 1980
         Vol. I pp. 309-319
         Berlin 1981

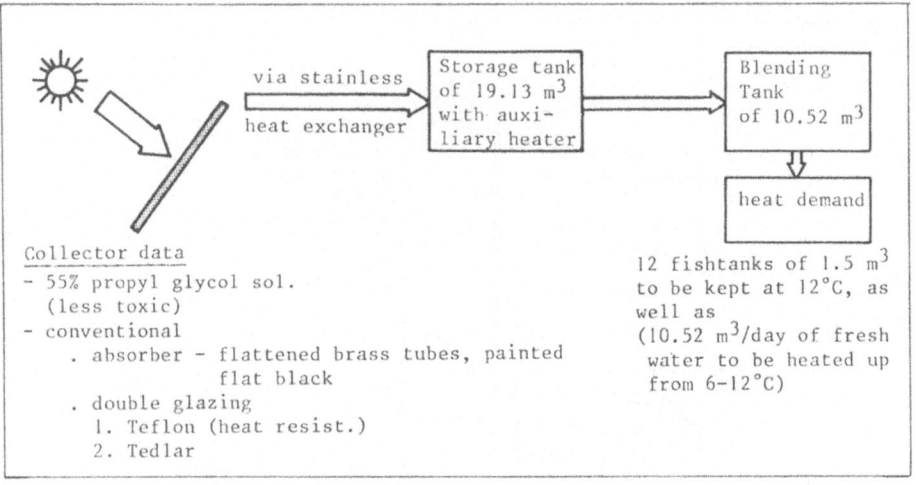

Collector data
- 55% propyl glycol sol.
  (less toxic)
- conventional
    . absorber - flattened brass tubes, painted
              flat black
    . double glazing
      1. Teflon (heat resist.)
      2. Tedlar

12 fishtanks of 1.5 m^3
to be kept at 12°C, as
well as
(10.52 m^3/day of fresh
water to be heated up
from 6-12°C)

Fig. 2'1 : Data on the installation in Winnipeg.

2. GENERAL CONCLUSIONS FOR SOLAR HOT WATER PRODUCTION

2.1. Discussion of the individual applications

2.1.1. Solar heating of a manure digester for production of biogas

The idea of heating a biogas digester by means of solar energy is relatively new and research in this area still scarce.

The option chosen in B.D. nr. 2 and North American example 1' *(1)* consists in raising the temperature of the <u>otherwise unheated</u> digester by solar energy whenever desirable and possible. Indeed, as shown on figure 2, p. 325, a slight rise in digester temperature can have a significant impact on biogas production.
Given the relatively small temperature difference between unheated digester and environment, this option allows for the application of simple installations operating at relatively high efficiency rates, even during wintertime.

Such a solar solution obviously adds to the already existing temperature variations within the digester tank; the potentially undesirable effects of which should be carefully considered.
The encouraging yields obtained in the North American example constitute a basis for hope that within certain limits, temperature variations of this kind will not be prohibitive.

In order to obtain a significantly more rational biogas production, providing at least as much biogas during wintertime as over summer, in practice the digesters are usually kept at a constant temperature by the combustion of part of the produced biogas itself. (Cf. B.D. nr. 1 & 3). Given the greater energy demand on the farm during wintertime, this upkeep of optimal digester temperatures is especially crucial during that period and significant amounts of already produced biogas are sacrificed for this purpose.
Solar energy is sometimes proposed as a way to save part of the biogas otherwise wasted on digester temperature upkeep.
Given the consistently raised digester temperature, in this case the collector efficiency is significantly lower, <u>especially</u> during wintertime, just the period when a significant impact of the solar energy is desirable. Consequently, the solar option ends up as a rather ineffective tool.

Furthermore, as B.D. nr. 3 and reference 2 on p. 327 suggest, bringing about a <u>rational</u> biogas production is a delicate process requiring much knowledge and attention. One might question the inclination of farmers to add an extra (solar) system, involving additional skills and risks, to an already complex system.

(1) Since the experience with solar biogas digester heating is very limited within the EC, while in other countries significant research has already been done on the subject, a project from outside the EC is described on p. 380 by way of example.

2.1.2. Solar heating of fish farms

The difference in fish production mentioned in the project in the U.K. (cf. basic description (B.D.) nr. 6) is one of the many illustrations of the impact that water temperature can have.

An element worth considering is whether the use of solar energy could possibly save the heated fish farms from the status of "secondary" users of waste energy, with all its implications. Indeed, the day on which the waste energy producer – say, a power plant – decides to stop its operation for a while (possibly a whole season) because there is no electricity need or for repairs, the fish farmer – a non-essential and so "secondary" element in the producer's system – could be confronted by serious difficulties.

In this same context it would be interesting to obtain more experience concerning the capacity of the fish to cope with varying water temperatures. This would determine within what limits the growth rate of the fish could be improved when using heat sources with intensities varying over time (e.g. : simple solar systems without back-up heater). B.D. nr. 6 sheds some light on this question.

Practical experience acquired in B.D. nr. 5 brought to light a few potential problems with solar heating of recirculating fishwater :
- The need for specially built heat exchangers, since the excrement and other dirt in densely populated water causes problems when run directly through the collectors.
- The need for specific heat exchanger materials, which are not good heat conductors, since the fishwater should not be in direct contact with most metals. (exception : cast iron)
(compatibility problems – cf. fundamental criteria, p. 1)

Concerning the second problem, it should be mentioned that several existing applications of heated fishfarming even outside the solar heating realm can be found which bring copper and other metal pipes in contact with the fish water, without catastrophic results.
It would seem then, that both for solar as well as for other types of water heating, this problem should not be seen as an absolute one, but rather as one of finding the right balance between danger for the health of the fish (which is a function of concentration in the water of the leached metal in question) and economic feasibility of the installation itself. Practically speaking, the set-up should be such that the dissolved metal concentration does not exceed a critical value. (It is clear that, given the corrosive nature of salt water, the limitations on use of metals in the fishwater will be significantly greater for salt water than for fresh water fish.)

The problem as a whole can be avoided by solar heating of the inflowing fresh water only. In this case use of a heat exchanger is not essential any more, and if water renewal occurs at a constant flow rate, a storage tank is not needed either.
The resulting system is very simple and yields high efficiency values (low temperature operation). Its impact on the total heat demand for fish water temperature upkeep is however rather limited. (The solar energy input to help cover the significant heat demand of the recirculated fish water is nil.)

Based on the results of a simulation model as well as on some simplifying assumptions *(1)* , the payback period for the solar experimental set-up described in B.D. nr. 5, adjusted to heat only the directly inflowing fresh water (at a constant flow rate of 2.1 m³/day) was estimated to be around 10 years.
Optimization of the component parts, reduction of conception, installation and component costs, etc... can further significantly reduce this figure.

The estimation of this payback period is however based on a comparison with oil- or gas-heated cultures.
These reference solutions themselves are actually economically feasible only for small scale experimental set-ups, or very <u>intensive larvae hatcheries.</u>
In general, low cost waste heat is used for normal size heated fish cultures, and solar energy will find it more difficult to compete with this source.

Note : Given the fact that research in the area of solar heating of fish cultures is so recent within the EC, while in other countries significant experience has already been gained, an example of a further developed research program from Canada is given on p. 380.

2.1.3. Solar assisted floor heating in farrowing houses.

In B.D. nr. 7 the performance of a solar system for floor heating of a farrowing house was simulated and an optimized version calculated. The payback period corresponding with the optimized model was calculated to be a little less than 10 years. *(2)*

The figure calculated following the procedure described in the note on p. 387 was a factor 3.7 higher for the installation described in B.D. nr. 8. This difference can probably be attributed to two main factors :
- choice of the collector : the performance characteristics of the simple collectors are not well adapted to the temperature ranges in which they need to operate. (Another crucial element in solar system optimization !) This resulted in a very low yield for a relatively very large collector surface.
- Volume of the storage : as was found in B.D. nr. 7 for the given energy demand characteristics, a large storage volume is economically undesirable.
A problem with compatibility which is especially significant for pig farms is the corrosive nature of the environment causing rapid deterioration of the sensitive component parts (electronic regulation systems, insulation ...). Hence installation lifetime is significantly reduced,

(1) Estimations, based on the investment expenses given on p. 410 ; assuming heat exchanger costs at about 3/4 of the material costs for components other than the solar collector; following payback determination procedure described in the note on p. 387.
(2) The calculation of the payback period according to the procedure described in the note on p. 387 yielded 10.5 years. The small difference between the figures is mainly due to the fact that in this last case the conventional fuel cost was chosen at a level slightly lower than the one actually in practice.

and payback period should be further decreased to fall well within the limits of that lifetime.

In order to reach this goal, experiments are being done in the area of floor heating techniques, which would lower the current temperature level (45-55°C) of the floor-heating water (down to 32-42°C) and so improve the solar system efficiency, hence payback time (1) .

2.1.4. Solar heat for hot water production in dairy farms

The production of hot water for dairy farms seems quite comparable to that of domestic water heating except for the fact that higher temperature levels can be aimed at (desired water temperature is partly 40°C, partly 80°C), with inevitably lower system efficiency values.

Based on calculations for optimized operational set-ups, payback periods of 5 to 8 years are expected (depending on the type of fuel solar energy is replacing) (2)
However, on more and more farms where the milk is cooled to 4°C, the heat taken from the milk is "recovered" and applied for hot water preparation by means of a heat pump. Such heat recovery systems, with expected payback periods within the E.C. of 2-5 years on larger farms, leave little hope for the solar option, at least on farms with more than 15 cows. (cf. p. 352, ref. 5).

2.1.5. Solar heat for milk preparation on calf-rearing farms

Water heating for calf-rearing farms (with a desired water temperature of up to 80°C) could very well become the process within the solar assisted hot water production range with the highest potential of economic feasibility.

The system described in B.D. nr. 13 happens to make use of collectors with comparable performance characteristics as in B.D. nr. 7. Also the simulated annual net system energy yield per m^2 collector is situated in the same range. The payback period figure (29 years) calculated according to the procedures described in the note on p. 387, is however a lot higher than in B.D. nr. 7. The higher storage requirement has a role to play in this, but the extremely high collector costs constitute the main factor. Reducing the collector price to that found in B.D. nr. 7 results in a decrease of the estimated payback figure by more than 40 %!
This again stresses the importance in system optimization of the relationship : collector price versus performance, in relation to the required temperature level of operation.

The influence of the scale factor of the installation is expressed in B.D. nr. 14, a system designed in a way very similar to and applying several of the same component parts as in B.D. nr. 13.

(1) GOEDSEELS, V.; et al.; "Application of polymer materials in the floor construction of the pig house" (Paper submitted for the International Symposium on Agricultural Engineering, Pretoria, 20-24 January 1986.)
(2) G. Brouwer (Dutch action 2 participant, cf. p. XVI); personal communication; 1985.

The heat demand of the latter is however more than 15 times larger than that of the former.
The share of costs for conception, regulation and other devices per m^2 collector surface, as well as the heat losses per litre of heated water increase as system size decreases. Thus it is only by comparing the smaller system as an alternative to electricity (costing about 2.5 times more per unit of net energy yield than the energy price used in the payback time calculation procedure, cf. note on p. 387) that payback figures are obtained comparable with those of the larger system.

Another illustration of the influence of the scale factor is provided in the performance calculation of a solar system on an industrial calf-rearing farm with 6500 calf stands. After system optimization by means of an adapted version of the simulation model used in B.D. nr. 7, a payback period of 5 to 6 years was calculated[1].
Given the huge quantity of 58000 1 of water use per day (at 80°C) it clearly concerns a large scale operation.

A final element of importance is the temperature level of operation of the solar system. Installations aimed at providing the larger part of the energy by means of the solar system, and so shooting for higher temperatures (70-80°C), will have higher payback times than others which only preheat the water in a lower temperature range (e.g.: up to 40°C).

Taking all these factors into account, and incorporating them in optimized set-ups yielded expected payback periods of 5 to 8 years. [2]
A factor that might further contribute to the success of this solar application could be the absence of strong competitive alternatives.

Note : Calculation procedures for payback period estimation.

The main focus of the research projects described lies in the area of efficiency assessment and optimization so as to ultimately reach a maximum value of energy savings per solar system cost.
The payback period provides a quantitative expression of this ratio.
(The payback period of an instrument is defined as the ratio of the total costs of the instrument (investment and actualized working costs) over the average value during the payback period of the annual earnings yielded by the instrument, actualized to the moment when the investment was made.)

Characteristic of techniques in their earlier stages of development are usually low efficiency rates, due to e.g. initially poor system optimization and component effectiveness. Cost factors on the other hand are high, since e.g. new materials are used, which are conceived and produced on an individual as opposed to large scale basis.
At this early stage, payback period figures for such systems may unwisely be interpreted as prohibitive even though further development may later yield very desirable options from an economic point of view.

Furthermore, economic variables can change abruptly in unexpected directions, completely altering the perspectives of a system in development.

(1) Cf. ref. 6 on p. 360.
(2) G. Brouwer (Dutch action 2 participant, cf. p. XVI). Personal communication; 1985.

Finally, elements like reduction of dependence on imported fuels by means of solar energy, desirability from the environmental point of view, etc... are not at all reflected in these payback period figures.

However, if used wisely the given figures do provide a valuable indication of how different applications compare with one another in this area, as well as of the range in which they can be situated. (Ranges resulting form differences in system conception, local differences in climate, prices, etc...)

For this purpose an estimation was made of the payback period for a few of the above installations, according to the following procedures :
- The payback period corresponds to the moment when the "present worth" becomes zero.
- The present worth was calculated as :

$$PW = - Iv + \sum_{i=1}^{n} \frac{F (1 + b)^n}{(1 + i)^n} - \sum_{i=1}^{n} \frac{M}{(1 + i)^n} - \sum_{i=1}^{n} \frac{O (1 + e)^n}{(1 + i)^n} + RV.$$

with PW = Present worth
      Iv = Investment costs
      F = expected annual fuel cost savings
           (for fuel costs uniformly fixed at 8.25×10^{-3} ECU/MJ)
      b = expected inflation free annual cost increase of fuel
      c = expected inflation free annual cost increase of electricity
           (both fixed at 0.02 (2 %/yr))
      i = expected inflation free interest rate
           (fixed at 0.05 (5 %/yr))
      n = number of years after investment for which the PW is calculated
      M = annual maintenance costs (its inflation free annual cost increase was estimated to be 0 %)
      O = annual operating costs (electricity for operation)
           (The annual working costs M + O were fixed at 1.5 % of Iv, with M = 0.85 (M + O))
      RV = Residual value of the installation
           (fixed at zero)

2.2. Discussion of some practical and specific issues in the solar water-heating sector

2.2.1. The investigation of system efficiency

Collector efficiency is the primary requirement for system efficiency. In this context the use of heat exchangers was kept to a minimum, since they raise the collector temperature and hence, reduce its efficiency. In a few cases, a drain-down frost protection was used (i.e. the liquid is drained from the collector circuit when frost danger occurs). This can eliminate the need for anti-freeze liquids and allows for direct heat transfer through one and the same liquid. (no need for a heat exchanger). This solution unfortinately has some undesirable implications, like e.g. the requirement of a large capacity of the collector circuit pump, as well as of a highly reliable temperature control system.

Under the headings of "acquired experience/problems arising in the course of the technical research activity", certain rather obvious, but often forgotten aspects of system efficiency were frequently mentioned.

- For applications where originally cold water is heated to high temperature levels (e.g. hygienic water for cows, milk preparation for calves) the solar system efficiency (net solar system yield over solar radiation input) decreases as the desired share of the solar system in the total heat supply increases.
- The system efficiency decreases greatly with decreasing compactness. If the system components are far removed from one another, a good part of the heat is lost. (This can be a significant problem when a solar system is being added to an existing building).
- Adequate insulation plays a crucial role in system efficiency.
- The use of a non-return valve in the collector circuit is often very important in order to prevent backward thermosiphonic flow at night.
- The criteria for when the collector pump needs to be switched on should be different from the switch-off criteria, in such a way that there is a logical link between the radiation intensities at which the pump is started and stopped. Inadequate regulation can result in either excessive use of energy for operation of pumps and valves, or in an insufficient recuperation of available solar energy, as well as in excessive wear and tear of the pumps.
- Certain aspects of the physical environment in which the solar system is placed can have a profound influence on system performance, efficiency and reliability.
  . Especially in pig houses significant problems are being experienced with corrosion and deterioration of materials. The corrosive air, as well as the heavy duty cleaning devices used (high pressure water cleaning), call for a secure packaging of electronic regulation components, and often lead to a high rate of deterioration of component parts (e.g. insulation ...). The predicted lifetime of the materials, based on normal operation conditions, sometimes far exceeds the actual lifetime in the farming set-up.
  . On farms where the ventilation air is very dusty, placing the collectors in the neighborhood of the ventilation shafts can significantly reduce the transmissivity of the collector covers, owing to dust build-up.
  . Certain antifreeze liquids can be toxic for animals. This factor has to be taken into account especially when the heated water is taken in by animals (e.g. calf-rearing). In this case so called "double" heat exchangers, placed around (instead of in) the hot water storage tank are often used (or even imposed by law) to provide extra security.

2.2.2. System economics

- Different collector types were chosen with different efficiency/cost ratios

      more expensive, and efficient collectors.
      E.g.: Coating the absorber with a spectral selective layer which allows for energy collection in some cases even in overcast weather conditions. These collectors are able to collect energy, even at higher temperatures, in a very efficient way.

      cheaper, less efficient collectors.
      E.g.: with plastic covers, the transmissivity of which is significantly reduced over time, especially when high

collector temperatures occur. This involves a lower efficiency as well as additional costs for replacement about every 6 years.

Some go even further than that by using an
air collector in combination with an air/water heat exchanger. This is done to avoid the complexity of liquid collectors (more expensive, more intricate, leaking danger ...) despite the fact that the heat exchanger lowers the system efficiency.

Which of the different possible approaches is best depends on the application itself. Where high water temperatures are the goal, efficiency of the collector tends to be of greater importance than in other cases.

- Installation of solar systems on <u>already existing farm structures</u> can pose some significant challenges :
  . The orientation, inclination and strength of the barn roof are factors that can significantly influence the cost and efficiency of the solar installation.
  In many cases, extra support structures with an orientation and inclination different from that of the barn roof will have to be provided.

  . Lack of space in existing farm buildings can jeopardise the system efficiency because components must then be scattered in different places (cf. need for compactness), or left outside (low temperature, higher heat losses).

2.2.3. Socio-economic acceptability

It is clear that the domain of solar hot water production research has some characteristics that clearly distinguish it from that of solar drying for instance. The described installations for water heating are usually of a higher degree of sophistication, due to the extra required care for water as a heat conducting medium, as well as the often more demanding requirements as to temperature levels and regulation of the applications in question. These aspects of solar hot water production, namely both the higher complexity of the system, and thus the tighter economic margin, call for a more rigorous approach, while at the same time often no immediately marketable results are to be expected. It is a natural consequence that the interest in this domain, as exemplified in the basic description overview, is more limited, but generally more rigorously scientific and from university, rather than industrial backgrounds.

3. PROSPECTS FOR FUTURE RESEARCH

It is impossible to arrive at definite conclusions for such varied processes, based on so few experiments, (many of which are still in full development). It would however seem fair to say that for most of the described set-ups, in their local climatic and economic setting, the potential for economic feasibility does not seem particularly bright at present. Further research and eventual mass production (leading to lower investment costs), as well as energy price increases might change the picture.

Going back to the fundamental criteria, (as explained on p. 1), and measuring up the current research results against those criteria, some aspects arise which would be worthy of further consideration.

- APPLICATION TRANSFER TO OTHER REGIONS: The climate was said to be a crucial factor both as to energy-supply and system efficiency.
  As can be seen from the overview, most of the research that has reached a minimal level of results is situated in the Northern part of Europe. Some of these studied applications could possibly be transferred to more sunny climates.
  This would be the case:
  - in the first place, when the energy demand is not too sensitive to the outside temperature so that it doesn't disappear in warmer climates.
  - in the second place, when the applied process is transferable or already exists in the socio-economic environment of Southern Europe.

For the applications which meet these requirements, it would certainly be worth while to apply the already acquired knowledge, and to simulate the solar system performance for the different climatic conditions and socio-economic situation of Southern Europe (different scales and different degrees of sophistication).
This added research could contribute significantly to the identification and development of applications which are economically feasible and marketable in the more sunny regions. The operational installations which might ensue could in turn lead to the acquisition of further experience, which could then lead to further optimization of the system ; at the same time a market could be created, which could bring about mass production of solar system components, thus leading to lower prices. Both these effects (efficiency increase - price decrease), could then possibly make the given application economically feasible even in the more Northern countries.

- LONG-TERM HEAT STORAGE: it is obvious that if effective methods for long term heat storage were found, the application of solar energy would be revolutionized not only in agriculture, but over the whole range of research on this subject.
  Little hope exists that practical solutions will be found in the near future. In any case, it is believed that this type of research would be best located in another domain, not in the agricultural context.

- AN IN-DEPTH APPROACH: As regards solar system efficiency, an essential step beyond the simple addition of a solar system to an existing process is the adaptation of the original process, to make it blend well with the solar installation. An example would be the adaptation of the floor heating technique for piglets in order to employ lower operating temperatures and consequently reach much higher efficiency values for the same solar installation. Another example would be the use of alternative milk powders for calves, requiring lower water temperatures, thus increasing the solar system efficiency.

Still another example would be the acquisition of more quantitative knowledge in the area of sensitivity of fish production to temperature levels and variations, so as to be able to design economically and energetically optimal fish production schemes.

To include the cost aspects in this same context, it would be worth while to develop quantitative models of the cost factors of important processes, and their evolution over time, so as to gain a better insight into which factors should be subsituted, and how much. This would make it possible to obtain a more adequate adaptation of the system as a whole to the new situations as they develop.

- COMPONENT MATERIAL ANALYSIS: Still in the realm of solar system cost, a primary requirement for economic feasibility is that the installation must have a lifetime that fully covers the payback period.
Research suggests that these payback periods are certainly not negligible at present.

Furthermore, experience shows that the lifetime which is normally expected of the equipment is significantly shortened by the corrosive atmosphere to which it is exposed in many of the agricultural applications. In this context, studies aimed at extending the lifetime of the equipment in corrosive environments, while minimally influencing system costs, would certainly not be out of place. In this same context, research in the area of minimization of dust build-up on the collector cover (influenced by - among others - collector cover characteristics, but also collector location) would be appropriate.

- OTHER APPLICATIONS:
  - In the area of heating and conditioning of buildings, it seems that other possible applications should be considered, as well as floorheating for farrowing houses and some of the currently studied applications in poultry houses.

  - It is interesting to note the scarcity of research projects in which the use of solar assisted heat pumps is tested. Obtaining more data in this area could be worth while.

  - Methods for storing solar heat during the day, or the adaptation of passive solar designs to help obtain acceptable environmental temperatures during cold nights should be studied.

  - Free cooling of animal housing during warm periods by means of a natural ventilation stimulated by a solar induced thermosiphon-effect should be promoted ; cf. the so called "cool roofs".
  Cool roofs are constructed in such a way as to allow for free movement of air between the roof cover and the insulation applied, 5-10 cm below the cover. During periods of intense solar radiation the roof cover can reach extremely high temperatures and so heat up the air underneath.

Because of the thermosiphon effect however, the heated air is swiftly moved up and out through the top of the roof, and is then replaced by colder air. Environmental temperatures in the building can be reduced significantly in this way.

APPENDIX TO PART III
Some specific research projects on solar hot water production

In this appendix, seven projects on solar hot water production are described in detail.
They represent each of the five applications studied in chapters one and two :

1. SOLAR ENERGY TO ACCELERATE CATTLE MANURE DIGESTION - p. 396
   (An Italian project on solar biogas digester heating)

2. THE USE OF SOLAR ENERGY FOR HEATING CIRCULATING WATER FOR EEL- AND CARP BREEDING - p. 404
   (A Dutch project on solar heating of a fish farm)

3. THE USE OF SOLAR ENERGY FOR FLOOR HEATING IN A FARROWING HOUSE - p. 414
4. SOLAR DOMESTIC HOT WATER PRODUCTION AND SOLAR HEATING OF FARROWING HOUSE - p. 431
   (Belgian and German projects on solar floor heating in farrowing houses)

5. THE USE OF SOLAR ENERGY FOR HOT WATER PRODUCTION FOR A DAIRY FARM - p. 442
6. APPRAISAL OF A SOLAR WATER HEATER FOR BRADMORES FARM DAIRY AT SEALE - HAINE COLLEGE - p. 448
   (Dutch and English projects on solar water heating for hygienic purposes in dairy parlours)

7. THE USE OF SOLAR ENERGY FOR HOT WATER PRODUCTION ON A CALF REARING FARM - p. 457
   (A Dutch project on solar water heating for milk preparation on calf rearing farms)

The reporting formats used to compile the detailed descriptions are explained in the addendum, cf. p. 467. The numbering of the sub-headings in the detailed descriptions corresponds to the reporting formats.

1. Solar energy to accelerate cattle manure digestion

Project location

LANDRIANO, PAVIA, ITALY

Study Center :
- Project leader : Prof. G. CASTELLI
- Institution :   Institute of Agricultural Engineering
                   Via Celoria 2
                   Milan
                   ITALY

Started in 1983

Project description

In Pavia, Italy, the contribution of solar collectors to biogas productivity of a batch manure digester is being studied.

The experimental set-up consists of :
- 2 similar batch digesters, the reference one unheated, the other one heated only by a very simple solar system. (Cattle manure capacity : Reference 30 m^3; solar 20 m^3).
- The solar system consists of 6 m^2 of collectors (rubber absorber, no cover, no insulation), which are laid out on top of the digester. The water heated in the collector is led through a 100 metre long

polyethylene tube (ϕ 25 mm), embedded in the digester floor.
- The produced biogas is collected in a 30 m³ rubber gas storage.
- The biogas is used by a dual fuel (petrol–biogas) 3.5 kW electricity generating set.

The aim of the experiment
- To assess the difference in yield, using a limited solar outlay, in comparison with the unheated reference.
- To verify reliability as well as economic feasibility of such a simplified system.
- The ultimate aim is to assess the possibility of freeing farms from their dependence on electric power plants, by means of a simple, reliable, low cost biogas installation.

Scheme

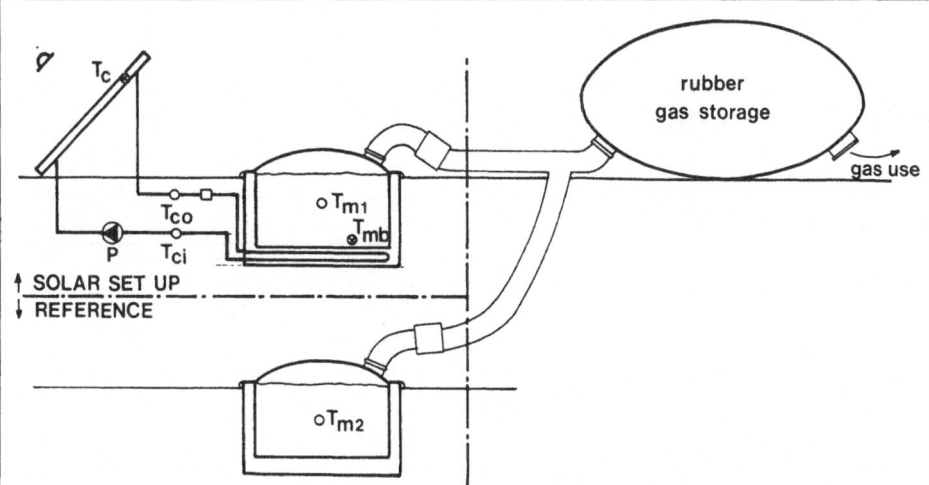

An explanation of the symbols used in the scheme is given on pp. 470–471.

Note : Another reference could be a conventional boiler, connected to the heat exchanger (e.g. a 10 kW gas fired unit).

Operating modes : If Tc > Tmb → P on
                  Tc < Tmb → P off

Conclusions

- Calculations predict an annual energy input from the solar system of 13 000 MJ/year, which corresponds to a system efficiency of 47%. (High, due to low temperature operation).
- During the days of fair solar radiation, the expected temperature increase of the manure mass should be of the order of 0.5 K, which at a manure temperature level around 18°C would yield a 7–8% biogas production rate increase. Under optimum weather conditions the plant could raise the manure temperature by 1.2 K, (assuming the heat is evenly distributed in the digester). The expected total production of biogas in the 20 m³ "solar" digester is about 3 700 m³/year.

- The given values are based on a calculation method described in section 5.1, which should be interpreted with caution, since very little practical experience has yet been gained with the system.

Experience acquired

- The efficiency of the heat exchanger is adversely affected by the need for it to be embedded in the digester floor, in order not to be destroyed when the batch digester is being filled and emptied by a mechanical spade.
- The most practical place for the collector is on top of the digester itself. (No extra space, little risk)
  The presence of the collectors does however require extra labour when the batch is filled or emptied (every 40 days in the warm season, max. every 120 days in the cold season. Also in this position venting problems occur.

PROJECT ANALYSIS

1. Site and climate

1.1. Site

Latitude :	45° N *(1)*
Longitude :	9° E *(1)*
Altitude :	115 m
Nearest main city :	Milan
Distance from main city :	10 km (as the crow flies)
Direction from main city :	Southeast
Obstructions :	none

(1) Approximate

1.2. Site micro-climate

The climate is typical of the Milan province.
Main characteristics :
- Poor insolation in the winter because of fog.
- Good insolation in summer. The diffuse component is constantly high; generally, the sun can be seen only for a few hours after sunrise even on a clear summer day.

1.3. Annual long-term averages

			Project	Country in general
1. Prevailing wind direction	W		E	
	S		NW	
3. Total precipitation		mm/yr	1 195	870
4. Absolute humidity	W	g/kg	3.8	
	S	g/kg	6.2	
5. Global irradiation on horizontal plane		MJ/(m².yr)	3 960	(1)
7. Degree days (Base temp.=18)		C days	2 340	
8. Hours sunshine		hrs/yr	1 848	
9. Ambient temperature		C	12.2	14.5
10. Average max. temp.(July/Aug)		C	27.4	27.8
11. Average min. temp.(January)		C	-4.6	3.0

(2)

```
12. Meteorological station : Milan Linate
13. Latitude :               45°28' N
14. Longitude :              9°12' E
15. Altitude :               120 m
16. Distance from site :     7.5 km (as the crow flies)
```
Note : W = Winter
S = Summer

(1) The map on page 5 gives an average annual solar radiation of $\pm$ 5 250 MJ (m² horizontal plane . yr) for the region in question.
(2) Referred to the common value of the plain.

2. Plant

2.3. Collector

1. Absorber : Rubber, no transparent cover or insulation
   Coolant : Water
   Working pressure : 50 kPa
   Collector fluid content : 3 1/m²
   Emittance of plate : 0.9
   Coefficient of absorption
       of visible light : 0.95
2. Dimensions :
   Absorber area : 6 m²
3. Flow rate : 28 x 10⁻⁶ m³/(s.m² collector surface)
4. Diagram of collector : cf. figure 2.3.4.

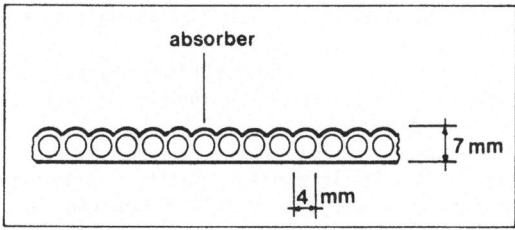

Fig. 2.3.4 : cross-section of collector

5. Collector performance :
   The instantaneous collector efficiency was assessed as

$$\eta_c = 0.75 - 20.3 \left(\frac{Tc - Ta}{I_c} \right)$$

based on the following measured values :

Ta (°C)	Tci (°C)	Tco (°C)	I_c (W/m²)	$\dot{V}$ (x 10^{-6} m³/s)	T' (°Cm²/W)	η_c (decimal)
25.0	30.9	37.6	860	92	0.007	0.55
22.0	32.4	38.1	710	89	0.015	0.55
25.0	33.4	38.1	850	96	0.010	0.41
22.0	33.7	39.6	710	89	0.016	0.49
24.5	36.4	40.3	850	94	0.014	0.34
22.0	37.6	40.4	725	89	0.022	0.27
22.0	38.9	42.0	750	89	0.023	0.28
22.0	41.2	43.5	780	93	0.025	0.21
22.5	42.5	44.3	740	100	0.027	0.19
25.5	43.3	45.6	750	97	0.028	0.23
25.0	51.6	53.0	900	97	0.030	0.12
25.0	52.1	53.4	880	99	0.031	0.11
25.5	52.1	53.7	875	94	0.030	0.13

Tc   = average collector temperature (°C)
Tci = temperature at collector inlet (°C)
Tco = temperature at collector outlet (°C)
I_c = insolation on the collector (W/m²)
Ta   = ambient temperature (°C)
$\dot{V}$ = flowrate through the collector (x 10^{-6} m³/s) *(1)*
$T' = \frac{Tci - Ta}{I_c} \quad (\frac{°Cm^2}{W})$
η_c = collector efficiency (decimal)

*(1) The flowrate of the tests does not fully correspond with the flow
rates apparently used in normal operation.
(compare with sections 2.3.3 and 5.6.4) (N.O.T.E.)*

2.5. Installation parts and measuring devices

1. Technical description of installation parts
   - 0.3 kW pump -flow rate 0.7 m³/h
   - heat exchanger -> 100 m of φ 25 mm PE tubing, embedded in
     digester floor.
2. Technical description of the sensors *(1)*
Ambient temperature : Pt 100 thermoresistor, to the above standard. DIN
43760 (tolerance 0.2°C).

Manure temperature : Custom-built Pt 100 thermoresistor, to the
standard. At present, 4 probes (2 per digester)
are installed.

Solarimeter : Kipp & Zonen, Model CM5

Heat energy meter : Model WM22 by Cazzaniga, Milan ; composed of an
electronic unit, two Pt 100 thermoresistors and a
turbine flow meter *(2)*
Precision 2%

Gas volume meter : Model Ariete by SIM Brunt, Milan

Precision 2%
Digital tester for data reading.

(1) Installed at present
(2) Also used to monitor collector flow rate.

---

4. Economic aspects

Cost basis 1983

4.1. Economic environment

	Solar only	Reference *(1)* solution
1. Energy price of gas ECU/m³	0.29	0.29
2. Energy price of electricity ECU/kWh	0.075	0.075
5. Interest rate of loans for farm equipment (rate without subsidy) %/yr	22	22
6. Subsidies available for the solar and competitor farm equipment (as capital grant or interest rate reduction) %	30–50 *(2)* 50–60 *(3)*	50–60 *(3)*

(1) The reference solution considered here is not the one given in the experiment. It applies to an assumed 20 m³ digester equipped with a 10 kW gas fired conventional boiler, which is operated in such a way as to yield the same biogas production as the solar digester.
(2) Principal
(3) Interest

4.2. Investment expenses

	Solar only	Reference solution *(1)*
1. Total investment budget (equipment only) ECU	1 280	1 900
2. Cost of equipment :		
– cost of collectors per m² ECU	45	not applicable
– expenses for buildings ECU	75	226
– cost of regulation equipment ECU	150	377
– other material costs ECU	332	829
3. – Time of personnel for the conception, installation and start-up of equipment Hours	37	45
– Cost of personnel for the conception, installation and start-up of equipment ECU	452.2	467 *(2)*
4. Lifetime of the equipment years (evaluation)(circulat.pump excluded)	15	10

(1) cf. footnote (1) in section 4.1
(2) It is not clear why the labour cost per hour is different in the 2 situations (N.O.T.E.)

4.3. Operating expenses

	Solar only	Reference solution *(1)*
1. Energy cost ECU/yr	45	225
4. Manpower cost ECU/yr	75	150

(1) cf. footnote (1) in section 4.1

4.4. Hidden cost and benefits *(1)*

1. Production obtained *(2)*	3 700 m³/year
2. Average sales price per unit produced *(3)*	0.136 ECU/m³
3. Total income obtained (4.4.1 x 4.4.2)	503 ECU/year
4. Storage costs due to the installation *(4)*	1 500 ECU/yr

(1) Given the chosen assumptions this section yields identical figures for both solar and reference solutions
(2) Depending on digester temperatures, the daily biogas yield ranges from 0.05-0.65 m³ per m³ digester
(3) Based on commercial methane gas prices pro rata of their energy content.
(4) Refers to the biogas storage tank for the 2 digesters in the actual experimental set-up.

5. Results

5.1. Calculation method

Since the experiment is still at an early stage, some tentative results were calculated in the following way.

1) The collector efficiency in function of the circumstances was :

$$\eta = 0.75 - 20.3 \left(\frac{Tc - Ta}{I_c} \right) \text{ (cf. section 2.3.5)}$$

2) Since the limited collector surface does not raise the digester temperature significantly (max. 1-2 K), the year-long performance of the solar system was calculated, using previously measured climatological data (average monthly radiation, ambient temperature, and corresponding unheated digester temperature) in combination with the collector efficiency formula (assuming a water temperature in the heat exchanger equal to the digester temperature + 5 K. The useful amount of solar energy extrapolated in this way is in the neighbourhood of 13 000 MJ/year, which would correspond to an overall solar system efficiency of 47%.

3) The electric consumption was calculated based on the number of probable hours of daily operation of the solar plant.

5.2. Weather conditions and energy

1.	1/1	1/2	1/3	1/4	1/5	1/6	1/7	1/8	1/9	1/10	1/11	1/12		from :	PERIOD	
2.	31/1	28/2	31/3	30/4	31/5	30/6	31/7	31/8	30/9	31/10	30/11	31/12		till reference year		
3.	0.4	6.0	10.7	12.9	20.1	24.1	25.5	24.1	22.1	14.5	7.8	2.8	°C	average ambient TEMP(1)		
5.	18	38	69	95	131	113	110	110	90	54	28	45	MJ/day	average SOLAR RAD on collector(2)		
6.	–	11.5	32.6	52.5	89.1	76.8	73.0	46.8	24.7	10.6	9.2	–	MJ/day	input SOLAR energy		solar
10.	–	11.5	32.6	52.5	89.1	76.8	73.0	46.8	24.7	10.6	5.8	–	MJ/day	USEFUL energy		
14.	–	3.7	5.5	6.7	7.8	8.6	9.8	8.8	6.5	4.2	2.0	–	MJ/day	ELECTRICITY consumpt.(3)		

(1) During the day
(2) Daily average over 6 m², actually measured in the year considered
(3) Daily average

5.6. (Pre)-heated water

1.	1/1	1/2	1/3	1/4	1/5	1/6	1/7	1/8	1/9	1/10	1/11	1/12		from :	PERIOD
2.	31/1	28/2	31/3	30/4	31/5	30/6	31/7	31/8	30/9	31/10	30/11	31/12		till	1980
3.	12	12	13	14	16	21	28	33	27	23	17	14	°C	average manure temperature(1)	
4.	0.16	0.16	0.16	0.16	0.16	0.16	0.16	0.16	0.16	0.16	0.16	0.16	l/s	average flow rate	
6.	17	17	18	20	24	26	33	38	32	28	22	19	°C	output temperature required in solar plant	(2)

(1) Actual readings in the unheated digester
(2) The water temperature is assumed 5 K higher than the manure.

2. The use of solar energy for heating circulating water for eel and carp breeding

Project location

BARNEVELD, NETHERLANDS

Study Center :
- Project leader : Ing. G. BROUWER
- Institution : Van Heugten, Consulting Engineers
  P.O. Box 305
  6500 AH Nijmegen
  THE NETHERLANDS

Running since September 1983

Project description

At the experimental station in Barneveld, the Netherlands, a simple solar collector of 15 m² was installed on top of a heated fishbreeding facility, where the water is kept at a temperature of 26.5°C.
- Given such a low temperature requirement, a low investment installation was chosen (collectors without glazing or insulation, basins not insulated).
- Indirect heat exchange was necessary between collector and storage, since the fishwater is too polluted with faeces to flow through the collector.
- The instantaneous heat demand is unpredictable, since for reasons of

evaporation and water quality requirements, a considerable amount of fresh water at 10-13°C is added by hand in an irregular way : an average of about 2.1 m³ water per day added into a total water volume of 16 m³ of the fishbreeding plant. This obviously creates some irregular peak values of the heat demand.

Goals of the experiment :
- To gain experience with this technique.
- To optimize the component capacities, as well as to predict the results in bigger installations using a simulation model that is being developed and validated by means of this set-up.
- To determine its economic feasibility.

Scheme (1)

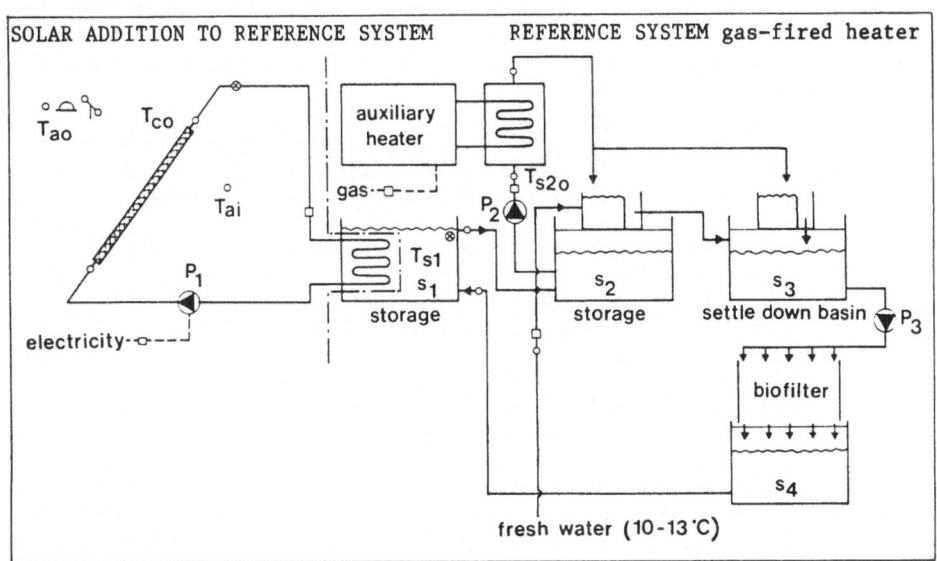

- P_3 constantly circulates water at a rate of 6 m³/h
- Operating modes
  Frost protection control drain down system
  If Ts1 > 35°C -> P1 off
  If Tco > Ts1 + 5.5°C and Ts1 < 35°C -> P1 on
  If Tco < Ts1 + 2°C -> P1 off
  If Ts2o < 26.5°C -> auxiliary heater on
  If Ts2o > 26.5°C -> auxiliary heater off

NOTE : the weather data come from a meteorological station 30 km away, except in the short periods in which a data acquisition system is used.

(1) An explanation of the symbols used in the scheme is given on pp. 470 and 471.

Conclusions

- Using the set-up as a reference, a simulation model was developed and validated.
  Running this model over a climatological reference year over hourly calculation steps, the net heat production of the system was calculated and found to be very low : 500 MJ/(m².year). (This corresponds to a solar system efficiency of less than 15%. The solar system then only contributes 2.8% of the total heat demand). The weak point in the set-up is the heat exchanger, which needs to be used, since the fishwater is too polluted to flow through the collector itself.
  Due to the fact that the fishwater should not be in contact with copper, iron or zinc, the heat exchanger was made in plastic, which, together with its special construction in order to avoid plugging up with dirt, and to be cleaned easily, contributes to a very low heat transfer capacity.
  Hence the temperature level of the collector is raised and consequently its efficiency is lowered.

  Constructing the heat exchanger in stainless steel (the only metal the fish tolerate) would yield 850 MJ/m² collector per year in useful energy. This type of heat exchanger is very expensive however.

  Other possible ways to influence the existing system efficiency are :
    - Covering the collector with a light-transmitting plate.
    - Adding the fresh tap water directly into the storage tank containing the collector circuit heat exchanger. (permits functioning of collectors at lower temperature, and promotes stability of fishpond temperature).
    - further sophistication of the heat exchanger
    - combination of collectors with heat pump.

  Still another possibility would be to directly heat up the fresh tapwater, rather than the recirculating water. This would eliminate the need for the heat exchanger. Supposing 600 l of fresh water are used 4 times per day, the existing collector system, combined with a 1 m³ storage of fresh water being heated up by the collectors, would yield 830 MJ/m² collector per year according to the simulation model.

  In larger fishfarms, the addition of fresh water to the system could be less irregular. Assuming a constant flow of 240 l/hr of fresh water into the system over a period of 10 hours during the day, circulating that water through the 15 m² of collectors first, and without use of the above mentioned fresh water storage tank, would yield 1 250 MJ/m² collector per year, again according to the simulation model.

  The simulation model was validated using a short measuring period. Further validation as well as further acquisition of experience should make it possible to draw conclusions of a less provisional nature.

- The pollution in the water (slime and faeces) as well as the fact that according to the involved researchers no copper, iron or zinc should be in contact with the fishwater should be borne in mind when designing systems or measuring devices.
  e.g. : flowmeters need to be made in plastic (less accuracy), and need

regular cleaning.

- Future research
  More experimentation is planned with the above described alternatives,
  as well as with an installation of greater volume.

PROJECT ANALYSIS

1. Site and climate

1.1. Site

Latitude :	52°10' N
Longitude :	5°10' E
Altitude :	< 20 m
Nearest main city :	Amersfoort
Distance from main city :	15 km
Direction from main city :	West
Obstructions :	None

1.2. Site micro-climate

See photograph
The buildings are situated in a forest region.

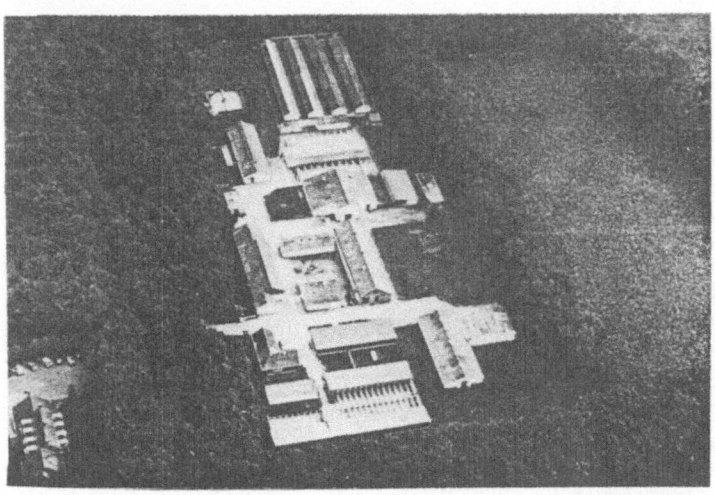

1.3. Annual long-term averages

			Project
1. Prevailing wind direction	W		SW
2. Average wind speed	W	m/s	4.0
	S	m/s	2.8
3. Total precipitation		mm/yr	795
4. Absolute humidity	W	g/kg	4
	S	g/kg	7
5. Global irradiation on horizontal plane		MJ/(m²yr)	3 530
6. Diffuse proportion		%	60
8. Hours sunshine		hrs/yr	1 506
9. Ambient temperature		C	9.2
10. Average max. temp. (July/Aug)		C	21.1
11. Average min. temp. (Jan)		C	3.5

12. Meteorological station : De Bilt
13. Latitude : 52°20'N
14. Longitude : 4°50 E
15. Altitude : < 20 m
16. Distance from site : 30 km

    Note : W = Winter
    S = Summer

2. Plant

2.3. Collector (Flat-Plate collector)

1. Absorber
   - Coolant : water
   - Material : modified ABS (Acrylonitril Butadieen Styreen)
   - Aperture area : 15 m² λ = 0.027 W/mK
   - Absorber area : 15 m²
   - Collector fluid content : 3.5 1/m²
   - Working pressure : 200 kPa
   - Flow rate : 17 x 10^{-6} m³/(s.m²) collector surface)
   - Emittance of plate : 0.95
   - Coef. of absorption of visible light : 0.95
2. Glazing : none
3. Insulation : none
4. Mounting
   - orientation : South
   - tilt : 20°
5. Collector performance (η_c) :

$$\eta_c = 0.9 - 18 \; \frac{Tc - Ta}{I_c}$$

   with Tc = collector temperature (°C)
        Ta = ambient temperature (°C)
        I_c = incident radiation on the collector (W/m²)

2.4. Heat storage

1. Medium of heat storage : water
2. Description, location in the system : see technical scheme.

The heat storage is part of the conventional breeding system.
(settle down basin)
3. Insulation : no insulation
4. Dimensions : 2 tanks of 3.6 x 0.7 x 0.6 m = 3 m³
5. Thermal capacity : 12.6 MJ/K
6. Overall heat loss coefficient : 60 W/K

Note : - pipes of the collector circuit are insulated
  - the traditional fishfarm components are not insulated.

2.5. Installation parts and measuring devices

1. Technical description of the installation parts
   - Circulation pump of the collector circuit 1.5 m³/h, 75 to 120 W (variable)
   - Heat exchanger, 12 m² modified ABS, suspended in the storage basins. This material is the same as that of the absorber.
   The heat exchanger can easily be removed in order to enable cleaning of its surface.
   - Control

type :	Resol, El
switch-on temperature difference :	variable, 2-16 K
switch-off temperature difference :	ca. 2 K below switch-on temp. difference
temperature range :	0-110°C
sensors :	N.T.C.-resistors

2. Description of the sensors

Type of sensor	Temperature sensors	Flowmeter	Heatflow meters	Pyranometer
Principle	platinum 100 Ω resistance	float meter, with built in magnet	multiplication and integration of flow and temp. diff.	sternpyrano-meter (Dirmhirn)
Precision	0.5°C	2.5%	2% - 5%	
Measuring range	0 - 100°C	different types	ΔT : 80 K different types	1 260 W/m²

(Periodically a data acquisition system is used.)

4. Economic aspects

Incl. V.A.T. 1983

4.1. Economic environment

1. Energy price of gas	0.24 ECU/m³
2. Energy price of electricity	0.12 ECU/kWh
3. Energy prices evolution prospects	
- of gas	+ 10%/yr
- of electricity	+ 10%/yr
5. Interest rate of loans for farm equipment (rate without subsidy)	9%/yr

4.2. Investment expenses

1. Total investment budget	2 999 ECU
2. Cost of equipment :	
- cost of collectors per m^2	34 ECU
- expenses for buildings	75 ECU
- cost of regulation equipment	67 ECU
- other material costs	1 315 ECU

3. - time of personnel for the con-
    ception, installation and start-up
    of equipment
	50 hours

- cost of personnel for the con-
    ception, installation and start-up
    of equipment
	1 032 ECU

4. Lifetime of the equipment (evaluation)	15 years

4.3. Operating expenses

2. Spare parts	20 ECU/yr
4. Manpower cost	cleaning of heat exhanger

5. Results

5.1. Method of calculation used in table 5.2.1.

Measured are mainly :
- The useful energy given off by the solar system (Qsi)
  (This corresponds to the input energy into the solar storage tank, since
  the storage tank is part of the fishbreeding system)
- The useful energy produced by the gas heater (Qag)
- The necessary energy input in order to heat up the incoming fresh water
  to the desired temperature level (Qdl).

These energy flows are assessed indirectly by measuring the water flow ($\dot{V}$)
and the water temperature difference (ΔT) (i.e., before and after the heat
transfer) over time (t).
The energy flow Q is then :

$$Q = \int_{to}^{t1} (c_w \, \rho_w \, \dot{V} \, \Delta T) dt$$

with c_w the specific heat of the water ($4.18 \frac{kJ}{kgK}$)

      ρ_w the specific mass of the water (1000 kg/m^3)
      $\dot{V}$ the flow rate of the water (m^3/s)
      ΔT the temperature difference (K)
      t the elapsing time (s)
      Q given in kJ

The heat losses (Q_{d2}) from the breeding system can be calculated using an
energy balance equation :

$$Q_{d1} + Q_{d2} = Q_{si} + Q_{ag}$$

For the period 1-8 September 1983 this yields the energy flows given in
figure 5.1.

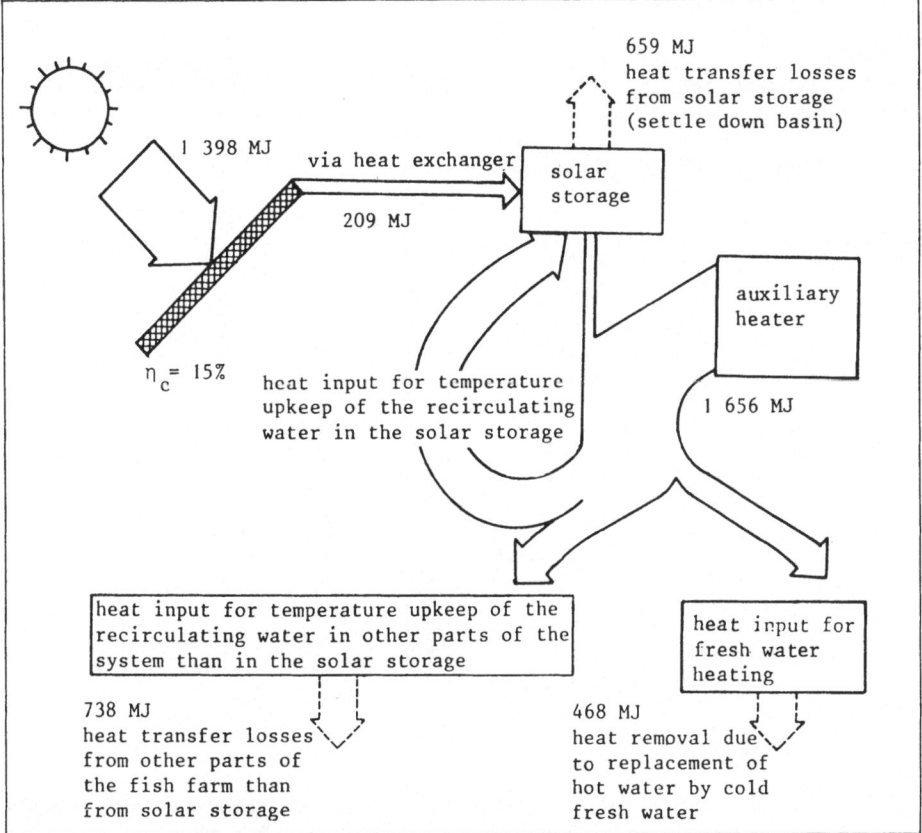

Fig. 5.1 : Energy flow scheme over the measuring period of 1-8 September 1983.

As regards the simulation program P 635, to be validated for the given set-up :
The program is meant to be useable for most space or tapwater heating solar installations.
It is based on an hourly calculation of the energy balance :
1) The energy demand, which is a function of
   - the climatological data : a reference year for the ambient temperature and the solar radiation was developed, based on 10 years of data from the Royal Meteorological Institute of the Netherlands.
   - the characteristics of the system itself :
     + for the case where cold tapwater is directly heated : e.g. the draw-off characteristics
     + for the case where the water flowing through the solar storage tank is being heated (as with experimental set-up) :
       e.g. - stable temperatures
            - heat losses in function of building components
            - heat to be added to incoming water
2) The energy to be gained from the collectors, which is a function of :
       e.g. : - The climatological data
              - Collector characteristics

These two calculated energy values are then compared to find the amount of heat to be added by an auxiliary heater or the amount of heat to be stored.
For 2 different set-ups the sum of these hourly calculated energy flows over a whole year is given in sections 5.2.2. and 5.2.3..

5.2. Weather conditions and energy

1. Measured values

1.	1-9-83		from : PERIOD	
2.	8-9-83		till	
3.	15	°C	average ambient TEMP	
4.	4	m/s	average WIND	
5.	1398	MJ	average SOLAR RAD on collector	
9.	24	°C	average TEMP storage	
10.	209	MJ	USEFUL energy	
11.	1656	MJ	AUXILIARY HEATING energy	solar
12.	a+2.5	MJ	energy for OPERATION	
13.	2365	MJ	GAS consumption η=0.7	
14.	a+2.5	MJ	ELECTRICITY consumption	
15.	1865	MJ	HEATING energy	
16.	a	MJ	energy for OPERATION	reference
17.	2664	MJ	GAS consumption	
18.	a	MJ	ELECTRICITY consumption	

a = unknown

2. Results of the actual experimental set-up extrapolated to a whole year. (weather conditions according to the "Van Heugten" Standard year)

	1/1	1/2	1/3	1/4	1/5	1/6	1/7	1/8	1/9	1/10	1/11	1/12	from : PERIOD	
1.	1/2	1/3	1/4	1/5	1/6	1/7	1/8	1/9	1/10	1/11	1/12	1/1	till reference year	
3.	1.7	2.0	5.0	8.5	12.4	15.5	17.0	16.8	14.3	10.0	5.9	3.0	°C average ambient TEMP	
5.	1300	1945	4280	6330	7268	7513	6862	5969	4417	2633	1274	860	MJ average SOLAR RAD on collector	
9.	18.1	18.1	18.1	18.3	18.7	19	20	20	18.5	18.2	18.1	18.1	°C average TEMP storage	
10.	0	0	0	650	1595	1674	1515	1285	766	0	0	0	MJ USEFUL energy	
11.	40435	37854	33198	23805	14152	6700	3567	4349	9824	20350	29830	35517	MJ AUXILIARY HEATING energy	solar
12.	a	a	a	a+15	a+36	a+44	a+41	a+34	a+26	a	a	a	MJ energy for OPERATION	
13.	57765	54077	47426	34007	20218	9570	5096	6212	14035	29072	42614	50739	MJ GAS consumption η=0.7	
14.	a	a	a	a+15	a+36	a+44	a+41	a+34	a+26	a	a	a	MJ ELECTRICITY consumpt.	
15.	40435	37854	33198	24455	15747	8374	5083	5634	10591	20350	29830	35517	MJ HEATING energy	
16.	a	a	a	a	a	a	a	a	a	a	a	a	MJ energy for OPERATION	reference
17.	57765	54077	47426	34936	22495	11986	7262	8048	15130	29072	42614	50719	MJ GAS consumption	
18.	a	a	a	a	a	a	a	a	a	a	a	a	MJ ELECTRICITY consumpt.	

a = unknown

3. Simulated results for the case where fresh-water supplied at a continuous rate of 0.24 m³/h (10 hours/day) would be directly warmed up by the collector (no solar storage or heat exchanger) (Weather conditions according to the "Van Heugten" standard year.

	1/1	1/2	1/3	1/4	1/5	1/6	1/7	1/8	1/9	1/10	1/11	1/12		PERIOD	
1.	1/1	1/2	1/3	1/4	1/5	1/6	1/7	1/8	1/9	1/10	1/11	1/12		from :	
2.	1/2	1/3	1/4	1/5	1/6	1/7	1/8	1/9	1/10	1/11	1/12	1/1		till reference year	
3.	1.7	2.0	5.0	8.5	12.4	15.5	17.0	16.8	14.3	10.0	5.9	3.0	°C	average ambient TEMP	
5.	1300	1945	4280	6330	7268	7513	6862	5969	4417	2633	1274	860	MJ	average SOLAR RAD on collector	
10.	54	248	1163	2394	3424	3801	3690	3366	2365	944	274	65	MJ	USEFUL energy	solar
11.	40381	37606	32035	22061	12323	4573	1393	2268	8226	19356	29556	35452	MJ	AUXLIARY HEATING energy	
12.	a	a	a	a	a	a	a	a	a	a	a	a	MJ	energy for OPERATION	
13.	57687	53722	45764	31516	17604	6532	1990	3240	11751	27651	42222	50646	MJ	GAS consumption η=0.7	
14.	a	a	a	a	a	a	a	a	a	a	a	a	MJ	ELECTRICITY consumpt.	
15.	40435	37854	33198	24455	15747	8374	5083	5634	10591	20350	29830	35517	MJ	HEATING energy	reference
16.	a	a	a	a	a	a	a	a	a	a	a	a	MJ	energy for OPERATION	
17.	57765	54077	47426	34936	22495	11963	7262	8048	15130	29072	42614	50739	MJ	GAS consumption	
18.	a	a	a	a	a	a	a	a	a	a	a	a	MJ	ELECTRICITY consumpt.	

a = unknown
Collector pump is not needed

6. Remarks

The use of gas as a heating source is practically non-existent in normal fishfarms.
The heat demand values calculated over the reference year for the studied set-up were about 16 700 MJ per year and per m³ water in the system. This would amount to a tremendous annual heating cost of ± 150 ECU/m³ water holding capacity of the system if one used a heating system of 70% efficiency and gas of about 30 MJ/m³ heat value costing 0.194 ECU/m³.
Therefore the use of conventional heaters is restricted in practice to fish larvae breeding places, which are even more intensive than the real fishfarms, or experimental stations where economic feasibility is not a primary concern.

3. The use of solar energy for floor heating in a farrowing house

Project location

SCHRIEK, BELGIUM

Study Center :
  - Project leader : Prof. V. Goedseels
  - Institution :   Laboratory for Agricultural Buildings Research
                    Catholic University of Leuven (K.U.L.)
                    Kardinaal Mercierlaan 92
                    3030   Leuven
                    Belgium
  - Industrial cooperator :
                    Belgische Boerenbond

Running since 1981

Project description

In a farrowing house in Schriek, Belgium, a small scale solar collector system (4 m^2 collector surface) was set up to contribute to the floor heating used all year round to create the right climate for the piglets.
- The collector is coated with a spectral selective layer.
- The antifreeze fluid of the collector circuit is separated from the heating system by a heat exchanger in the storage tank.
- Two storage tanks of different sizes (100 1/300 1) are built in, with

the possibility of experimenting on both sizes individually or combined.
- The required temperature of the heating water varies between 40°C in summer and 55°C in winter.
- The average primary heat demand in this <u>small scale experiment</u> (involving 6 farrowing sow stands) varies over the year between 600 – 1 100 Watt. (For a heating system efficiency of 70 %.)
- Every 2 minutes a whole set of data is measured and recorded by data-log.

Goals of the experiment :
- To select a set of typical weather-days, and use the measured values to evaluate a simulation model that is being developed.
- To gain experience with the specific application.
- Using the simulation model, to optimize the component parts in order to reach a maximum efficiency/cost ratio for different situations.
- To assess the economic feasibility.

Scheme (1)

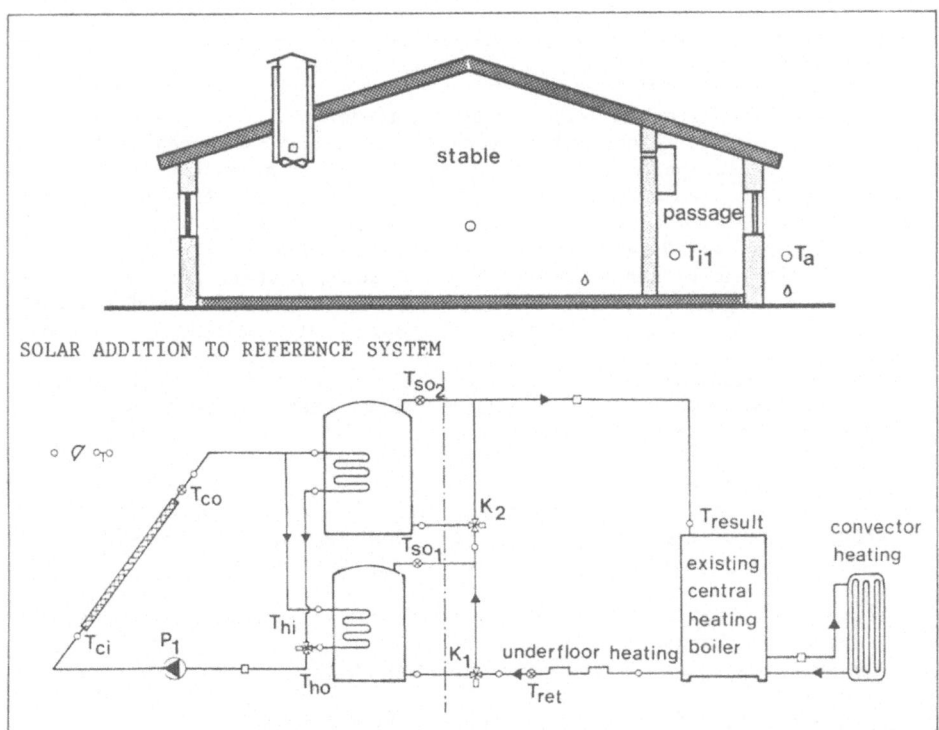

(1) An explanation of the symbols used in the scheme can be found on pp. 470-471.

Operating modes (e.g. : in case only storage 2 is used)

In installation :

if $T_{co} > T_{so2} + X$ -> P_1 on

if $T_{co} < T_{so2} + X$ -> P_1 off

if $T_{co} > T_{so2} + Y$ -> return water thru storage $(K_2$ ⟶✕▢ $)$

if $T_{so2} < T_{ret} + Y$ -> no return water thru storage $(K_1$ ✕▢ $)$

$(X$ and Y are set values)

In simulation model :

if $T_{co} > T_{so2} + X^{on}$ and $T_{so2} < 85°C$ → P_1 switches on

if $T_{co} < T_{so2} + X^{off}$ or $T_{so2} > 98°C$ → P_2 switches off

$\quad X^{co} \neq X^{so2}$

if $T_{so2}^{on} < T_{ret}^{off} + Y$ -> $(K_2$ ✕▢ $)$

if $T_{ret} + Y < T_{so2} < T_{demand}$ $\quad (K_2$ ⟶✕▢ $)$

if $T_{so2} > T_{demand}$ -> K_2 is automatically adjusted so $T_{result} = T_{demand}$

Note : The existing central heating boiler functions as auxiliary heater, and adjusts the water temperature to make it correspond to the desired value.

Conclusions

- A simulation model was developed and evaluated.
- Using the model, the optimal component dimensions have been calculated for a full scale operation, based on a climatological reference year. These dimensions were found to be 18 m² collector surface with a storage tank of 100 l for a net heat input into the floor of 5.6 - 10.4 kW corresponding to 80 farrowing stands for a sow and her piglets.
- Based on the energy input coming from the solar system, and the current energy and component prices, the payback period was calculated for this optimal system. This was about 10 years (cf. calculation method, section 5.1.).
- Given the experience acquired, which would predict a short lifespan of the installation, it was concluded that this application is not economically feasible.

- Experience acquired :
  + The atmosphere in and around a pig farm is very agressive (corrosive, dusty). This -together with the fact that regular cleaning with water under high pressure occurs - posed significant problems with the installation.
    - adequate protection of the components is essential (especially electronic regulation devices)
    - regular technical maintenance and supervision is necessary to ensure reliability.
    - degeneration of the materials occurs very quickly.
    - given the dusty environment, the collector needs to be regularly cleaned, if collector efficiency is to be kept at a reasonable level.
  + In the operating modes a fixed X value was found to be inadequate. A different X value for the switch-on and switch-off position was established. For full scale application additional operation modes are required :

if $T_S > T_{demand}$ -> K_x should act as a mixing valve in order to establish the desired floorheating water temperature. Furthermore, the occurrence of excessive storage water temperatures needs to be avoided.
+ The importance of a non-return valve in the collector circuit was noted.
- The measurements were chosen from data recorded over the period between April 1981-April 1982. Collecting valid data was a significant problem due to the aggressivity of the environment and the complexity of the activities on the pig farm. A regular and strict supervision program was required.
- The global efficiency of the simulated full scale installation was calculated as 32%.
  Research on development of a different floor heating technique which would lower the required temperature level, and improve the system efficiency, is now under way.
  Since warm ventilation air is always available, the feasibility of using a heat pump as an alternative energy source is also being assessed.

Further references

GHIJS, M. : "Onderzoek naar de integratiemogelijkheden van Zonneenergie in kraamstallen." (1981)
VAN DER STUYFT, E. : "Rekenmodel voor energiebesparingsberekeningen bij vlakke plaat watercollectoren in kraamstallen." (1982)
VAN DUYSE, N. : "Technico-economische beschouwingen van een zonne-installatie in een mestkalverij bij middel van een simulatie-model." (1983)

The given references are dissertations, made by engineering students at the "Laboratory for Agricultural Buildings Research".

PROJECT ANALYSIS

1. Site and climate

1.1. Site

Latitude :	51°02' N
Longitude :	4°42' E
Altitude :	
Nearest main city :	Leuven
Distance from main city :	18 km
Direction from main city :	North
Obstructions :	none

1.3. Annual long-term averages (for country in general)

1. Prevailing wind direction	W	SW
	S	
2. Average wind speed	W m/s	4.2
	S m/s	3.4
3. Total precipitation	mm/yr	835
5. Global irradiation on horizontal plane	MJ/(m².yr)	3514
6. Diffuse proportion	%	59
7. Degree days (base temp.= 15°C)	C days	2087
8. Hours sunshine	hrs/yr	1555
9. Ambient temperature	C	10.1
10. Average max. temp. (July)	C	22.7
11. Average min. temp. (January)	C	-0.3

12. Meteorological station :   K.M.I. Ukkel
13. Latitude :   50°48'N
14. Longitude :   4°21' E
15. Altitude :   100 m
16. Distance from site :   35 km

Note : W = Winter
S = Summer

2. Plant

2.3. Collector (Flat-plate collector)

1. Absorber
   - Coolant : 50% PKL 100 (anti-freeze) + 50% water
   - Material : Aluminium
   - Aperture area : 2 x 2.071   m²
   - Absorber area : 2 x 1.978   m²
   - Collector fluid content : 0.302   l/m²
   - Working pressure : 180600   Pa
   - Flow rate : 17 x 10^{-6} m³/sm²
   - Emittance of plate : 0.30   decimal
   - Coef. of absorption of visible light : 0.96

2. Glazing
   - emittance of glass : 0.88
   - number of glass covers : 1
   - material : glass
   - thickness : 4 mm

3. Insulation
   - Polyurethane 3 cm
   - Rockwool 5 cm

4. Mounting
   - Orientation : South
   - Tilt : 51°

5. Collector performance : cf. figure 2.3.5.

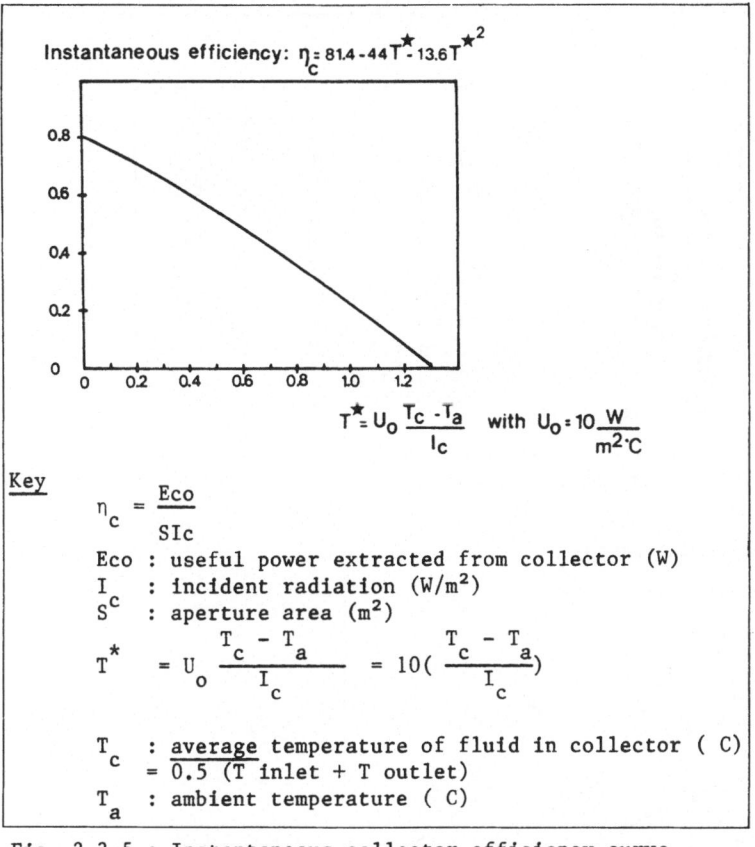

Key

$$\eta_c = \frac{Eco}{SI_c}$$

Eco : useful power extracted from collector (W)
I_c : incident radiation (W/m²)
S : aperture area (m²)

$$T^* = U_o \frac{T_c - T_a}{I_c} = 10(\frac{T_c - T_a}{I_c})$$

T_c : average temperature of fluid in collector (C)
$= \overline{0.5 (T \text{ inlet} + T \text{ outlet})}$
T_a : ambient temperature (C)

Fig. 2.3.5 : Instantaneous collector efficiency curve.

2.4. Heat storage

1. Medium of heat storage : Water
2. Description, location in the system : 2 Boilers : sheet-iron E24.1
   one 100 1, the other 300 1
3. Insulation : Polyurethane 5 cm (top and bottom)
   10 cm (other parts)
4. Dimensions : see diagram below
5. Thermal capacity : 0.418 and 1.26 MJ/K
6. Overall heat loss coefficient : 0.468 and 0.974 $\frac{W}{K}$
7. Diagram : cf. figure 2.4.7.

2.5. Installation parts and measuring devices

1. Technical description of the installation parts :
   - Pump : GRUNDFOSS UPS 20-45 (cf. figure 2.5.1.1.
     connections : 1"
     max. pressure : 10 bar
     max. temperature : 120 C
     weight : 3.9 kg
     position 1 : 1600 r.p.m. : 42-45 W
           2 : 2000   : 55-65 W
           3 : 2400   : 80-90 W

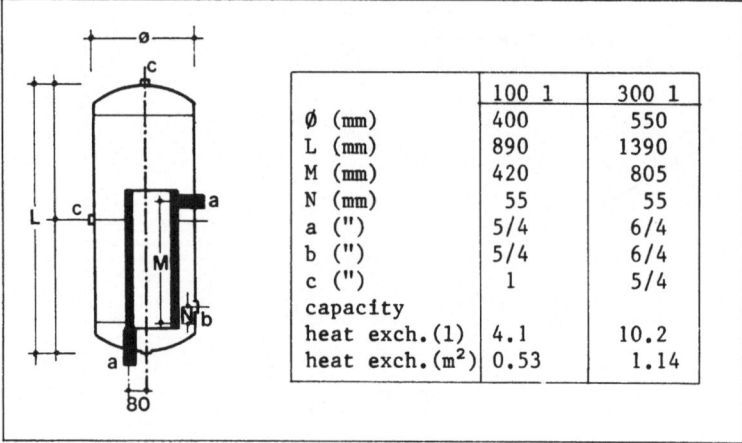

Fig. 2.4.7 : Diagram of storage tank.

	100 l	300 l
Ø (mm)	400	550
L (mm)	890	1390
M (mm)	420	805
N (mm)	55	55
a (")	5/4	6/4
b (")	5/4	6/4
c (")	1	5/4
capacity heat exch.(l)	4.1	10.2
heat exch.(m²)	0.53	1.14

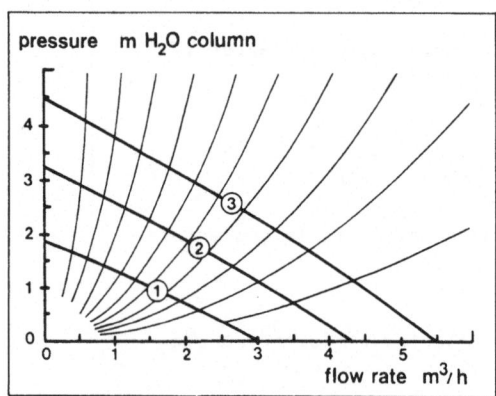

Fig. 2.5.1.1 : Pump performance data.

- Floor heating : (cf. figure 2.5.1.2)

2. Description of the sensors
   - Temperature : - Analog Device 590 M
                   - precision : $\pm$ 0.3 K
                   - measuring range : - 55 ... 150 C
                   - reference : legally stamped thermometer
   - Flow rate (liquid flow) : - CONTEURO mechanical water meters
                   - read-out is a function of the fluid flow
                   - precision : 0.1 litre
                   - minimum 50 l/hr
          (air flow) : - propeller 21281 Young Company on DC-motor MBLE 980412001806
                   - measuring range : 0.1-30 m/s
   - Solar radiation : - Solar cell
                   - precision $\pm$ 5%
                   - measuring range : 0 - 1 000 W/m²
                   - reference : in laboratory

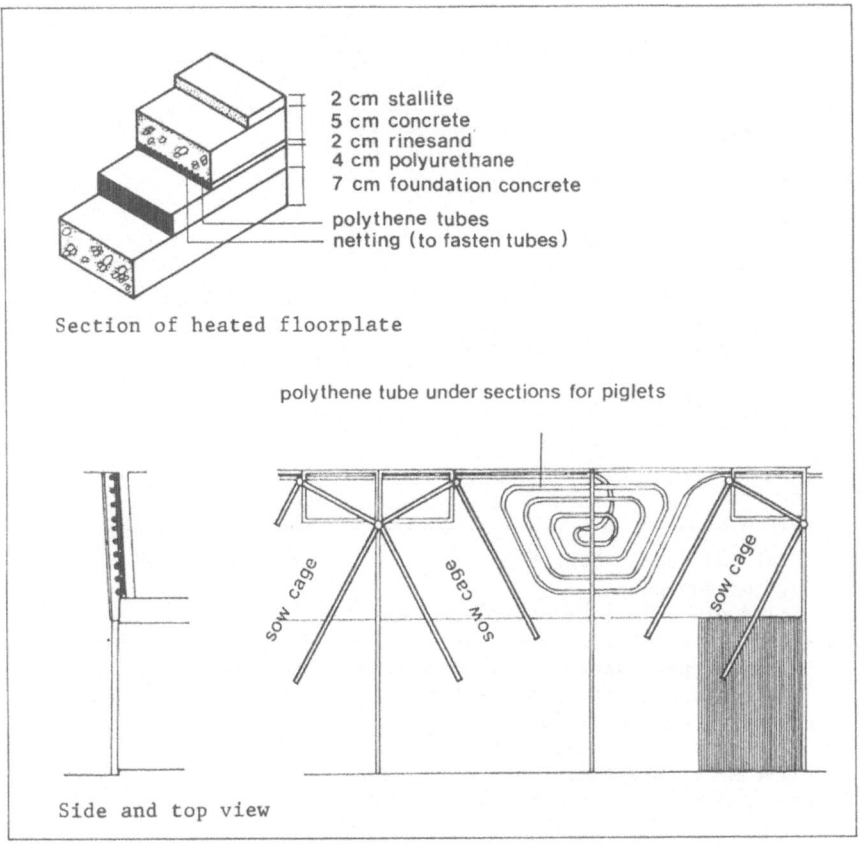

2 cm stallite
5 cm concrete
2 cm rinesand
4 cm polyurethane
7 cm foundation concrete
— polythene tubes
— netting (to fasten tubes)

Section of heated floorplate

polythene tube under sections for piglets

sow cage

sow cage

sow cage

Side and top view

Fig. 2.5.1.2 : Scheme of floorheating system.

- Wind : - contact-anemometer : Fluess Berlin - Steglitz
    nr. 92 b
       - Chronograph : Fluess Berlin - Steglitz E 9095/
                                                  E 2669
- Humidity : - Capacity humidity sensor Philips (Valvo)H1
         - Precision : 5% using a temp. dependent
           electronic circuit.

	Measuring frequency	Storage	Units
Temperatures	Every 2 min	tape	C
Liquid flow rate	Reading at beginning and end of period	manual	litre
Air flow rate	Every 6 sec	average of 20 measurements on tape = every 2 min.	m^3/s
Radiation	Every 6 sec	average of 20 measurements on tape	W/m^2
Wind	Reading of average every hour	manual (reading from paper)	m/s
Humidity	Every 2 min	on tape	-

Humidity and air flow rate have been measured only during a short period, in order to calculate the heat balance and energy demand.

4. Economic aspects

1982 prices.

4.1. Economic environment

1. Energy price of oil	0.31 ECU/1
2. Energy price of electricity	0.074 ECU/kWh
3. Energy prices evolution prospects *(1)*	
– of oil	2%/yr
– of electricity	2%/yr

(1) inflation free

4.2. Investment expenses *(1)*

Solar only

1. Total investment budget	2225 ECU
2. Cost of equipment :	
– cost of collectors per m^2	78 ECU
– cost of regulation equipment	378 ECU
3. – Time of personnel for the conception, installation and start-up of equipment	30 hr
– Cost of personnel for the conception, installation and start-up of equipment	445 ECU
4. Lifetime of the equipment (evaluation)	15 years
5. Residual value of the investment	0 ECU

(1) In this and the next paragraphs all data are given for an assumed collector area of 17.82 m².

4.3. Operating expenses *(1)*

Solar only

1. Energy cost	22 ECU/yr
2. Maintenance cost (replacement of material or components)	
– spare parts	
– salaries (internal-external)	111 ECU/yr
4. Other manpower costs	45 ECU/yr

(1) Cf. footnote under section 4.2.

4.4. Hidden cost and benefits *(1)*

Solar only

1. Production obtained : energy : hot water	21425 MJ/yr
2. Average sales price per unit produced	0.012 ECU/MJ
3. Total income obtained (4.4.1. x 4.4.2.)	263 ECU/yr

Note : 3 diagrams, showing the calculated present worth of the optimized solar installation for different situations, can be found in section 5.1.

(1) cf. footnote under section 4.2.

5. Results

5.1. Method of calculation

The measurements were made for an installation with 3.96 m^2 of collector area and a heat storage tank of 100 or 300 1. The data were recorded using a data-acquisition system :

+ Data-logger and data-storage :
  - Mess & System Technik MDL 500
    50 channels max.
  - Data storage on small cassette :
    . media : ECMA-34/ANSI compatible Philips Cassette, 2 tracks
    . format : Phase Encoded, 800 bits/inch, single track recording
    . code : ECMA-34
    . tape speed : 15 inch/s

+ Data treatment : (cf. figure 5.1.1)
  - The data on the small cassette were rewritten on a DC 300 A cassette by a Tektronix computer. The program examined the data to determine certain errors.
  - The development of the simulation model was done on the Tektronix computer.
  - Calculations based on the reference year were made at the computer centre of the university.

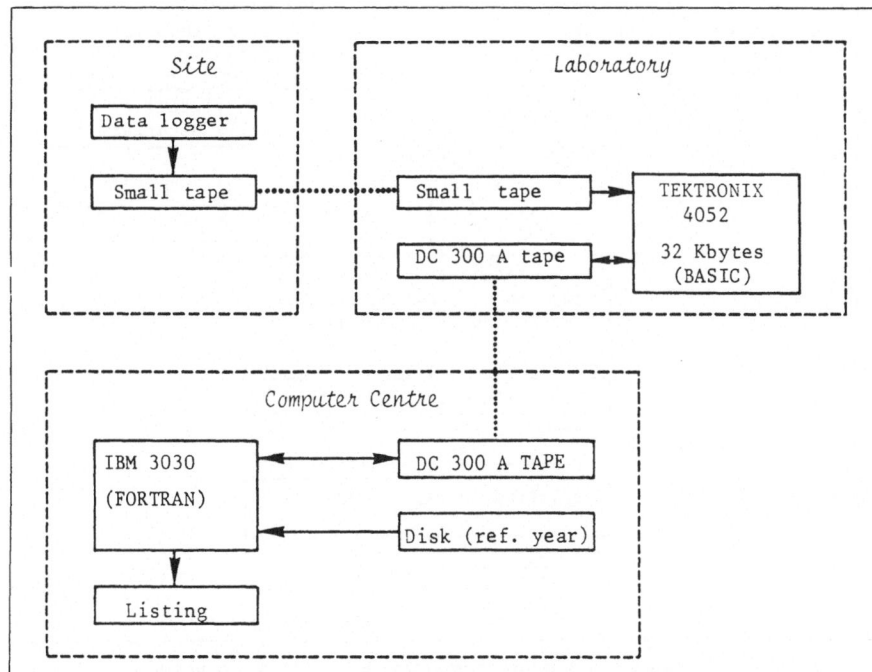

Fig. 5.1.1 : Computer configuration.

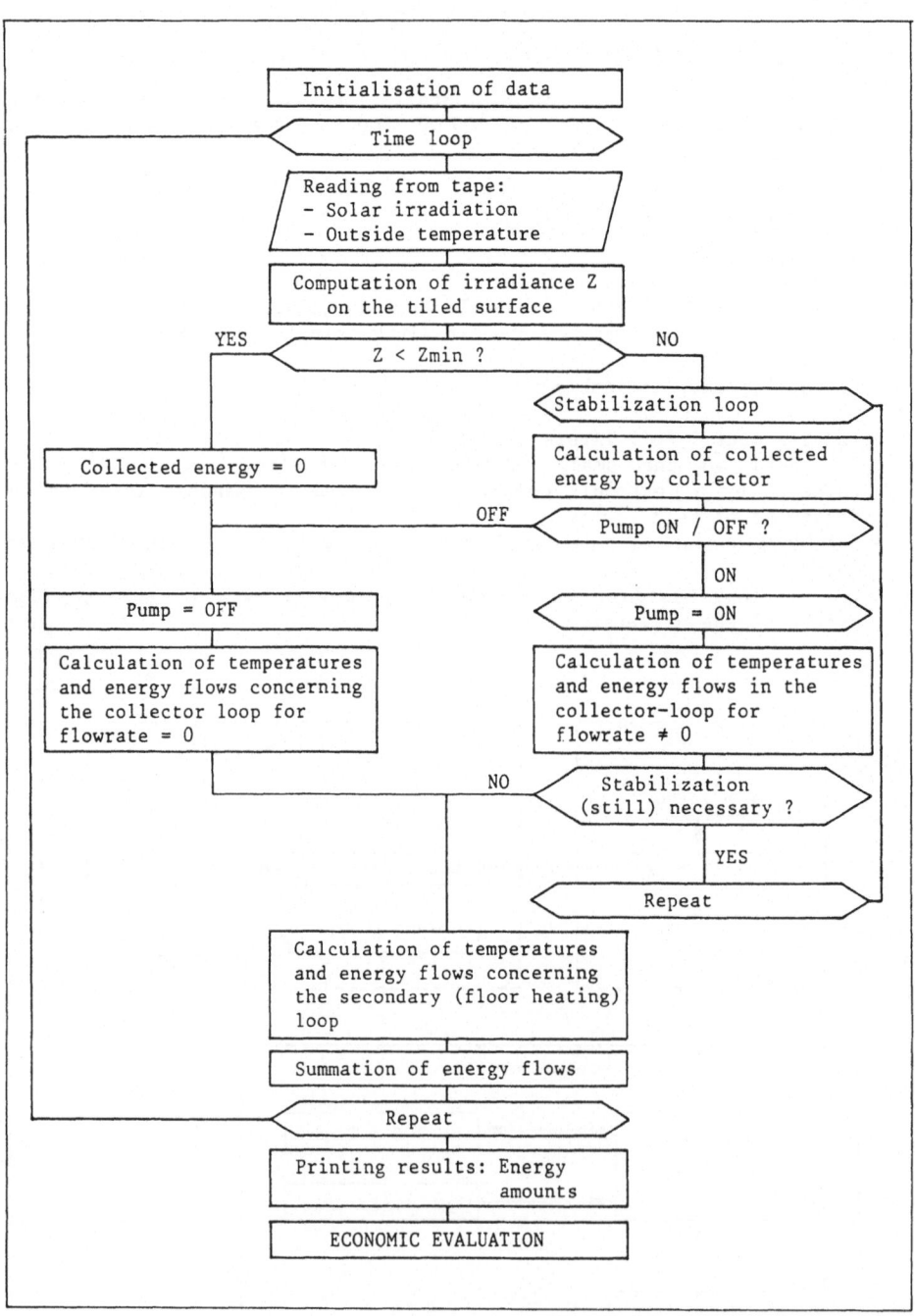

Fig. 5.1.2. : Stream diagram of computer simulation program.

+ A mathematical model was developed to determine the influence of certain
  parameters, such as e.g. collector area, storage, ..., (figure 5.1.2).

To be able to investigate different kinds of flat plate collectors, a
rather general expression for the energy losses has been used. This
implied the necessity to assume a quasi-static situation.   The
storage tank is described by a one node model.
To calculate the energy which can be saved, the outside weather
conditions from a reference year are read from a tape. Using the
irradiance data on a horizontal plane, the radiation energy on the
tilted collector surface was calculated.

If the radiation intensity is too low, no energy can be collected and
the pump will be off. Otherwise, the pump can be on or off,
depending upon the different temperatures and the characteristics of
the control unit.
Because of the static description of the collector, it is possible
that the results of the calculation should be corrected in a
stabilization loop. This occurs e.g. when the pump starts working.
On the basis of the calculated energy savings, an economic evaluation
is possible.

Figure 5.1.3 illustrates the simulated variation of installation
efficiency with collector surface and storage volume. A maximum
value ($\pm$ 34 %) is obtained for 10 m^2 collectors and 100 1 storage
capacity.

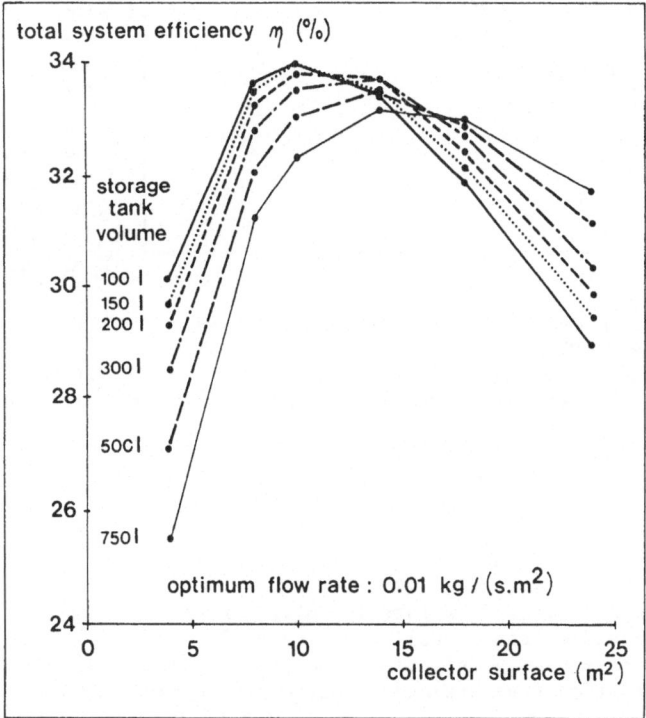

Fig. 5.1.3. : Installation efficiency as a function of storage tank
              capacity and collector surface

+ To find the economic optimum, the Present Worth (for definition, cf. p. 428) has been calculated for different collector surfaces and storage volumes. A 17.82 m² collector, together with a storage tank of 100 1 seems to be the most promising combination. In that case, the payback period is about 10 years (cf. figures 5.1.4 and 5.1.5). Also the influence of other variables can be investigated. E.g. the inflation free increase of the fuel price (0-2-4-6%). This results in a payback time between 8-11 years.

Finally, the price of the collector is very important. The results for a collector price of 2 000/2 500/3 000/3 500/4 000 Bfr./m² are given in figure 5.1.6 (1 000 Bfr = 22.3 ECU).

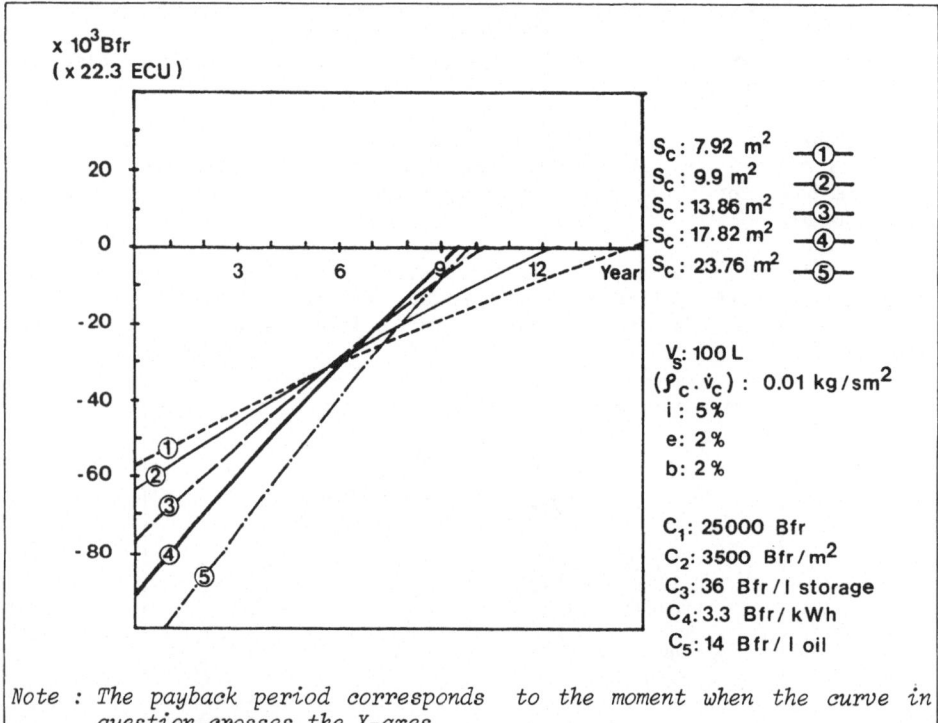

Note : *The payback period corresponds to the moment when the curve in question crosses the X-axes.*

Fig. 5.1.4 : Present worth as a function of time of different collector surfaces

The key for the 3 diagrams on present worth of the installation is given below.

S_c = collector area (m²)
V_s = volume of storage tank (1)
$(\rho_c \dot{v}_c)$ = specific mass flow rate in collector circuit (kg/(s.m² collector))
i = inflation free actualisation rate (%)
e = average yearly inflation free price increase of electricity (%)
b = inflation free average yearly price increase of the fuel (%)

The C factors are cost factors (in Bfr.)

C_1 = a fixed investment cost for any installation dimension
C_2 = the cost per m^2 collector surface
C_3 = additional costs (exceeding a fixed amount for any storage volume, already included in C_1) per litre of storage

with Total investment cost $C_I = C_1 + C_2 S_c + C_3 V_s$

C_4 = price per kWh of electricity
C_5 = price per l of oil

Maintenance costs (C_M) are given as a function of collector surface and investment cost

$$C_M = 500 + 50 \times S_c + 0.005 \times C_I$$

To convert to ECU, multiply the C values by 0.0223

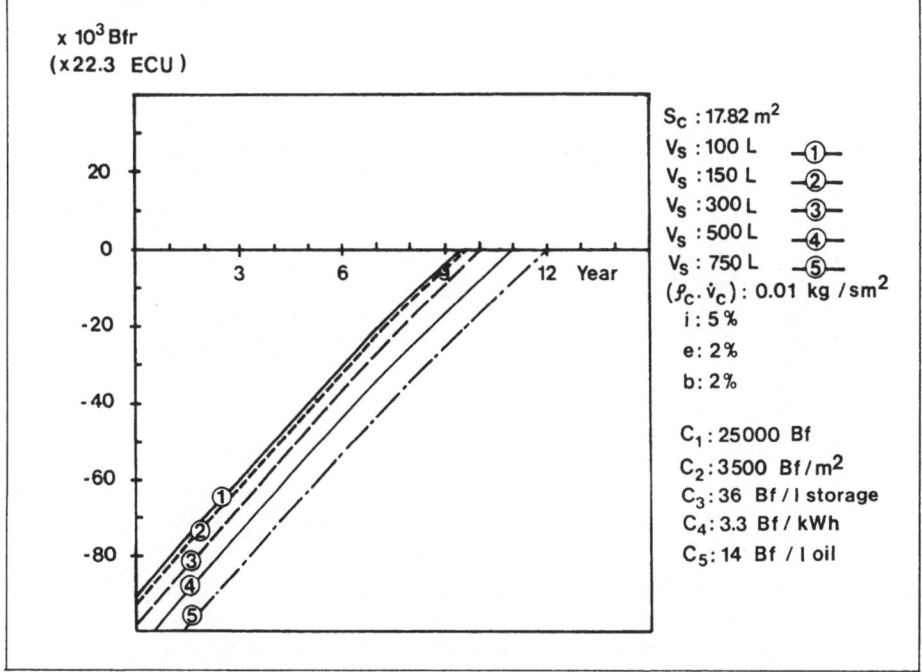

Fig. 5.1.5 : Present worth as a function of time for different storage volumes.

Note : Care obviously needs to be taken that, in case $T_{storage}$ exceeds 98°C, the liquid in the collector circuit that is being switched off does not start boiling.

The instantaneous collector efficiency curve given in section 2.3 would predict a maximum collector temperature of about 160°C in extreme situations :

$Ta_{max} = 35°C$; $I_{cmax} = 1\ 000$ W/m^2 collector

(The instantaneous efficiency curve yields $\eta_c = 0$ for $T^* = 1.26$

Solving the equation $1.26 = \dfrac{Uo\ (T_c^c - Ta_{max})}{Ic_{max}}$

with the above values yields $T_c = 160°C$)

The mixture of 50% PKL – 100 with 50% water used in the experiment should be changed to 90% PKL – 100 with 10% water in order to safely withstand the high collector temperature of 160°C without starting to boil.
A collector circuit pump should likewise be chosen that can withstand this temperature.

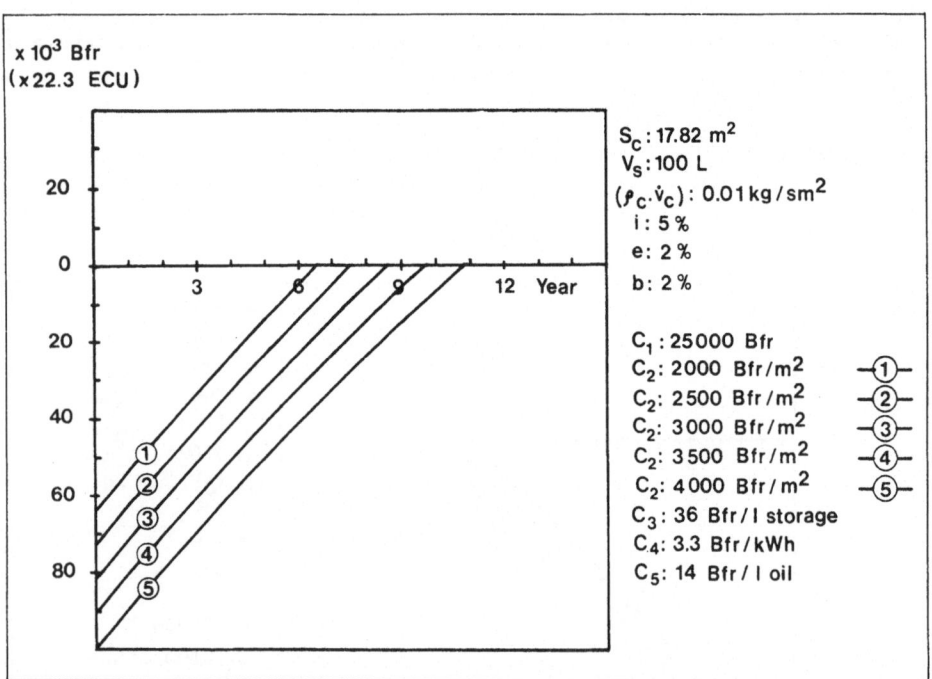

Fig. 5.1.6. : Present worth as a function of time for different collector prices

In order to avoid having to stop the collector at periods of high insolation, a storage tank of a least 300 l can be chosen instead.
(The storage temperatures calculated over the reference year never exceeded the boiling point of water for a 300 l storage).

Note :
The PRESENT WORTH at a given time is the sum over the period from operation start to the time in question of all net cash flow figures, discounted at a specific rate to the starting date.
At the starting point of operation, the present worth equals the negative value of the investment cost.
By the time which corresponds to the payback period, the present worth equals zero.

5.2. Weather conditions and energy (simulation results)

1.	Jan	Feb	Mar	Apr	May	Jun	Jul	Aug	Sep	Oct	Nov	Dec		PERIOD	
2.	31	28	31	30	31	30	31	31	30	31	30	31	days/month	reference year	
3.	3.8	3.2	6.6	9.1	11.9	16.7	16.1	17.1	16.1	11.0	6.1	3.1	°C	average ambient TEMP	
4.	3.8	5.1	4.4	4.7	3.4	3.3	3.2	3.5	3.6	3.6	3.8	3.6	m/s	average WIND	
5.	1905	3274	5313	6442	8089	9398	7849	7823	7593	3977	1993	1377	MJ	average SOLAR RAD on collector	
6.	197	547	1350	2122	3079	4441	3225	3376	3292	1090	265	90	MJ	input SOLAR energy	solar (17.82 m²) (100 l)
7.	188	525	1310	2072	3015	4378	3170	3322	3235	1061	254	85	MJ	input energy STORAGE	
8.	147	493	1276	2043	2988	4355	3145	3294	3205	1030	223	57	MJ	output energy STORAGE	
9.	40.9	44.5	44.1	41.8	39.8	38.7	38.3	39.8	42.3	42.3	42.0	41.3	°C	average TEMP storage	
10.	147	493	1278	2019	2867	3970	3004	3169	3168	1030	223	57	MJ	USEFUL energy	
11.	27.5	23.2	22.7	18.6	16.1	12.8	14.4	15.5	17.2	22.8	25.2	27.6	x 10^3 MJ	AUXILIARY HEATING energy (non-solar)	
12.						(1)							MJ	energy for OPERATION	
13.						(2)							MJ	FUEL consumption	
14.	3.4	7.6	16.4	24.7	37.1	43.5	39.2	37.5	33.1	13.6	4.4	1.7	MJ	ELECTRICITY consumption(3)	
15.	27.6	23.7	24.0	20.6	18.8	16.8	17.4	18.7	20.4	23.8	25.4	27.6	x 10^3 MJ	HEATING energy	reference
16.						(4)							MJ	energy for OPERATION	
17.						(5)							MJ	FUEL consumption	
18.						(6)							MJ	ELECTRICITY consumption	

Note : A diagram of the variation of installation efficiency as a
       function of storage tank capacity and collector surface, is
       given in section 5.1.

(1) Energy for operation : the energy for operation in the solar
    system is the same as in the reference solution, only some extra
    energy for the pump in the primary collector loop is needed. (See
    (3))
(2) Fuel consumption : the fuel consumption is proportional to 5.2.11
    but dependent on the efficiency of the heating system.
(3) Electricity consumption : in this column only the extra energy
    which has been used for the pump in the primary collector loop has
    been mentioned.
(4) Energy for operation : has not been measured.
(5) Fuel consumption : is proportional to 5.2.15 (heating energy) but
    dependent on the efficiency of the normal heating system.
(6) Electricity consumption : has not been measured.

5.6. (Pre-)heated water

1.	Jan	Feb	Mar	Apr	May	Jun	Jul	Aug	Sep	Oct	Nov	Dec		from : PERIOD
2.														till
3.	49.7	48.2	45.7	42.6	39.9	38.3	38.2	39.8	42.4	45.5	48.2	49.7	°C	average input water temp
4.	0.48	0.48	0.48	0.48	0.48	0.48	0.48	0.48	0.48	0.48	0.48	0.48	x10^{-3}m³/s	average flow rate
5.	49.7	48.3	45.9	43.0	41.4	41.4	38.8	40.4	43.0	45.7	48.3	49.7	°C	average (pre-)heated water temperature
6.	54.8	53.1	50.2	46.5	43.4	41.5	41.4	43.3	46.3	49.9	53.1	54.8	°C	required output temp

5.7. Data

In figure 5.7 an example of the main measurements is given for April 17 th, 1981.

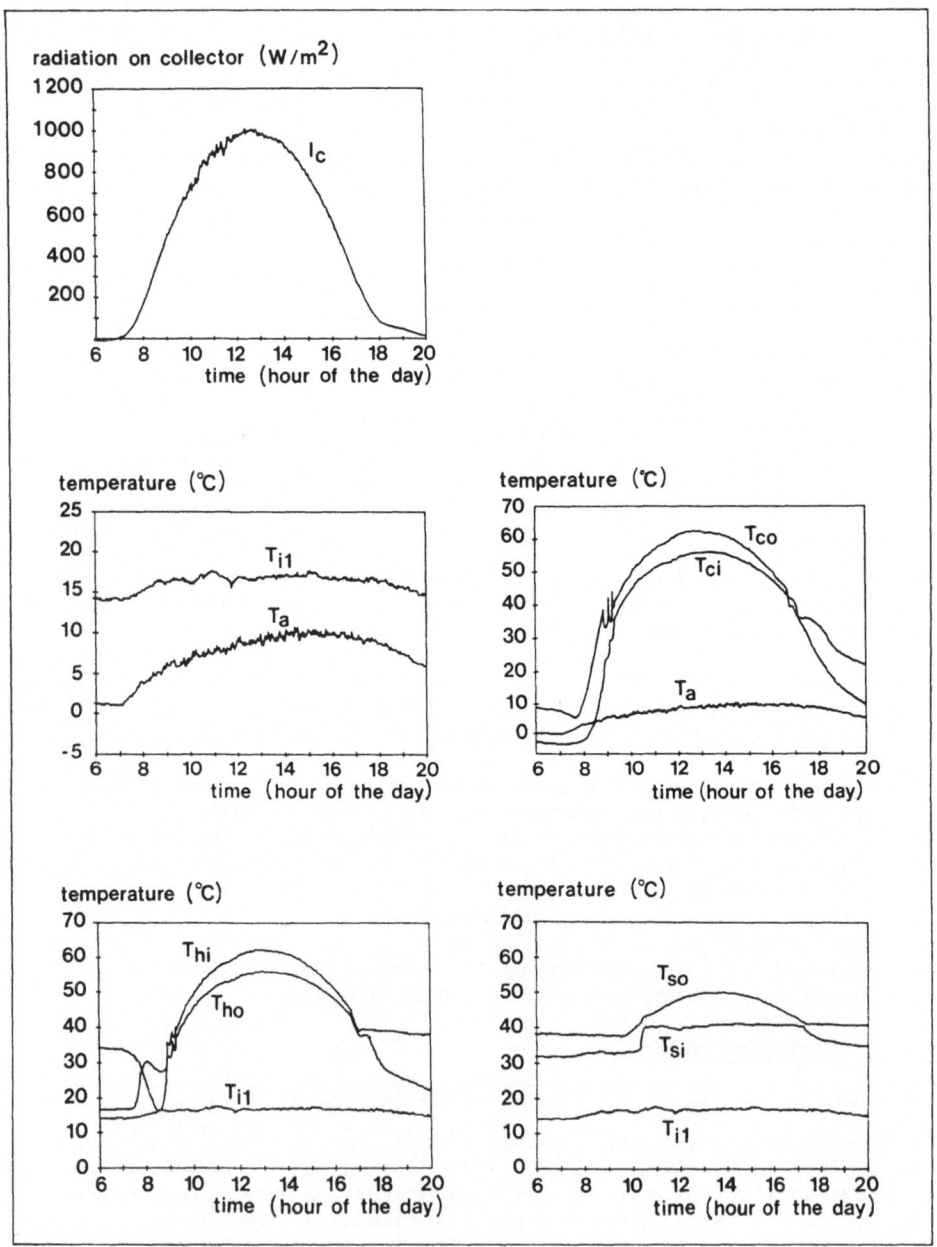

Figure 5.7 : Examples of measured data.

Note : for explanation of the symbols, cf. "scheme"-section, p. 415.

4. Solar domestic hot water production and solar heating of farrowing house

Project location

BUCHHOFEN, BAYERN, WEST GERMANY

Study Center :
- Project leader : Dr. H. SCHULZ; K. MEUREN
- Institution :   Bayerische Landesanstalt fur Agrar Technik
                   Vöttinger Strasse 36
                   D-8050 Freising-Weihenstephan
                   West-Germany

Project monitored since 1983/Set-up partly exists since 1981.

Project description

On the experimental farm in Buchhofen, West-Germany, 2 self-made collector types were placed on top of the roof of a pig house, in order to help meet the heat demand of the farrowing house heating devices, as well as that of tap-water heating in stable and domestic house.
The 2 collector types were placed in parallel :
- 70 m^2 surface of type I ; a "herring-bone type". (Installed in 1981)
- 54 m^2 surface of type II ; a "serpentine type". (Installed in 1983)
The water heated in the collectors flows into a 6 000 1 storage tank.

If the storage temperature is high enough, water from this tank is circulated to the low temperature heating system of the stable (floorheating, radiators, convectors, connected in series with one another).
If not, an auxiliary (oil) heater is used instead.
Water from the same storage tank can also circulate to a heat exchanger placed in the lower part of a boiler of the domestic dwelling house. This is the case if the storage tank is at a higher temperature (T_{s1}) than the temperature level in this lower part of the boiler (T_{s2}).
(T_{s2} can be far below the set-point value of the electric heater in the boiler due to the phenomenon of temperature stratification, promoted as it is by the positioning of the cold water inlet in the lower, and of the electric heater and hot water outlet in the upper part of the boiler).
Finally, stable tapwater is heated by running it through a heat-exchanger in the storage tank. Here no auxiliary heating is provided.

Goals of the experiment :
- To gain experience with the system.
- To build and validate a simulation model, which will help design systems with optimal component parameters, as well as optimal operating modes.

Scheme (1)

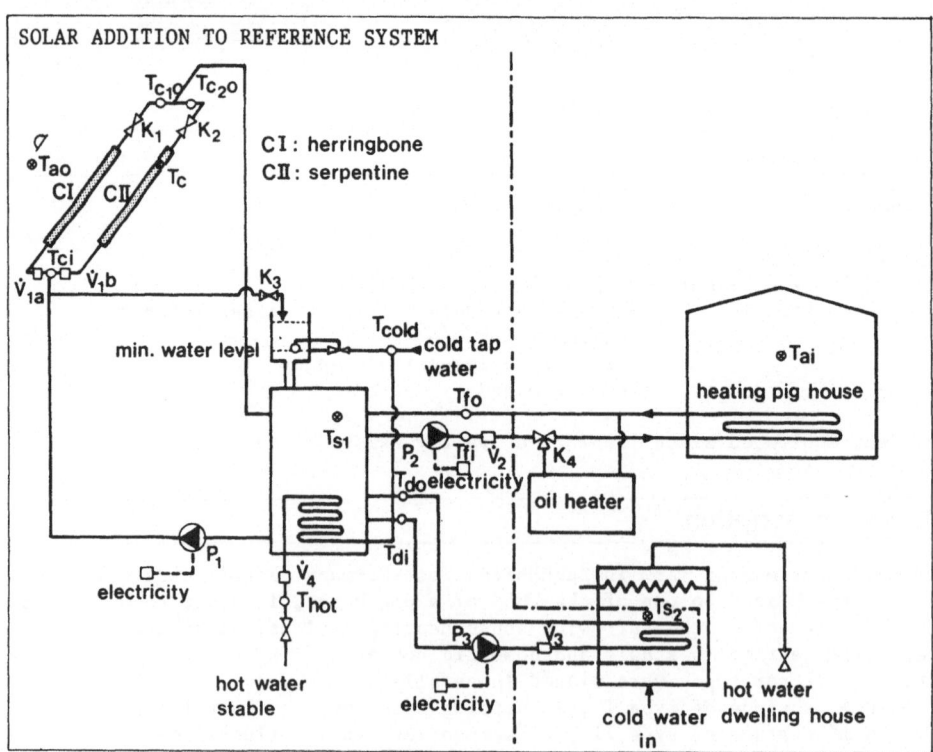

(1) An explanation of the symbols used in the scheme can be found on pp. 470-471.

Operating modes

- if $T_c > T_{s1} + \Delta T \rightarrow P_1$ on.
  $T_c < T_{s1} + \Delta T \rightarrow P_1$ off.
- Anti-frost protection :

  if $T_{ao} < 3°C \rightarrow K_1$, K_2, K_3 are positioned so that the water content
  drains into the storage tank.

  if later $T_{ao} > 3°C$ and $T_c > T_{s1} + \Delta T \rightarrow P_1$ fills the collectors again.
  K_1 and K_2 purge the air from
  the system.

- A protection against collector overheating is also provided.
- Heating of the stable :

  if $T_{s1} > 33°C \rightarrow K_4$ in position ⟶⫟-
  . if $T_{ai} < 25°C \rightarrow P_2$ on (= solar heating of stable).
  if $T_{ai} > 25°C \rightarrow P_2$ off.
  if $T_{s1} < 33°C \rightarrow K_4$ in position ⫟-
  . if $T_{ai} < 25°C \rightarrow$ auxiliary heater on.
  if $T_{ai} > 25°C \rightarrow$ auxiliary heater off.

- Heating of the domectic tapwater :

  if $T_{s1} > T_{s2} + 4K \rightarrow P_3$ on (= solar heating of domestic tapwater)
  $T_{s1} < T_{s2} + 4K \rightarrow P_3$ off

  (At any time the temperature in the upper part of the boiler is kept
  at a minimum level by means of an electric heater)

Conclusions

- Measurements over the period of September 1983 to August 1984 yielded
  an annual average collector efficiency of 16.2 %.
  Of this collected energy 18 % was lost from the storage tank; 32 %
  was used for stable heating; the remaining 50 % was routed to the
  domestic tapwater boiler.
  The energy demand for hot tapwater in the stable was negligible.
  The electrical energy input into the solar system corresponded with
  3 % of the net energy output from the solar storage tank.
- The energy demand was nearly fully covered over the months of August
  to Oktober.
  During the winter months (December to February) the solar
  contribution was however practically nil.
- Acquired experience :
  - Problems with thermosiphonic flow were encountered due to lack of
    a non-return valve.
  - Where the covering of collector 1 makes contact with the absorber
    material, light transmissivity decreases. This will eventually
    necessitate replacement of the cover by sheets of polycarbonate
    instead of PVC.

- The amount of heat lost in the long pipes between solar storage and domestic boiler was about as great as what could be delivered to the boiler, (cf. importance of adequate insulation and compactness of installation).
- Further experimentation is planned with other applications : preheating of inlet air into the stable ; heating of food for a liquid feeding system.

PROJECT ANALYSIS

1. Site and climate

1.1. Site

Latitude :	48°45' N
Longitude :	12°50' E
Altitude :	330 m
Nearest main city :	Deggendorf
Distance from main city :	25 km
Direction from main city :	S
Obstructions :	None

1.3. Annual long-term averages

			Project west
1. Prevailing wind direction			
2. Average wind speed	W	m/s	3
	S	m/s	2
4. Absolute humidity	W	g/kg	1–2
	S	g/kg	6–10
5. Global irradiation on horizontal plane		MJ/(m².yr)	(1)
6. Diffuse proportion		%	60
8. Hours sunshine		hrs/yr	1 700–1 800
9. Ambient temperature		C	7.7
10. Average max. temp. (Jul/Aug)		C	17
11. Average min. temp. (Jan)		C	−2.4

Note : W = Winter
*   S = Summer*
(1) According to the map on p. 5 the average annual value would be
*   around 4 200 MJ/(m².yr) (N.O.T.E.)*

12. Meteorological station :	Weihenstephan/Freising
13. Latitude :	48°24' N
14. Longitude :	11°43' E
15. Altitude :	490 m
16. Distance from site :	100 km

2. Plant

2.3. Collector (flat-plate collector)

1. Absorber
   - Coolant :   clear water
   - Material :   Polypropylene
   - Aperture area :   I. 70 m^2
                                II. 54 m^2
   - Absorber area :   I. 66.3 m^2
                                II. 45 m^2
   - Collector fluid content : 6.0 $1/m^2$
   - Working pressure :   40 000 Pa (static pressure)
   - Flow rate :   I. 29 x 10^{-6} $m^3/(sm^2)$
                                II. 18 x 10^{-6} $m^3/(sm^2)$
2. Glazing
   - Emittance of glass : 0.75
   - Number of covers: 2
   - Material : 70 m^2 PVC (Palram[R]), 54 m^2 Polycarbonate both in combination with Polyester foil (Hostaphan[R])
   - Special characteristics : cf. figure 2.3.2.

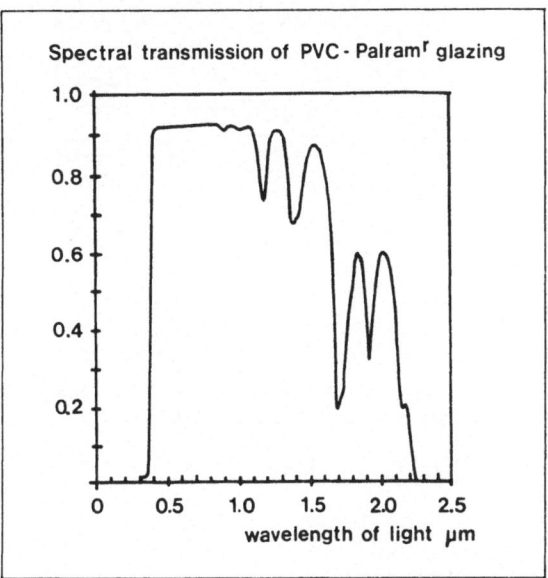

Fig. 2.3.2 : Initial spectral transmission of the glazing of collector I.

3. Insulation : Heraklit 15 mm + 40 mm Styropor
4. Mounting - Orientation : 180°
              - Tilt : 35°

Diagrams of absorbers for large "Rippenrohr" (corrugated tube) collectors (20 - 100 m²). (cf. fig. 2.3.4.1 & 2.3.4.2)

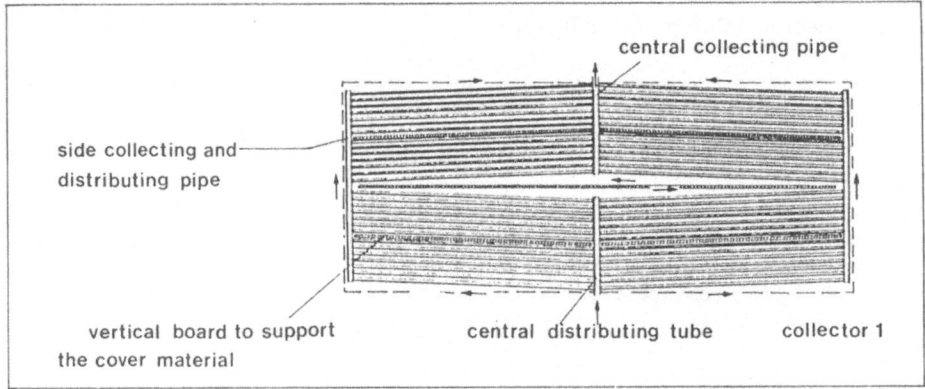

Fig. 2.3.4.1 : Herringbone-type (for installations without heat exchanger and anti-freeze liquid : drain back system).

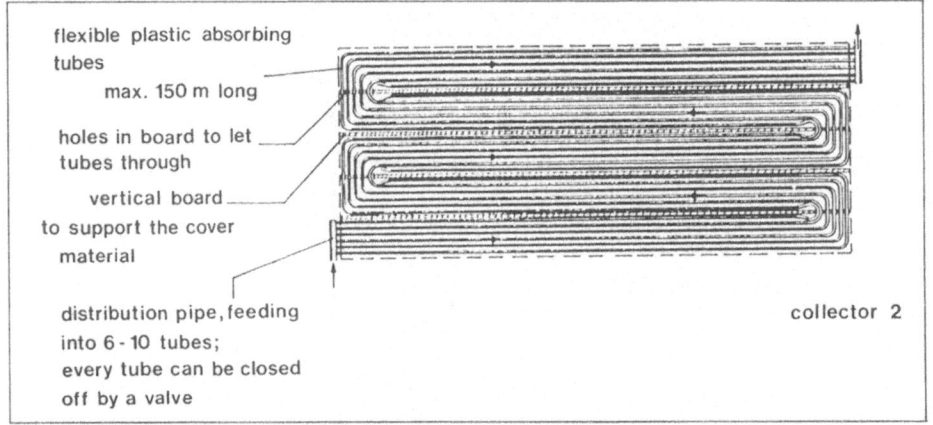

Fig. 2.3.4.2 : Serpentine type (preferably used in installations with heat exchanger and antifreeze liquid).

5. Collector performance (cf. fig. 2.3.5)

2.4. Heat storage

1. Medium of heat storage : clear water
2. Description, location in the system : Pressureless steel tank located in the heating room below the collectors
3. Insulation : 10 cm of rock-wool
4. Dimensions : 6 m³, horizontal
5. Thermal capacity : 25 MJ/K
6. Overall heat loss coefficient : 20 W/K

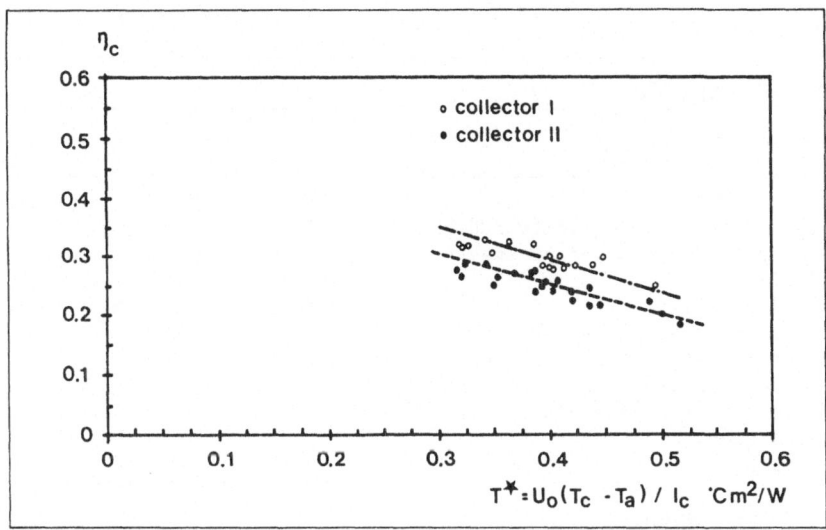

The instantaneous collector efficiency η_c is defined as $\eta_c = \dfrac{Eco}{S_c I_c}$

as a function of T^*

with Eco : the useful power extracted from the collector (W)

I_c : the incident solar radiation on collector (W/m²)

S_c : aperture area (m²)

$T^* = Uo \left(\dfrac{Tc-Ta}{I_c} \right) = Io \left(\dfrac{Tc-Ta}{I_c} \right)$ (decimal) since $Uo = 10$ W/(m² K)

With Tc average temperature of fluid in collector (°C)
Ta ambient temperature around collector (°C)

Fig. 2.3.5. : Instantaneous collector efficiency curve.

Note : Even though collector I is older and covered with a PVC plate of
already reduced transmissivity, its performance is still better
than that of collector II. This is due to the low ratio in
collector II of the absorber surface over the total aperture area.

2.5. Installation parts and measuring devices

1. Technical description of the installation parts :
   pumps : solar circuit (P1 and P2), 600 W, 10 m³/h,
   able to fill the system.
   stable heating circuit (P3) : 3 x 110 W, each 2.8 m³/h
   charging pump for boiler in dwelling house : 110 W,
   3 m³/h
2. Description of the sensors :
   - Temperature sensors : Pt 100, 1/3 Din
   - Mass-flow sensors : water meters with read contacts,
   + 1% precision
   - Solar radiation : Pyranometer CM 11 (Kipp & Zonen).

4. Economic aspects

4.1. Economic environment

1. Energy price of oil	ECU/1	0.31
2. Energy price of electricity ECU/kWh	0.11	
3. Energy prices evolution prospects		
– of oil	%/yr	10
– of electricity	%/yr	10
5. Interest rate of loans for farm equipment (rate without subsidy)	%/yr	8
6. Subsidies available for the solar and competitor farm equipment (as capital grant or interest rate reduction)	%	50

4.2. Investment expenses

1. Total investment budget 54 + 70 m^2 collectors	ECU	10 250
2. Cost of equipment :		
– cost of collectors per m^2	ECU	44
– cost of regulation equipment	ECU	701
– other material costs	ECU	1 985
3. – Time of personnel for the conception, installation and start-up of equipment	hr	$\pm$ 400
– Cost of personnel for the conception, installation and start-up of equipment, estimations	ECU	2 103
4. Lifetime of the equipment (years) (evaluation)		10
5. Residual value of the investment storage + equipment	ECU	2 191

4.3. Operating expenses

1. Energy cost 1 000 kWh el/yr	ECU/yr	110

5. Results

5.1. Method of calculation

1. Measured data : (cf. technical scheme, p. 432)
   - Temperatures : mean values of 15 min (6 000 measurements per 15 min).
   - Mass-flow data : integration of impulses, integration time 15 m^{2}n. 1 and 1.5 litres per impulse.
   - Solar radiation : integration of 15 min.
2. Calculated values, based on the measured data.
   - Average solar radiation on collector (E_{ci} in W) :
   $$E_{ci} = I_c (S_{c1} + S_{c2})$$

with S_{c1} : area of collector I (m^2)
S_{c2} : area of collector II (m^2)
I_c : radiation on collector (W/m^2)

- Input energy in storage :

$$Q_s = c_w \int_{t=0}^{t=15 \text{ min}} \rho_w \dot{V}_{1a} (T_{c1o} - T_{ci}) \, dt + c_w \int_{t=0}^{t=15 \text{ min}} \rho_w \dot{V}_{1b} (T_{c2o} - T_{ci}) \, dt$$

with : c_w = specific heat of water (J/(kg K))
ρ_w = specific mass of water (kg/m^3)
t = time (s)
for other symbols, cf. technical scheme (Temperature differences given in K; flowrate in m^3/s; energy input in kJ).
- Output
  - Stable heating (Q_{no1}) :

$$Q_{no1} = c_w \int_{t=0}^{t=15 \text{ min}} \rho_w \dot{V}_2 (T_{fi} - T_{fo}) \, dt$$

  - Domestic hot water (Q_{no2})

$$Q_{no2} = c_w \int_{t=0}^{t=15 \text{ min}} \rho_w \dot{V}_3 (T_{di} - T_{do}) \, dt$$

  - Stable hot tapwater (Q_{no3})

$$Q_{no3} = c_w \int_{t=0}^{t=15 \text{ min}} \rho_w \dot{V}_4 (T_{hot} - T_{cold}) \, dt$$

  Losses of the storage (Q_L)
$$Q_L = Q_{S_i} - (Q_{no1} + Q_{no2} + Q_{no3})$$

5.2. Weather conditions and energy

The energy flows measured during one year for the studied installation are given in figures 5.2.1 and 5.2.2.

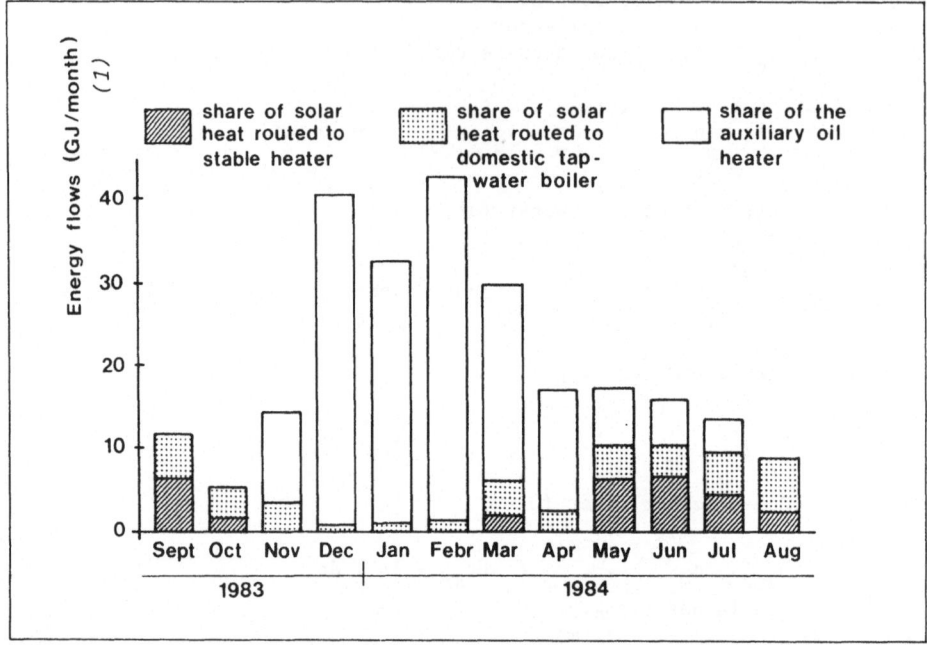

Fig. 5.2.1. : Monthly values of the distribution of some of the energy flows in the studied installation.

(1) electrical energy for domectic water heating not measured.

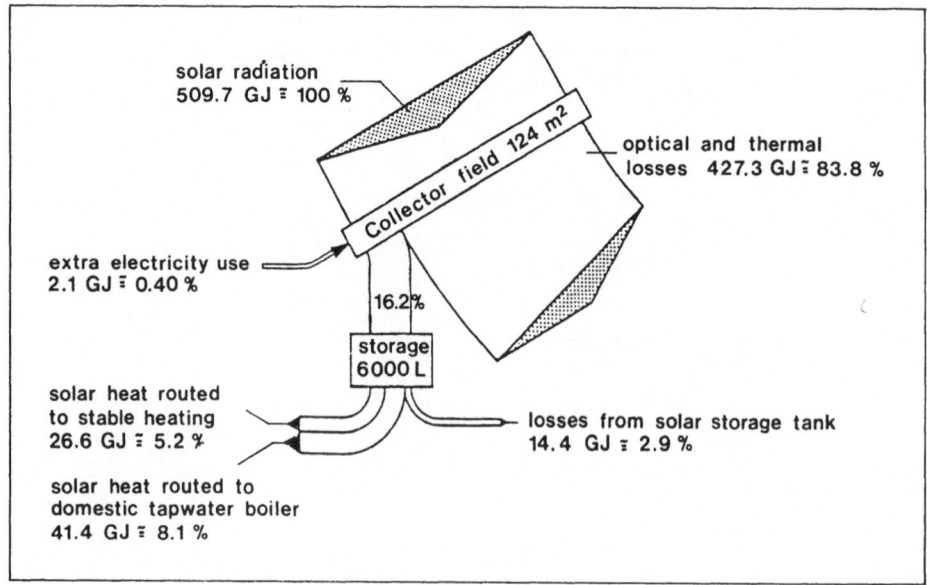

Fig. 5.2.2. : Global energy flow chart for the whole measuring year 1.9.83 - 31.8.84.

5.7. An example of some measured data (cf. fig. 5.7)

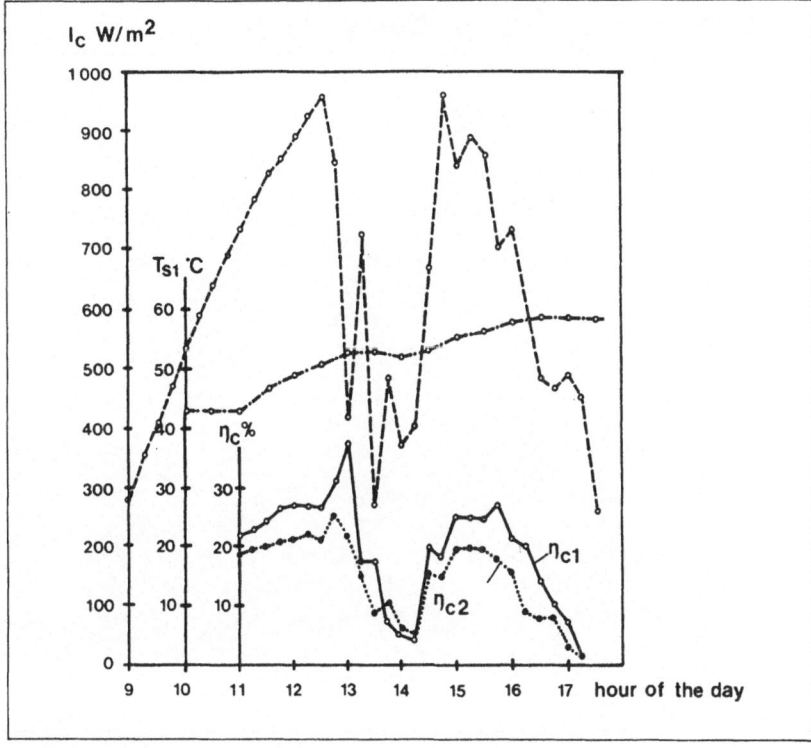

Fig. 5.7 : Most important measured and calculated data for one day
(7/9/1983)

5. The use of solar energy for hot water production for a dairy farm

Project location

DUIVEN, NETHERLANDS

Study Center :
- Project Leader : J.M. Lange
- Institution :   Institute for Mechanization, Labour and
                  Buildings. (IMAG)
                  Mansholtlaan 10-12
                  6708 PA Wageningen
                  The Netherlands
Running since 1977

Project description

On the IMAG experimental farm in Duiven, 12 m² of collectors were installed to heat up tap water in the farm.

- The collector has an absorber coated with a spectral selective layer.
- The anti-freeze fluid that flows through the collectors transfers its heat to the water through a heat exchanger in a storage tank of 1 m³.
- The water is put to use twice during the day and the demand totals 250 litres per day at 70°C.

Goals of the experiment :

- To measure the energy that can be extracted using the collectors.
- With these data to estimate the economic feasibility of such a set-up.
- To stimulate the integration of active solar systems in agriculture.

Scheme *(1)*

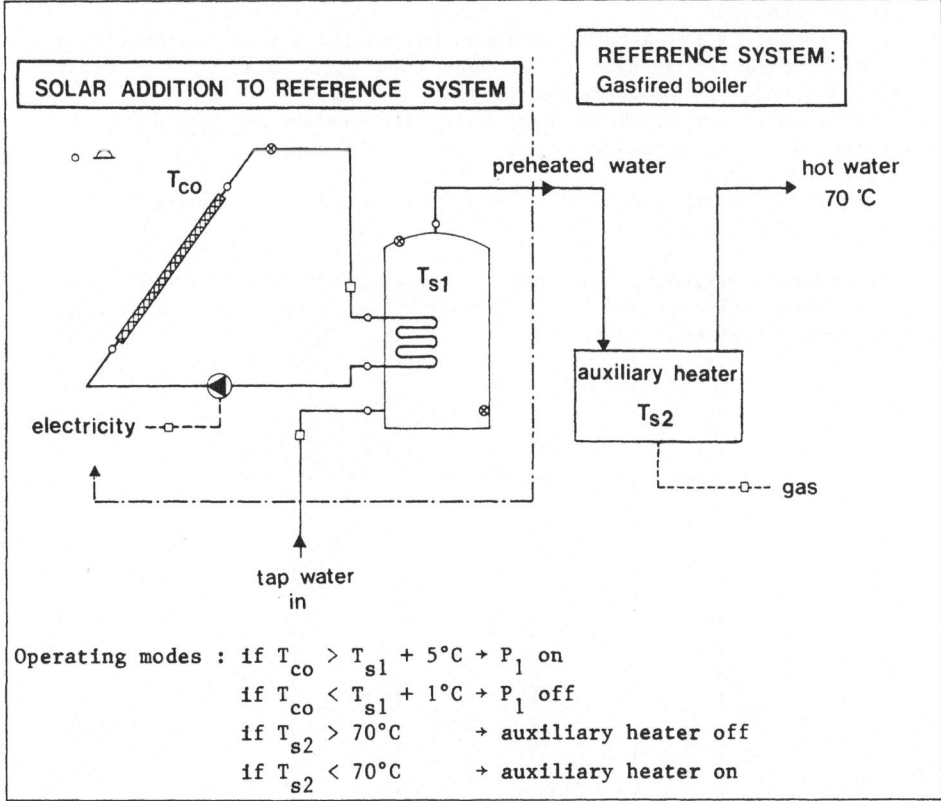

Operating modes : if $T_{co} > T_{s1} + 5°C \to P_1$ on

if $T_{co} < T_{s1} + 1°C \to P_1$ off

if $T_{s2} > 70°C \qquad \to$ auxiliary heater off

if $T_{s2} < 70°C \qquad \to$ auxiliary heater on

Conclusions

- The application is not only attractive but also economically feasible. *(2)*
- Experience acquired :
  + Even in the case of diffuse solar radiation a significant amount of energy could be secured, due to the use of an absorber coated with a spectral selective layer.

(1) An explanation of the symbols used in the scheme can be found on pp. 470-471.
(2) For an evaluation of this statement, cf. section 6.

+ Compactness of the system (i.e. distances between collector, storage tank and auxiliary heater) and insulation of storage were unsatisfactory, which caused a significant loss of energy. (cf. fig. c on page 444)
(Distance collector-storage tank was 30 m.)
+ One should be aware of the potential of intoxication by the antifreeze liquid.
- The conclusions are based on an experiment running between June 1977 and June 1978. The values measured corresponded well with the previous computations.
- Projected future research would focus on feasibility of the heat pump as an alternative system. (The heat pump would heat up the water with the energy it recovers during milk cooling.)
It would be interesting to know how these systems compare with one another, and if it would make sense to mix the systems.
- Further references :
  + "Zonneenergie als hulpmiddel bij verwarming van tapwater" J.M. Lange, IMAG Wageningen, Publ. 132, 1980.

  + "Warmwatervoorziening op het melkveebedrijf door middel van zonnewarmtekollektoren" G. Brouwer (Van Heugten, Consulting engineers) Nijmegen 1976.

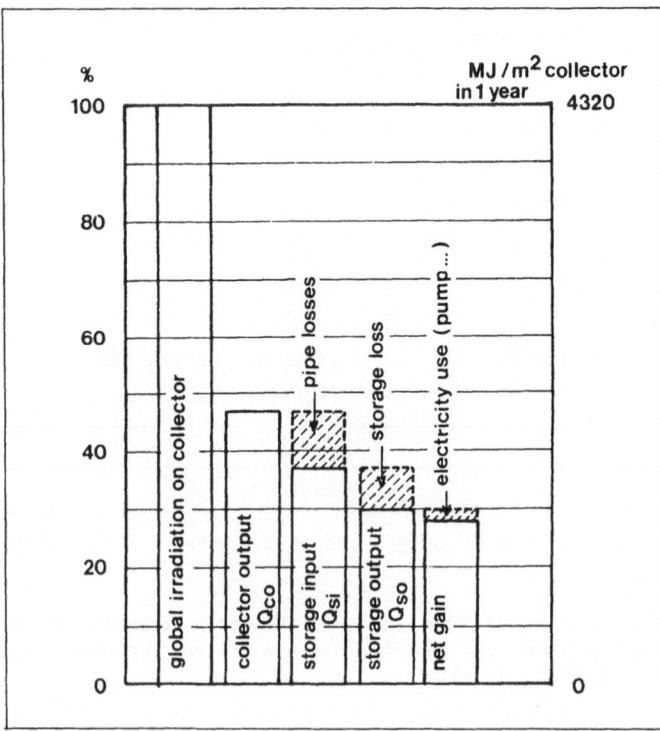

Fig. c : Energy performance throughout the whole solar system (average over 1 year).

PROJECT ANALYSIS

1. Site and climate

1.1. Site

Latitude :	52°20' N
Longitude :	6° E
Altitude :	< 20 m
Nearest main city :	Arnhem
Distance from main city :	10 km
Direction from main city :	East
Obstructions :	none

1.2. Site micro-climate

The solar collectors are situated on an inclined roof.

1.3. Annual long-term averages

3. Total precipitation	(mm/yr)	750
5. Global irradiation on horizontal plane	(MJ/(m²yr)	3 400
6. Diffuse proportion	(%)	55

2. Plant

2.3. Collector : (Flat plate collector)

1. Absorber
   - Coolant :   anti freeze
   - Material :   aluminium
   - Aperture area :   12.8 m²
   - Absorber area :   12 m²
   - Collector fluid content :   0.7 1/m²
   - Flow rate :   16.7 x 10^{-6} m³/(sm²)
   - Emittance of plate :   0.1
   - Coef. of absorption of visible light :   0.9
2. Glazing
   - Emittance of glass :   0.9
   - Number of glass covers :   1
   - Material :   glass
   - Special characteristics :   4 mm
3. Insulation   5 cm rockwool
4. Mounting
   - Orientation :   200° (South = 180)
   - Tilt :   52° (horizontal = 0)

5. Collector performance

$$\eta_c = 0.7 - 5.83 \frac{T_c - T_a}{I_c}$$

with I_c = collector temp. (°C)
T_a = ambient temp. (°C)
I_c = solar irradiation on collector (W/m²)

2.4. Heat storage

1. Medium of heat storage :
   Water

2. Description
   - heat exchanger in double wall of the storage tank
   - at a distance of 30 m, situated near the heater

3. Insulation
   40 mm Rockwool, 15 mm wood

4. Dimensions
   1.30 x 1.0 m²

5. Thermal capacity
   4.2 MJ/K

6. Overall heat loss coefficient
   6.0 W/K

4. Economic aspects

4.1. Economic environment (prices 1982, incl. VAT)

1. Energy price of gas
   0.187 ECU/m³

2. Energy price of electricity
   day :   0.082 ECU/kWh
   night : 0.067 ECU/kWh

3. Energy prices evolution prospects
   - of gas   + 10 %/yr
   - of electricity + 10 %/yr

5. Interest rate of loans for farm equipment.
                (rate without subsidy)
   9 %/yr

6. Subsidies available for the solar and competitor farm equipment
   (as capital grant or interest rate reduction) :
   none

4.2. Investment expenses

(for solar part only, since it is an addition to
the existing system)

1976 excl. V.A.T.

1. Total investment
 1 880 ECU
 2. Cost of equipment
   - Cost of collectors per m²
   - Expenses for storage
   - Cost of regulation equipment
   - Other material costs
	
	51 ECU
	1 082 ECU
	68 ECU
	118 ECU

4.3. Operating expenses

	Solar only
1. Energy cost	18.7 ECU/yr.

5. Results

5.2. Weather conditions and energy

1.	1/1	1/2	1/3	1/4	1/5	1/6	1/7	1/8	1/9	1/10	1/11	1/12		from: PERIOD
2.	31/1	28/2	31/3	30/4	31/5	30/6	31/7	31/8	30/9	31/10	30/11	31/12		till: reference year
3.	3.0	1.0	5.0	7.0	12.0	14.5	16.8	16.0	13.0	11.0	6.5	4.5	°C	average ambient TEMPERATURE
5.	1469	2549	3197	6826	6394	5486	6566	5530	4838	3586	1901	1426	MJ	average SOLAR RAD on collector
6.	600	1080	1680	3024	3720	3324	3240	2292	1896	1344	816	600	MJ	input SOLAR energy
7.									total per year		18313		MJ	input energy STORAGE
10.									total per year		74764		MJ	USEFUL energy
11.										total	10977		MJ	AUXILIARY HEATING (non-solar)
12.										total	820		MJ	energy for OPERATION

The reference solution would use about the sum of 5.2.10 + 5.2.11 as
gas consumption, and a lower electricity consumption, since in that
case pumps and other electricity consuming parts of the added solar
system (adding up to 820 MJ) are non-existent.

6. Remarks

Some rough calculations tend to make us suspicious of the conclusion that
this application is "economically feasible". If the energy saved over the
test year (about 14 700 MJ) were provided by natural gas (at 0.187 ECU/m³
and given the average value of 30 MJ/m³) in a traditional heater with η =
0.7, this would cost about 130 ECU. This is a very small figure compared
to the investment expenses (1 878 ECU) for the solar system, and would
suggest a long payback period. (Optimization of the system parameters
might however brighten the picture).

6. Appraisal of a solar water heater for Bradmores farm dairy at Seale - Hayne College

Project location

NEWTON ABBOT, UNITED KINGDOM

Study Center :

  - Project Leader : J.L. CARPENTER
  - Instituttion : Seale-Hayne College
                    Newton Abbot
                    Devon, TQ 12 6NQ
                    UNITED KINGDOM

- Supporting organisations
    . Agricultural Research Council
    . Dept. of Energy (E.T.S.U.)
    . Electricity Council (Farm Electric Centre)
    . Ministry of Agriculture, Fisheries and Food (A.D.A.S.)
    . National Institute for Research in Dairying
    . Science and Engineering Research Council
    . South Western Electricity Board.

Running since 1980.

Project description

On the dairy farm of Seale-Hayne College in Devon, a solar collector was installed to provide hot water for plant cleaning and cow hygiene purposes.

- The used collector surface can be adjusted in steps of 2 m² up to a total of 10 m².
- 3 different storage tanks can be put to use alternatively. Their capacities are 190 l, 280 l and 360 l respectively.
- The heat from the collector is transferred to the tap water in the storage tank through a heat exchanger.
- The water demand is simulated. (tested draw-off quantities vary from 146 - 243 l/day) of which 60 l are requested at 82°C, and the rest at 40°C.
- The dairy parlour caters for a herd consisting of up to 100 cows.

Scheme (1)

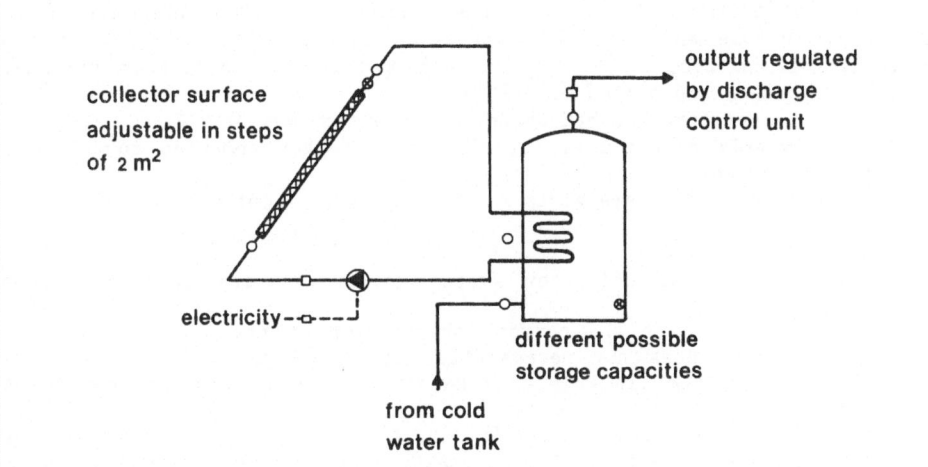

Simulated draw-off quantities were :

AM	PM
97 l	47 l
46 l	97 l
81 l	122 l

In practice, the solar system could be a supplement to the reference system between the cold water supply and the traditional water heater.

(1) For an explanation of the symbols used in this scheme, cf. pp. 470-471.

Goals of the experiment :

The experiment is part of an overall program which aims at establishing
- through mathematical modelling and the use of practical surveys - a
system for milking parlours with minimal energy requirements (possibly
using a mix of different energy supply solutions).
This part of the program involves a long term assessment of the solar
system
- to establish the annual amount of energy to be saved
- to assess the variation of the system efficiency with varying
  component capacities and draw-off characteristics.

Conclusions

+ The measurements were made during an exceptionally sunny summer
  (1983), so further monitoring is necessary to achieve more "normal"
  results.
  Even given the exceptional year the economic feasibility is
  disappointing compared to other alternatives, like heat recovery from
  milk cooling, especially if the current low prices for night tariff
  electricity are maintained.
- System efficiency η^*_{system} increased from 9.9-26.8 % (1) roughly as
  storage volume and draw off quantities increased in relation to the
  collector surface. (η^*_{system} = 26.8 % for 6 m^2 collector surface,
  280 l storage, and a load of 243 l)
- Maximum temperature reached by the solar system was 65.9°C.
- Maximum weekly contribution of the solar system amounted to 60 % of
  the heat demand.
- Some measured values are given in the tables in section 5.7.

+ Experience acquired :
- Installing a solar collector system in an existing structure poses
  problems :
  - in connection with the correct orientation of the collectors, since
    the building does not necessarily correspond to what is ideal for
    the solar system (In this case an extra support structure had to be
    built).
  - in connection with available space for storage tanks in the
    building e.g. possibility of a compact system is often jeopardized.
  - in connection with the available head provided by the pump in the
    existing heating system.
    (In this case a stronger pump had to be installed to provide a
    sufficient head to run the heating water through the storage tank
    in the loft).
- No mechanical failure was experienced with the system over 3 years.
  However, deterioration of the insulation material around the storage
  tank, caused by mealworms, was observed.
- The importance of a non-return valve to the efficiency of the solar
  loop was noted. (Adding such a valve to the system increased the
  recorded daily flow through the collector by 50 %).

(1) For the correct interpretation of the system efficiency values, cf.
    section 6.

+ The conclusions are based on measurements running from June 8th, 1983 – November 15th, 1983.
- Over the 3 years after installation (in 1980) a sophisticated measuring set-up using a data logger failed to produce reliable information. This system was then discontinued and a simpler approach, using chart recorders, was implemented.
(Where rapid results are required, a more limited approach to monitoring may produce substantial data at much lower costs).
- Due to poor water quality, filters had to be used to protect measuring devices.

+ Future work :
- to continue monitoring the system, in order to establish a 'normal' annual energy delivery prospect.
- to continue analysis of the relationship between efficiency, the component volumes and the heat draw characteristics.

+ Further references

Carpenter, J.L. et al. "The viability of the use of solar water heaters in milking parlours in the U.K. "Energy conservation and the use of solar and other renewable energies in agriculture, horticulture and fish culture. Ed. Vogt. F. Pergamon Press. 1981.

Norman, A.J. et al. "Solar water heating for the farm dairy in England". World Solar Forum, Brighton. 1981.

Sudman, C.J. "A once through system for farm dairies in New Zealand using solar heated water". J. agric Engng Res, 24, 149–156. 1979.

PROJECT ANALYSIS

1. Site and climate

1.1. Site

Nearest main city :	Plymouth
Distance from main city :	51 km
Direction from main city :	North East

1.2. Site micro-climate

Partially sheltered, low-lying area, exposed to South Westerly winds and within 9 km of the coast. The area has a relatively high level of radiation when not subject to the influence of nearby upland moors.

1.3. Annual long-term averages

		Project	Country in general
3. Total precipitation	mm/yr	1 036	912
5. Mean irradiation on a horizontal plane for the year	MJ/(m^2.yr)	4 015	3 470
10. Average max. temp. (July/Aug.)	(°C)	16.2	14.7
11. Average min. temp. (Jan)	(°C)	6.6	4.0

2. Plant

2.3. Collector (: Flat-Plate collector)

'Solacyl' New Senior panels by Don Engineering
(Now no longer manufacturing)

1. Absorber
   - coolant : Water + Fernox A/F at 3:1
   - material : Copper with selective coating
   - aperture area : 10 x 1 m^2
   - absorber area : 10 x 1 m^2
   - working pressure : 1.1 bar
   - flow rate : 7 x 10^{-6} m^3/(sm^2)

2. Glazing
   - material : Tedlar film – one layer
   - special characteristics : Unknown grade

3. Insulation : 50 mm of expanded polyurethane backing

4. Mounting
   - Orientation : 17°E of South – 163°
   - Tilt : 40° from horizontal
   Above cow entry point at South end of parlour

2.4. Heat storage

1. Medium of heat storage : Water
2. Description, location in the system :
   In food storage loft above parlour, in separate compartment with additional 25 mm of Dow Styrofoam.
3. Insulation :
   25 mm of "spray-on' urethane foam (Herculage)
4. Dimensions :
   a) 1.2 m x 0.5 m
   b) 1.2 m x 0.6 m   3 tanks installed
   c) 1.5 m x 0.6 m
5. Thermal capacity : 0.79/1.17/1.50 MJ/K
7. Diagram : cf. figure 2.4.7

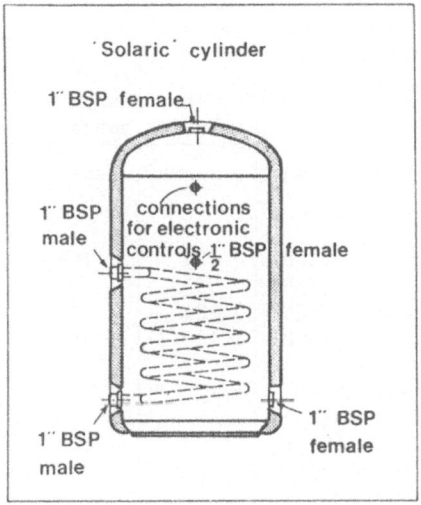

Fig. 2.4.7 : Heat storage tank.

2.5. Installation parts and measuring devices

1. Technical description of the installation parts :
   Pump - 'Solacyl' bronze pump, 1" B.S.P. conn.
   Control system - 'Solacyl' No. 1   1° min. differential
                                       temp. setting.

4. Economic aspects

4.1. Economic environment

2. Energy price of electricity	0.092 ECU/kWh day 0.027 ECU/kWh night
3. Energy prices evolution prospects	stable (perhaps +4-5 % in 1984)
4. Manpower cost (relevant for the needs of functioning of the solar installation)	Installer 9.7 ECU/h Maintenance 6.5 ECU/h
5. Interest rate of loans for farm (%) equipment (rate without subsidy)	18 %/yr unsecured 15 %/yr secured
6. Subsidies available for the solar and competitor farm equipment (as capital grant or interest rate reduction)	Nil

4.2. Investment expenses

		Solar only	Solar + support energy	Reference solution
1. Total investment budget	(ECU)	4 680	4 930	385
2. Cost of equipment :				
- cost of collectors per m²	(ECU)	184	184	0
- expenses for buildings	(ECU)	384	384	0
- cost of regulation equipment	(ECU)	284	368	42
- cost of other materials	(ECU)	1 120*(1)*	1 287	292
3. - Time of personnel for the conception, installation and start-up of equipment	(man hour)	105	105	10
- Cost of personnel for the conception, installation and start-up of equipment	(ECU)	1 053	1 053	50
4. Lifetime of the equipment (evaluation)	(years)	10	10	15
5. Residual value of the investment	(ECU)	0	0	0

(1) 1 tank only included in costs shown.
Total for 3 tanks was 2 673 ECU plus 300 ECU for installation.

4.3. Operating expenses

		Solar only	Solar + support energy	Reference solution
1. Energy cost	ECU/yr	8	420	477
			Estimate for full year in parlour	

No other operating expenses were incurred over the testing period.

5. Results

5.7. An example of some measured values (cf. tables 5.7.1 & 5.7.2).

Table 5.7.1 : Output from a solar panel system designed for a farm
dairy at Seale-Hayne College

Week com-men-cing	Max & Min afternoon storage temperature achieved during week (°C)	Energy output of the collectors (Q_{co}) (MJ equiv/wk)	Pump MJ used/wk	Net solar contribution MJ/wk	Solar contribution as a % of total draw-off at 82 °C	Solar contribution as a % of supplement to raise 60 l x 2 to 82 °C	Draw-off per day am/pm (l)	Panel area in use (m^2)
8.6	58 – 43	126^E	14	112	35.3	40.2	97 + 49	10
15.6	64 – 38	150	14	136	43.6	48.7	97 + 49	10
22.6	59.8 – 31.5	127	12	115	38.1	42.3	97 + 49	10
29.6	61.3 – 47.5	152	16	136	46.3	51.2	97 + 49	6
6.7	59.8 – 35.5	123	15	108	38.0	42.8	97 + 49	6
13.7	62.5 – 44.2	225	18	207	44.5	60.6	146 + 97	6
20.7	52.8 – 31.5	142^E	11	131	31.0	43.0	81 + 122	6
27.7	65.9 – 35.2	170	14	156	38.0	50.5	81 + 122	6
3.8	50.3 – 39.6	147^E	13	134	32.4	44.3	81 + 122	6
10.8	63.7 – 43.7	206	17	189	46.6	59.0	81 + 122	6
17.8	52.5 – 28.0^E	135	10	125	30.4	42.3	81 + 122	6
24.8	60.5^E– 42.3	187	16	171	43.1	54.4	81 + 122	6
31.8	49.9 – 30.6	136	10	126	29.7	41.4	81 + 122	6
7.9	44.4 – 19.7	70	9^E	61	9.5	21.2	81 + 122	6

Solar storage tank capacity used was 280 l throughout period.
E – Indicates the inclusion of an estimated value in sequence of
observed readings.

Table 5.7.2 : Performance characteristics of the experimental solar
heating system for water used in a farm dairy at
Seale-Hayne College, over periods of 5 days throughout
the summer of 1983.

Date (beginning date of the considered 5 day period)	Mean and range of maximum daily temps. reached by draw-off water (°C)	Quantity of water drawn-off (l) am and pm per day	Storage capacity (l) and panel area (m^2)	Mean system efficiency η*system (%) See note as well as section 6
15.6	38.0 (50.7) 64.0	97 and 49	280 – 10	9.4
21.6 (4 days only)	31.5 (47.5) 60.0	97 and 49	280 – 10	10.2
26.6 (3 days only)	35.8 (49.9) 59.8	97 and 49	280 – 10	10.2
30.6	50.3 (55.2) 61.3	97 and 49	280 – 6	14.1
5.7	51.0 (56.8) 60.5	97 and 49	280 – 6	13.7
14.7	47.8 (54.0) 57.8	146 and 97	280 – 6	26.8
25.7	44 (52.9) 65.9	81 and 122	280 – 6	20.6
30.7	35.2 (47.0) 62.4	81 and 122	280 – 6	21.0
12.8	43.7 (55.9) 63.7	81 and 122	280 – 6	22.2
17.8	28.0 (42.7) 52.5	81 and 122	280 – 6	22.7
24.8	42.3 (53.7) 60.5	81^E and 122^E	280 – 6	25.1

E – Estimated due to meter failure in water supply during the period.

Note : System efficiency obtained by comparing daily output (MJ/m²)
with the daily radiation received on a horizontal surface at a
nearby site monitored by research workers from the University of
Bath.

6. Remarks

The system efficiency in this report is expressed in

$$\eta^*_{system} = \frac{\text{Net heat gain per m}^2 \text{ collector surface}}{\text{Radiation per m}^2 \text{ \underline{horizontal} surface at 10 km distance}}$$

These values do not correspond with the efficiency as defined in this book ($\eta_{system} = \frac{\text{Net heat gain per m}^2 \text{ collector surface}}{\text{Solar radiation per m}^2 \text{ \underline{collector} surface}}$).

In order to accurately transpose these values to the definition used here, the course of the sun in function of time should be known. In an experiment in the Netherlands over this same time period, the radiation received on 1 m^2 of the collector (at 52° from horizontal instead of 40° as is the case here) was average 17 % more than the radiation at the same time on a 1 m^2 horizontal plane. This figure was used to give a rough estimate of what η_{system} could be :

$$\eta_{system} \cong \frac{\eta^*_{system}}{1+0.17*\frac{(4)}{5}}$$

$$\eta_{system} \cong 0.88 \, \eta^*_{system}$$

7. The use of solar energy for hot water production on a calf rearing farm

Project location

EEN, NETHERLANDS

Study Center :
- Project Leader : Ing. G. Brouwer
- Institution :   Van Heugten Consulting Engineers
                   P.O. Box 305
                   6500 AH Nijmegen
                   THE NETHERLANDS

Running since 1982.

Project description

On a calf rearing farm with 800 calf stands in Een, 120 m² of collectors were integrated in the roof structure to heat up tap water for purposes of milk preparation, drinking, and equipment cleaning in the farm.

- The collector has a single cover and a spectral selective coated absorber.
- Water (without antifreeze) flows through the collectors and directly heats up a 10 m³ water storage tank.
  Twice during the day the solar preheated water is pumped into a consumer storage tank, where it is postheated to 80°C by a gasfired

auxiliary heater. The 10 m³ storage is then supplied with fresh (cold) water (Via Level Controls).
- The hot water demand amounts to 2 m³ twice a day.

Goals of the experiment :

- To determine the possibilities of the use of solar energy in this application field.
- To measure the heat gain and the temperature levels of operation.
- To validate the computer simulation model using the measured results.
- To develop and to give advice for the construction of optimal installations in this field using the acquired experience and the simulation model.

Scheme (1)

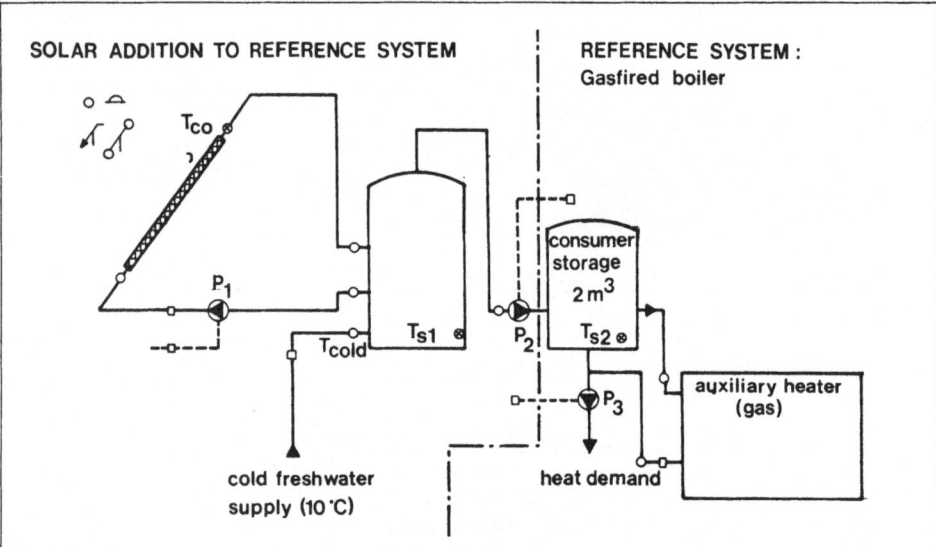

Note :
1. Weather data from meteorological station nearby (except for short periods of automatic data acquisition)
2. Temperature sensors and flow meters are connected with heat flow meters (integrators) (Q_{si}, Q_{so}, Q_{ag})

<u>Operating modes</u>
if $T_{s1} > 80°C \rightarrow P_1$ off and collector drains down
if $T_{s2} < 80°C$
 and if $T_{co} > T_{s1} + 5°C \rightarrow P_1$ on
  if $T_{co} < T_{s1} + 2°C \rightarrow P_1$ off and collector is drained

if $T_{s2} > 80°C \rightarrow$ aux. heater off
if $T_{s2} < 80°C \rightarrow$ aux. heater on

(1) for an explanation of the symbols used in the scheme, cf. pp. 470-471.

Conclusions

- Over the period 11/6/83 until 5/10/1983 the solar system provided 49.5 % of the total heat demand of 123 GJ over the period. Its system efficiency was 20-25 %.
  The gas savings over this period amounted to 3 200 m³ gas (for natural gas with a lower heating value of 31.7 MJ/m³ gas and heater efficiency of 60 %).
- The net energy input from the solar system, extrapolated using the simulation model, and given some minor changes in the installation, (*) is about 143 GJ/year (solar system efficiency = 37 %), which would correspond to a gas saving of 6 500 m³ of natural gas (for η of the conventional heater = 70 %).

  (*) example of changes :
    - change of ON-OFF criterion
      ΔT_{off} = 0.5 - 1 K instead of 2 K
    - better pressure loss distribution over the collector and distribution pipes
    - better position of the collector temperature sensor.

- The average final temperature of the solar storage before transfer to consumer storage was 43°C over the measuring period and 33.4°C over the reference year.

- Experience acquired

  + Over the 8 months of system operation, no malfunctions occurred.
  + High temperatures (70 to 85°C) of fresh water cause internal calcareous deposits on installation components.
    This means the measuring and control sensors have to be regularly cleaned.
    This factor will also influence maintenance requirements and installation lifetime in parts where such high temperatures are reached.
  + During the measuring period, the heat storage tank of the solar system with connection pipe was located outside. The heat losses were fairly high.
  + Even though the collectors are usually easily integrated into the existing roof structure, in this case the roof structure had to be straightened and strengthened in order to adequately support the collectors.
  + The drain-down frost protection control of the collector eliminates the need for a heat exchanger in the solar storage tank (since no antifreeze is needed in the solar circuit). Direct heating in the collector of the water to be used keeps the collector temperature as low as possible. This implies a higher collector efficiency value.

- Future work : the whole system is broken down into 2 independent identical parts. Comparative experiments will be done to help optimize operating modes, flow rates, as well as to assess influence of transparency of the cover (in relation to dust buildup ...). Comparison of efficiency of different collector-types for different working temperatures is also envisaged.

PROJECT ANALYSIS

1. Site and climate

1.1. Site

Latitude :	53°0' N
Longitude :	6°30' E
Altitude :	0 m
Nearest main city :	Assen
Distance from main city :	15 km
Direction from main city :	NW
Obstructions :	none

1.2. Site Micro-Climate

The long-term weather conditions from the meteorological station at Eelde
(about 12 km away) are used.
The farm is situated in a flat area of arable land.

1.3. Annual long-term averages

			Project	Country in general
1. Prevailing wind direction	W		SW	SW
2. Average wind speed	W	m/s	5.0	4.0
	S	m/s	3.5	2.8
3. Total precipitation		mm/yr	781	795
4. Absolute humidity	W	g/kg	4	4
	S	g/kg	7	7
5. Global irradiation on horizontal plane		$MJ/(m^2yr)$	3 600	3 530
6. Diffuse proportion		%	60	60
8. Hours sunshine		hrs/yr	1 450	1 506
9. Ambient temperature		C	8.6	9.2
10. Daily average max. temp. (July/Aug)		C	20.7	21.1
11. Daily average min. temp. (Jan)		C	3.8	3.5

12. Meteorological station	de Bilt
13. Latitude	52°20' N
14. Longitude	4°50' E
15. Altitude :	< 20 m
16. Distance from site :	140 km

Note : W = Winter
    S = Summer

2. Plant

2.3 Collector : (flat-plate collector)

1. Absorber
   - Coolant :   water
   - Material :   copper pipes with aluminium plate
   - Aperture area : $2 \times 60 \ m^2$
   - Absorber area : $2 \times 60 \ m^2$
   - Collector fluid content : $0.4 \ 1/m^2$
   - Working pressure :   100 kPa
   - Flow rate : $8.33 \times 10^{-6} \ m^3/(sm^2)$
   - Emittance of plate :   0.15
   - Coef. of absorption of : 0.95
     visible light
2. Glazing :
   - Emittance of glass :   0.9
   - number of glass covers : 1
   - material :   glass
   - special characteristics : low-iron content
3. Insulation : 4 cm PIR (polyisocyanuraat) with reflective coating
   ($\lambda = 0.2 \ \dfrac{W}{mK}$)
4. Mounting
   - Orientation : West south west
   - Tilt : 25°
   The collectors are integrated in the roof structure.
5. Collector performance (cf. figure 2.3.5).

2.4. Heat storage

1. Medium of heat storage : water
2. Description, location in the system :
   $10 \ m^3$ water storage horizontally situated outside the stable in a
   separate housing
4. Dimensions : 1 = 3.2 m d = 2.0 m
5. Thermal capacity : 42 MJ/K
6. Overall heat loss coefficient : 10.5 W/K

2.5. Installation parts and measuring devices

1. Technical description of the installation parts :
   - 2 circulation pumps collector circuit
     $3 \ m^3/h$, Euramo SB-50, 75 W
   - 2 controls
     manufacturer F.B.S.
     temperature difference control + 5, + 2 K
     maximum temperature control 80°C
   - sensors NTC semi-conductors
   - 1 circulation pump consumer circuit
     1 600 1/hour Euramo, SB-25, 72 W
   - level control Danfoss

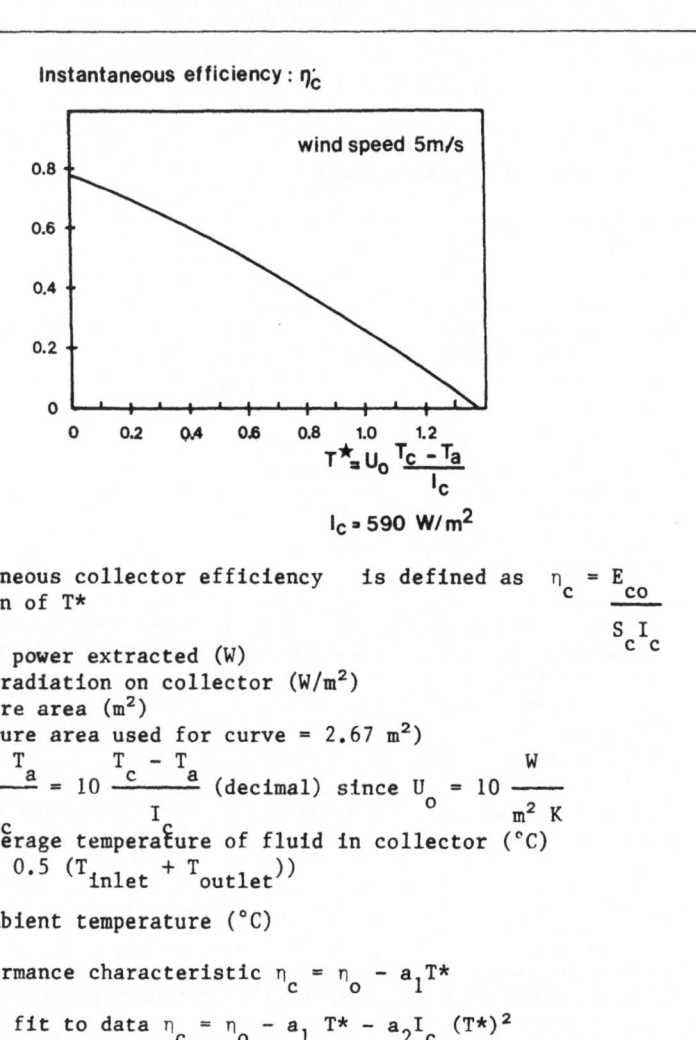

The instantaneous collector efficiency is defined as $\eta_c = \dfrac{E_{co}}{S_c I_c}$ as a function of T*

E_{co} : useful power extracted (W)
I_c : solar radiation on collector (W/m^2)
S_c : aperture area (m^2)
    (Aperture area used for curve = 2.67 m^2)

$T^* = U_o \dfrac{T_c - T_a}{I_c} = 10 \dfrac{T_c - T_a}{I_c}$ (decimal) since $U_o = 10 \dfrac{W}{m^2\,K}$

with T_c = average temperature of fluid in collector (°C)
    (= 0.5 ($T_{inlet} + T_{outlet}$))

    T_a = ambient temperature (°C)

Linear performance characteristic $\eta_c = \eta_o - a_1 T^*$

Second order fit to data $\eta_c = \eta_o - a_1 T^* - a_2 I_c (T^*)^2$

$\eta_c = 0.777 - 0.378\ T^* - 2.2 \times 10^{-4} I_c (T^*)^2$ for $I_c = 590$ W/m^2

Figure : 2.3.5 : Instantaneous collector efficiency curve

2. Description of the sensors

Type of sensor	temperature sensor	flowmeter	heatflow meters	pyranometer
Principle	platinum 100 Ω resistance	Float meter, with built in magnet	multiplication and integration of flow and temp. diff.	stern-pyranometer (Dirmhirn)
Precision	0.5°C	2.5 %	2 % – 5 %	
Measuring range	0 – 100°C	different types	T : 80°C diff. types	1 260 W/m^2

4. Economic aspects

Incl. VAT 18 % 1983.

4.1. Economic environment

1. Energy price of gas	0.24 ECU/m^3
2. Energy price of electricity	0.12 ECU/kWh

3. Energy price evolution prospects
   – of gas
   – of electricity
	
	+ 10 %/yr
	+ 10 %/yr

5. Interest rate of loans for farm equipment (rate without subsidy)	9 %/yr.

4.2. Investment expenses

1. Total investment budget	26 400 ECU

2. Cost of equipment :
   – cost of collectors per m^2
   – expenses for buildings
   – cost of regulation equipment
     and other materials
	
	142 ECU/m^2
	2 384 ECU
	
	5 563 ECU

3. Cost of personnel for the conception, installation and start-up of equipment	1 391 ECU
4. Lifetime of the equipment (evaluation)	15 to 20 years
5. Residual value of the investment	0 ECU

4.3. Operating expenses

	Solar + support energy	Reference Solution
1. Energy cost (gas & electricity) ECU/yr	3 600	5 100
2. Maintenance cost ECU/yr	40	

5. Results

5.1. Method of calculation

1. For data from measuring period.
   - most of the given values were measured directly or (in the case
     of heat flow meters) indirectly. (For location of the measuring
     points, cf. scheme).

- the average temperature of the incoming cold water was estimated.
- the average preheated water temperature was calculated as

$$T_{s1} = \frac{Q_{no}}{M \times c_w} + T_w$$

With Q_{no} the net energy input from the solar system over period x (Joule)

     M   the total amount of heated water used over period x (kg)

     c_w the specific heat of water $\left(\frac{J}{kg\ K} \right)$

     T_w the temperature of the incoming fresh water (°C)

- The required output temperature was calculated as :

$$T_{s2} = \frac{Q_d}{M \times c_w} + T_w$$

with Q_d the total heat demand (Joule)

2. For the simulation results
   - An adapted version of the program described in the fish farm project in Barneveld, Netherlands.

5.2. Weather conditions and energy

a) measured

1.	11/6	18/6	24/6	1/7	28/7	26/8	30/8	20/9	5/10	24/10	11/6		from:	PERIOD	
2.	18/6	24/6	1/7	28/7	26/8	30/8	29/9*	5/10*	24/10	20/10	5/10		until:	1983	
3.	13.4	18.5	15.0	18.4	17.5	-	14.1	14.0	-	-	-	°C	average ambient TEMPERATURE		
4.	4.7	3.5	4.5	3.0	3.5	-	-	-	-	-	-	m/s	average WIND		
5.	18.1	18.9	15.9	66.2	56.2	-	35.5	35.3	-	-	246.1	GJ	average SOLAR RAD on HORIZONTAL SURFACE of 120 m²		
6.	8.1	8.2	5.4	20.4	22.4	1.1	7.9	8.4	5.5	6.0	81.7	GJ	input SOLAR energy		solar
7.	8.1	8.2	5.4	20.4	22.4	1.1	7.9	8.4	5.5	6.0	81.7	GJ	input energy STORAGE		
8.	7.0	6.9	4.7	15.9	11.2	. -	7.8	7.4	4.8	5.6	60.9	GJ	output energy STORAGE		
9	7.0	6.9	4.7	15.9	11.2	-	7.8	7.4	4.8	5.6	60.9	GJ	USEFUL energy		
11.	5.9	2.6	3.4	8.1	12.4	-	16.4	13.7		41.3	62.2	GJ	AUXILIARY HEATING energy		
12.	11/6 until 28/7: 0.608					28/7 until 5/10: 0.169				0.061	0.777	GJ	energy for OPERATION		
13.	9.8	4.3	5.7	13.5	20.7	-	27.3	22.8	41.2	68.9	103.7	GJ	GAS consumption η = 0.6		
14.	11/6 until 28/7: 0.608					28/7 until 5/10: 0.169				0.061	0.777	GJ	EXTRA ELECTRICITY consumption		
15.	12.9	9.4	8.1	24.0	23.6	-	24.2	31.1	29.5	46.9	123.1	GJ	HEATING energy		reference
17.	21.5	15.7	13.5	40.0	39.3	-	27.3	51.8	49.2	78.2	205.2	GJ	GAS consumption		
18.	0	0	0	0	0	0	0	0	0	0	0	GJ	EXTRA ELECTRICITY consumption		

Weather data from other station.
- data not available

b) simulated over reference year

1. 2.	Jan	Feb	Mar	Apr	May	Jun	Jul	Aug	Sep	Oct	Nov	Dec	Year	PERIOD reference year	
3.	2.3	2	6	9.5	14	17	18	17.5	15	11	7	4.5	9.3	average ambient TEMP	°C
4.	3	3	3	3	3	3	3	3	3	3	3	3	3	average WIND	m/s
5.	6.6	13.3	31.2	48.4	57.6	60.6	54.9	46.4	32.6	18.2	8.5	5.5	385.5	average SOLAR RAD on tilted collector	GJ
6.	2.3	4.4	12.1	19.1	22.7	23.9	22.4	19.5	13.9	7.3	2.7	1.2	151.6	input SOLAR energy	
7.	2.3	4.4	12.1	19.1	22.7	23.9	22.4	19.5	13.9	7.3	2.7	1.2	151.6	input energy STORAGE	
8.	2.2	4.1	10.7	17.4	20.9	22.2	21.0	18.6	13.7	7.8	3.1	1.5	143.1	output energy STORAGE	solar (GJ)
10.	2.2	4.1	10.7	17.4	20.9	22.2	21.0	18.6	13.7	7.8	3.1	1.5	143.1	USEFUL energy	
11.	34.1	28.7	25.7	17.8	15.4	13	25.3	17.7	21.5	28.5	32.1	34.8	284.7	AUXILIARY HEATING energy	
12.	0.02	0.04	0.12	0.19	0.23	0.24	0.23	0.20	0.15	0.09	0.03	0.02	1.56	energy for OPERATION	
13.	48.7	41	36.7	25.4	22	18.6	36.1	25.3	30.7	40.7	45.9	49.7	406.7	GAS consumption $\eta = 0.7$	
14.	0.02	0.04	0.12	0.19	0.23	0.24	0.23	0.20	0.15	0.09	0.03	0.02	1.56	EXTRA ELECTRICITY	
15.	36.3	32.8	36.3	35.2	36.3	25.2	36.3	36.3	35.2	36.3	35.2	36.3	427.8	HEATING energy	reference (GJ)
17.	51.9	46.9	51.9	50.3	51.9	50.3	51.9	51.9	50.3	51.9	50.3	51.9	611.0	GAS consumption	
18.	0	0	0	0	0	0	0	0	0	0	0	0	0	EXTRA ELECTRICITY	

5.6. (Pre-heated) water

a) over measuring period

1.	11/6	18/6	24/6	1/7	28/7	26/8	30/8	20/9	5/10	24/10	11/6		from:	PERIOD (1)
2.	18/6	24/6	1/7	28/7	26/8	30/8	20/9	5/10	24/10	20/11	5/10		until:	1983
3.	10	10	10	10	10	10	10	10	10	10	10	°C	average input water temperature	
4.	40	36	28	87	88	-	82	79	99	158	440	m³	average water demand	
5.	52	55.6	50	53.5	40	-	32.7	32.3	21.5	18.5	43	°C	average (pre-) heated water temp.	
6.	86.8	72.2	78.9	75.7	73.9	-	80.3	73.6	80.9	80.7	76.6	°C	output temperature required	

(1) total number of monitored days is 111.

b) over simulated reference year

1. 2.	Jan	Feb	Mar	Apr	May	Jun	Jul	Aug	Sep	Oct	Nov	Dec	year	PERIOD reference year	
3.	10	10	10	10	10	10	10	10	10	10	10	10	10	average input water temp.	°C
4.	124	112	124	120	124	120	124	124	120	124	120	124	1460	average water demand	m³
5.	14.2	18.7	30.5	44.6	50.3	54.3	50.5	45.8	37.3	25.0	16.2	12.8	33.4	average (pre-) heated water temperature	°C
6.	80	80	80	80	80	80	80	80	80	80	80	80	80	output temperature required	°C

6. Remarks

Given the fact that the value of the expected gas savings amounts to 1/17 th of the total investment expenses, a long payback period seems to be indicated.

However, on the basis of the expected significant energy price increases and cost price decreases of solar installations (cheaper techniques, mass fabrication, further optimizations ...), and given also the significant subsidies for such projects in the Netherlands, this application is described as "very promising" by the study center in Nijmegen.

ADDENDUM

ADDENDUM

REPORTING FORMATS

In each part of the book, chapter 2 as well as the appendix consist for the most part of formatted information :

- Basic descriptions of research projects within the EC.
(chapters 2)

- Detailed descriptions of a few of those projects.
(appendices to the parts I, II & III)

The format conception and content of these descriptions are explained below.

1. Basic descriptions

The basic descriptions of the research projects consist of 2 parts :

* a brief discussion of the project, containing the following elements :
  - Title
  - Location of the installation (also the year when the research was started).
  - Description
  - Conclusions (which, in many cases are still of a provisional nature)
  - Study Center - project leader
                - address

* a scheme of the installation, and of the energy flow through the system

On the scheme, the central element of the installation, namely the collector, is described in more detail.
Elements listed are : - the type of the heat transporting medium (air, water, ...)
- the nature of the collector itself
  1) collector-type (flat plate, greenhouse type ...)
  2) The nature of the different collector components (absorber, cover, insulation).
- a representative value of the cost for 1 m² of collector (Except where explicitly mentioned, the cost was assumed to refer to prices representative of the time when the investment was made when the research was started)
- a representative value of the long term collector efficiency in the given application.

$$(\text{collector efficiency } \eta_c = \frac{\text{energy extracted from the collector by the solar collecting fluid}}{\text{total amount of solar radiation on the collector plates}})$$

In chapters 2 this format was followed as much as possible.

In cases where the presentation of the material according to this strict format would not fully serve the purpose of the chapters in question - the purpose being to give as complete a view as possible of the current state of the art and the evolution of research in the considered area -, the description was slightly altered, in order to better serve the above mentioned purpose.

In this context we also refer to the "Methodology" section of the "Readers' Guide", cf. p. XXII, where an explanation is given of the way in which this and the other material in the book were conceived and developed, and of how they could best be interpreted.

2. Detailed descriptions

In the appendices to parts I, II and III, a detailed description is given of several research projects within the E.C.

These research project reports were written up, following a uniform pattern, with the intention of enabling the reader to quickly find the part that he is interested in, and even to consider one particular aspect of the research, across the board of the given descriptions.

Each description consists of 2 parts.
The first part aims at giving an introduction to the problem and its different aspects, leaving out the technical information that is not essential in this first stage.
In the second part, the technical data and scientific results are presented.

FIRST PART OF DETAILED DESCRIPTION

The first part, which could be called "the first level of interest", consists of 4 sections.

1. Project location   (and "study center")

in which useful identification data of the research program in question are given.

2. Project description

in which set-up and goals of the experiment are explained.

3. Scheme of the installation

In this scheme, the following elements are presented :
- A drawing of the solar installation, made with the intention of helping the reader to understand the system under consideration. In this drawing, all technical intricacies which do not directly contribute to this purpose (e.g. certain valves, expansion tanks, extra pumps for back up, etc. ...) were deleted.

Following symbols are used :

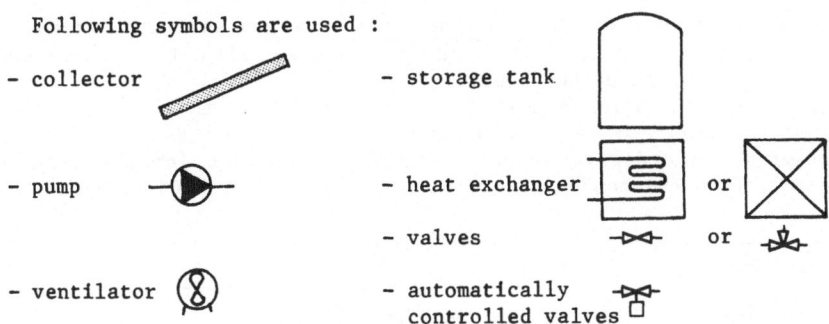

- collector

- pump

- ventilator

- storage tank

- heat exchanger   or

- valves   or

- automatically
  controlled valves

Where regulation of the valve postitioning is
shown, the direction of the flow that is blocked
is indicated by the solid black part of the
symbol.

E.g. the symbol (   ) means that flow is
possible only in the direction of the
dotted arrows.

- The location of the measuring points in the experimental set-up.

Following symbols are used :

solar radiation (if on a
horizontal surface)   (if on an inclined surface)

temperature   o

flow rate   □

relative humidity   ◊

moisture content   ✦

wind-speed

wind-direction

If a certain variable was measured only in a few places, a specific
symbol is used and explained in the scheme itself.

Where a certain value has also been measured for control or
regulation purposes, the same symbols are used, but with an x
inside them.
E.g. : the symbol ⊗ means : this temperature value is also
measured for system regulation purposes.
*Note : For practical reasons, the presentation of the location of
the measuring points in the appendix to part I (Greenhouses)
has been transferred to a later section (2.5.3) of the
detailed report.*
- The letter symbols : if unclear from the scheme, their explanation
can be found in the terminology section on p. XXIV.
- The operating modes of the installation.

- Description of the reference solution : a conventional set-up, the performance of which is normally compared with the experimental solution throughout the report.
  Sometimes the solar system is only an addition to the conventional installation. In that case a line (·—·—·—·) is added to the drawing, in order to separate the solar addition (on the left-hand side), from the conventional set-up (on the right-hand side).

4. Conclusions

The conclusions consist of the following elements.
- Some general information on the measured performance and expected feasibility of the experimental system.
- Possible elements which might help assess the reliability of the above conclusions. (duration and character of the monitoring period, difficulties with experiments ...)
- (In some cases the remarks of the editor, given in section 6 of the second part of the description, may throw some additional light on this subject).
- The acquired technical experience.
- Research activities planned for the future.
- Further references.

SECOND PART OF THE DETAILED DESCRIPTION

A list of the different items that are normally included in this second part is given below, together with some explanations in italics when the interpretation of the items in question is not self-evident.

1. Site and climate

1.1. Site

    Latitude :
    Longitude :
    Altitude : *(1)*
    Nearest main city :
    Distance from main city :
    Direction from main city :
    Obstructions : *(2)*

(1) Given in metres above sea level.
(2) Obstructions to sun and wind.

1.2. Site micro-climate :

Brief qualitative description of special conditions imposed by the local site, especially local variations from long-term averages presented in section 1.3.

1.3. Annual long-term averages

		Project	Country in general
1. Prevailing wind direction	W		
	S		
2. Average wind speed	W m/s		
	S m/s		
3. Total precipitation	mm/yr		
4. Absolute humidity	W g/kg		
	S g/kg		
5. Global irradiation on horizontal plane	MJ/(m².yr)		
6. Diffuse proportion	%		
7. Degree days (Base temp. = ...) (1)	Cdays		
8. Hours sunshine (2)			
9. Ambient temperature (3)			
10. Average max. temp. (Jul/Aug) (4)	C		
11. Average min. temp. (Jan) (5)	C		
12. Meteorological station : (6)			
13. Latitude :			
14. Longitude :			
15. Altitude :			
16. Distance from site :			

Note : W = Winter
S = Summer

(1) Degree days = $\sum\limits^{N} (Tb - T)^{+}$ *where :*
Tb = base temp. in °C
T = average daily temperature
N = number of days in the year
($^{+}$) Note : If Tb - T $\bar{\quad}$ 0, the value of (Tb - T), incorporated into the formula for that day is zero.

The base temperature is chosen at a critical level for a studied system. E.g. in the case of greenhouses, the outside temperature limit below which heating is needed. Thus in this example the degree day value provides an indication of the long term heating requirement of the greenhouse.

(2) Annual sunshine hours : $\sum\limits^{N} hr$ *where :*
hr : daily hours of sunshine
N : number of days in the year

(3) Annual average value

(4) Average max. temp. (Jul/Aug) = $\dfrac{1}{31} \sum\limits_{1}^{31} Tmax$ *for July or August.*
where Tmax = maximum daily temperature

(5) Average min. temp. (Jan) : as above, for Tmin in January

(6) Station where the given data were obtained.

2. Plant

2.1. Greenhouse

1. Dimensions - Floor area : m^2
                - volume : m^3
                - Glazing area : m^2
                - Wall area : m^2
2. Glazing :
3. Covers/screens :
4. Ventilation system :

2.2. Drying place

1. Drying system :
2. Materials :
3. Dimensions :
4. Capacity :
5. Ventilation - Power :   W
                - Flow rate : m^3/s
6. Conventional heating source :
                calorific power :   W
7. Diagram :

2.3. Collector

In the description of the collector 2 different formats were chosen

Format n^r 1

2.3. Collector

1. Absorber - Coolant : *(1)*
             - Material : *(2)*
             - Aperture area : *(3)* m^2
             - Absorber area : *(4)* m^2
             - Collector fluid content : *(5)* $1/m^2$
             - Working pressure :   Pa
             - Flow rate : m^3/sm^2
             - Emittance of plate : *(6)*
             - Coef. of absorption of
               visible light :
2. Glazing : - emittance of glass :
             - number of glass covers :
             - material :
             - special characteristics :
3. Insulation :
4. Mounting - Orientation : *(7)*
             - Tilt : *(8)*
5. Collector performance : *(9)*

*(1) Coolant : Heat transfer medium by which solar energy (heat) is removed
            from the absorber (e.g. water + 20 % glycol, air, ...)*
(2) Material : E.g. aluminium, copper, ...
(3) Aperture area : Area through which solar energy is admitted into the

solar collector (i.e. net area of glass)

(4) Absorber area : Total area of installed surface of absorber.

(5) Collector fluid content : Total coolant capacity of the absorber (in litres/m² of absorber area)

(6) Emittance of plate : Emissivity of the black absorber in the infra-red area.

(7) The orientation is expressed as the angle between the projection on a horizontal plane of the normal to the collector surface and the North direction (the angle is counted clockwise.)

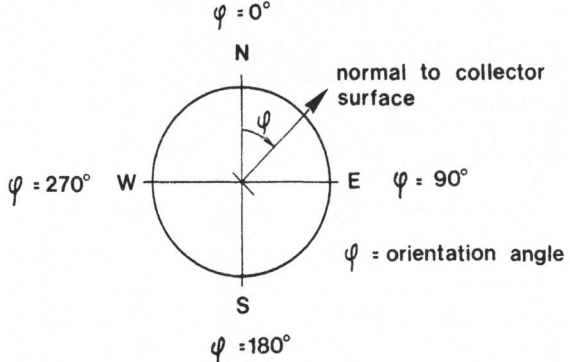

(8) Tilt is expressed as the angle between the collector surface and the horizontal plane.

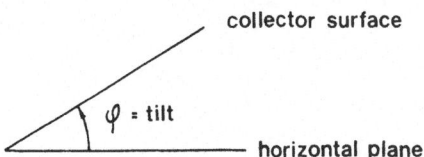

(9) At this point the instantaneous collector efficiency as a function of its operation temperature, ambient temperature and irradiation is given.

Format n^r 2

2.3. Collector

1. Material :
2. Dimensions :
3. Flow rate :
4. Diagram of collector :
5. Collector performance : *(1)*

(1) As in format n^r 1 (see footnote 9 above)

2.4. Heat storage

1. Medium of heat storage :
2. Description, location in the system :
3. Insulation :
4. Dimensions :

5. Thermal capacity : *(1)*   MJ/K
6. Overall heat loss coefficient : *(2)*   W/K
7. Diagram :

*(1) Thermal capacity : volume of the heat storage medium (m³) multiplied
     by the volumetric specific heat of the medium (MJ/m³K)*
*(2) Overall heat loss coefficient : Total storage heat loss coefficient,
     including thermal bridges.*

2.5. Installation parts and measuring devices

1. Technical description of the installation parts :
   (e.g. pumps, fans, radiators, heat exchangers, ...). Indication of
   type, dimensions, capacity, ...
2. Description of the sensors.
(3. Scheme of measuring points).

3. Products

3.1. Culture in greenhouse

1. Type of crop :
2. Date of sowing/planting :
                harvesting :
3. Growing system :
4. Required temp. regime : min. temp. :   C
                           max. temp. :   C
5. Required humidity regime : min. abs. humidity :   g/kg
                              max. abs. humidity :   g/kg
6. Required CO_2 regime : min. CO_2 :   %
                           max. CO_2 :   %
7. How sensitive is the crop to light ?
8. Other (related), crops which require an analogous inside climate and
   luminosity :

3.2. Drying-products

1. Designed for special products :
2. Required conditions of inlet air (flow rate, temperature, humidity,
   ...)
3. Properties of the products (quantity, temperature, humidity, ...)
4. Other (related) products which could be dried :
5. Any special treatment of the crop ? :

4. Economic aspects

*Note : 1. Where the 3 columns shown under sections 4.1. - 4.4. have been
          deleted, the given values apply to the solar system that is
          being tested.*
*   2. The monetary values, originally given in the local currency,
          were converted to ECU by the editor. In some instances
          however, the dates for which these prices are representative
          were not explicitly mentioned in the reports. In that case it
          was assumed that sections 4.1., 4.3. and 4.4. were priced at the*

time when the report was written (1983). Section 4.2. was assumed to refer to prices representative of the time when the investments were made.

4.1. Economic environment

		Solar only	Solar + support energy	Reference solution
1. Energy price of oil or gas	(ECU/1) (ECU/m³)			
2. Energy price of electricity	(ECU/kWh)			
3. Energy prices evolution prospects *(1)*				
- of oil or gas	%/yr			
- of electricity	%/yr			
4. Manpower cost (relevant for the needs of functioning of the installation)	(ECU/hr)			
5. Interest rate of loans for farm equipment (rate without subsidy) *(2)*	(%/yr)			
6. Subsidies available for the solar and competitor farm equipment (as capital grant or interest rate reduction)	(%, or ECU)			

(1) The given value includes inflation rate, even if this rate is mentioned separately.
(2) The interest rate is that which farmers would normally pay for a loan for such equipment, without subsidy.

4.2. Investment expenses

		Solar only	Solar + support energy	Reference solution
1. Total investment budget *(1)*	ECU			
2. Cost of equipment :				
- cost of collectors per m²	ECU			
- expenses for buildings *(2)*	ECU			
- cost of regulation equipment	ECU			
- cost of other materials	ECU			
3. - Time of personnel for the conception, installation and start-up of equipment	(time unit)			
- Cost of personnel for the conception, installation and start-up of equipment	ECU			
4. Lifetime of the equipment (evaluation)	(years)			
5. Residual value of the investment	ECU			

(1) Any rebate or difference from the market price should be mentioned.
(2) Where an existing building has been converted because of the solar system, the given figure represents the conversion costs.

4.3. Operating expenses

		Solar only	Solar + support energy	Reference solution
1. Energy cost	ECU/yr			
2. Maintenance cost (replacement of material or components)				
- spare parts	ECU/yr			
- salaries (internal-external)	ECU/yr			
3. Hire of surface for the installation (if any)	ECU/yr			
4. Other manpower cost (other than under 4.3.2.)	ECU/yr			
5. Insurance cost	ECU/yr			
6. Raw material cost (if any)	ECU/yr			

4.4. Hidden cost and benefits

		Solar only	Solar + support energy	Reference solution
1. Production obtained (dried substance, number of plants, ...	(kg, m^3, ./yr)			
2. Average sales prices per unit produced (1)	ECU/(kg, m^3 ...)			
3. Total income obtained (4.4.1 x 4.4.2)	ECU/yr			
4. Storage costs due to the installation	ECU/yr			

(1) The sales price is based on the quality of the obtained product, as well as on the time at which it becomes available for the market.

5. Results

5.1. Method of calculation

This part should normally consist of the following items :
- Justification of the applied method of calculation.
- Indication of the equations used and how the results are derived.
- Indication of total accuracy of the data and the results.

5.2. Weather conditions and energy

1.													from : PERIOD	
2.													till reference year	
3.												C	average ambient TEMP	
4.												m/s	average WIND	
5.												MJ	average SOLAR RAD on collector	
6.												MJ	input SOLAR energy	solar
7.												MJ	input energy STORAGE	
8.												MJ	output energy STORAGE	
9.												C	average TEMP. storage	
10.												MJ	USEFUL energy	
11.												MJ	AUXILIARY HEATING energy	
12.												MJ	energy for OPERATION	(1)
13.												MJ	OIL/GAS consumption	
14.												MJ	ELECTRICITY consumpt.	
15.												MJ	HEATING energy	Reference
16.												MJ	energy for OPERATION	(2)
17.												MJ	OIL/GAS consumption	
18.												MJ	ELECTRICITY consumpt.	

(1) Solar : In this part, the results of the solar installation are given.
(2) Reference : In this part, the results or approximations of the considered reference solution (cf. scheme) are given.

lines 1 & 2   : Time period over which the results in the given columns are obtained.
line 6   : Total solar energy collected by the collector and transferred to the other component parts of the system by the working fluid.
line 7   : That part of the solar energy that is transferred into the heat storage tank.
line 8   : The energy which is drawn from the heat storage tank.
line 9   : Average temperature of storage tank.
line 10   : That part of the solar energy which is effectively put to use.
line 11   : Where an auxiliary heater is used, its auxiliary energy input into the system.
lines 12 & 16 : The quantity of energy which is used for operation of e.g. pumps, fans, control units, etc...
lines 13 & 17 : The efficiency value of the oil/gas - fired heater is taken into account.

5.3. Inside climate of greenhouse

																from :		PERIOD	
1.																			
2.																till			
3.																C	min.		Solar
4.																C	average	TEMPERATURE	
5.																C	max.		
6.																g/kg	min.		
7.																g/kg	average	HUMIDITY	
8.																g/kg	max.		
9.																W/m2	Radiation		
10.																W/m2	PAR		
11.																%	min.		
12.																%	average	CO2	
13.																%	max.		
14.																C	min.		Reference
15.																C	average	TEMPERATURE	
16.																C	max.		
17.																g/kg	min.		
18.																g/kg	average	HUMIDITY	
19.																g/kg	max.		
20.																W/m2	Radiation		
21.																W/m2	PAR		
22.																%	min.		
23.																%	average	CO2	
24.																%	max.		

5.4. Crop results (greenhouse)

	Solar	Reference
1. Date of sowing/planting :		
2. Harvesting date :		
3. Flowering date :		
4. Flower number :		
5. Fruit number :		
6. Leaf area measurement : (m^2)		
7. Production obtained :		
8. Other differences : (in results, aspects, dimensions, ...)		

5.5. Drying results

1.																from :		PERIOD	
2.																till			
3.															m3/h	flowrate	⎫		
4.															C	temperat.	⎬ INLET AIR		
5.															%	humidity	⎭		
6.															%	moisture	⎫		Solar
7.															kg	useful quantity	⎬		
8.																quality	PRODUCTS		
9.															C	temperature	⎭		
10.															m3/h	flowrate	⎫		
11.															C	temperature	⎬ INLET AIR		Reference
12.															%	humidity	⎭		
13.															%	moisture	⎫		
14.															kg	useful quantity	⎬		
15.																quality	PRODUCTS		
16.															C	temperature	⎭		

Quality :

5.6. (Pre)-heated water

1.																from :	
																	PERIOD
2.																till	
3.															C	average input water temperature (1)	
4.															l/s	average flow rate (2)	
5.															C	average (pre-)heated (3) water temperature	
6.															C	output temperature required.	

(1) *Input temperature of the water that is to be heated.*
(2) *Flow rate of the water to be heated.*
(3) *Temperature of the same water after solar heating.*

5.7. An example of some measured data

6. Remarks

Editor's comments.